Lectures in Applied Mathematics

Proceedings of the Summer Seminar, Boulder, Colorado, 1960

VOLUME 1 LECTURES IN STATISTICAL MECHANICS
G. E. Uhlenbeck and G. W. Ford with E. W. Montroll

VOLUME 2 MATHEMATICAL PROBLEMS OF RELATIVISTIC PHYSICS
I. E. Segal with G. W. Mackey

VOLUME 3 PERTURBATION OF SPECTRA IN HILBERT SPACE
K. O. Friedrichs

VOLUME 4 QUANTUM MECHANICS
R. Jost

Proceedings of the Summer Seminar, Ithaca, New York, 1963

VOLUME 5 SPACE MATHEMATICS, PART 1
J. Barkley Rosser, Editor

VOLUME 6 SPACE MATHEMATICS, PART 2
J. Barkley Rosser, Editor

VOLUME 7 SPACE MATHEMATICS, PART 3
J. Barkley Rosser, Editor

Proceedings of the Summer Seminar, Ithaca, New York, 1965

VOLUME 8 RELATIVITY THEORY AND ASTROPHYSICS
1. RELATIVITY AND COSMOLOGY
Jürgen Ehlers, Editor

VOLUME 9 RELATIVITY THEORY AND ASTROPHYSICS
2. GALACTIC STRUCTURE
Jürgen Ehlers, Editor

VOLUME 10 RELATIVITY THEORY AND ASTROPHYSICS
3. STELLAR STRUCTURE
Jürgen Ehlers, Editor

Proceedings of the Summer Seminar, Stanford, California, 1967

VOLUME 11 MATHEMATICS OF THE DECISION SCIENCES, PART 1
George B. Dantzig and Arthur F. Veinott, Jr., Editors

VOLUME 12 MATHEMATICS OF THE DECISION SCIENCES, PART 2
George V. Dantzig and Arthur F. Veinott, Jr., Editors

Proceedings of the Summer Seminar, Troy, New York, 1970

VOLUME 13 MATHEMATICAL PROBLEMS IN THE GEOPHYSICAL SCIENCES
1. GEOPHYSICAL FLUID DYNAMICS
William H. Reid, Editor

VOLUME 14 MATHEMATICAL PROBLEMS IN THE GEOPHYSICAL SCIENCES
2. INVERSE PROBLEMS, DYNAMO THEORY, AND TIDES
William H. Reid, Editor

Proceedings of the Summer Seminar, Potsdam, New York, 1972

VOLUME 15 NONLINEAR WAVE MOTION
Alan C. Newell, Editor

MODERN MODELING OF CONTINUUM PHENOMENA

Summer Seminar on Applied Mathematics

Volume 16
Lectures in Applied Mathematics

Modern Modeling of Continuum Phenomena

Richard C. DiPrima, Editor

1977
American Mathematical Society, Providence, Rhode Island

The proceedings of the Summer Seminar were prepared by the American Mathematical Society with partial support from the following contracts and grants:

National Science Foundation under NSF grant MPS 75-00126,
Office of Naval Research under contract NOOO14-75-C0677.

Library of Congress Cataloging in Publication Data **CIP**

Summer Seminar on Applied Mathematics, 9th,
Rensselaer Polytechnic Institute, 1975.
Modern modeling of continuum phenomena.

(Lectures in applied mathematics; v. 16)
Includes bibliographies.
1. Continuum mechanics--Congresses.
2. Stochastic differential equations--Congresses.
3. Mathematical models--Congresses. I. DiPrima, Richard C. II. Title. III.
Series: Lectures in applied mathematics (Providence); 16.
QA808.2.S85 1975 531'.01'51 77-9041 ISBN 0-8218-1116-9

AMS (MOS) subject classifications. 73-02, 76-02, 86-02,
92-02, 34E15, 36B25, 60H10, 60H15.

Printed in the United States of America

Contents

Preface

The Ninth Summer Seminar on Applied Mathematics, sponsored jointly by the American Mathematical Society and the Society for Industrial and Applied Mathematics, was held at Rensselaer Polytechnic Institute from July 7 to July 18, 1975. The topic for the seminar was Modern Modeling of Continuum Phenomena. The program for the seminar was organized by George F. Carrier (Harvard University), Hirsh G. Cohen (Thomas Watson Research Center, IBM), Stephen H. Davis (Johns Hopkins University), Richard C. DiPrima, Joseph B. Keller (New York University), and Lee A. Segel (Weizmann Institute of Science and Rensselaer Polytechnic Institute).

The purposes of the seminar were (i) to introduce the participants to selected mathematical research areas of high current interest and relevance, (ii) to present the underlying fundamental laws of continuum model building, and (iii) to present selected mathematical topics particularly useful in solving modern mathematical problems of continuum phenomena. There were core series of lectures on continuum model building by Lee Segel, and on mathematical methods by Donald Cohen and George Papanicolaou, and in-depth case studies on problems of current interest by George Carrier, Garrett Odell, George Oster and Leon Knopoff. The present volume contains articles by all of the lecturers except George Carrier. In general, the written articles follow the pattern of the lectures, but are more detailed and include generalizations that could not be presented during the lectures.

I would like to express my appreciation to the Mathematics Branch of the Office of Naval Research and the Mathematics Division of the National Science Foundation for their financial support of the seminar, and to Rensselaer Polytechnic Institute for hosting the seminar. I am also grateful to Dr. Gordon L. Walker, Executive Director of the American Mathematical Society, and Ms. Hope Daly and Ms. Carol Kohanski of the American Mathematical Society for their generous help in planning and administering the seminar, and to Professor Lester Rubenfeld, Rensselaer Polytechnic Institute, for his efficient handling of local arrangements. I also wish to thank the speakers for their excellent lectures and articles and for their participation in the seminar. It is also appropriate to remember Mrs. Lillian Casey of the American Mathematical Society who was responsible for the initial administrative planning. Her death in February 1975 robbed the American Mathematical Society of a devoted worker and many of us of a dear friend.

Finally, during the seminar we learned of the death of Sir Geoffrey Ingram Taylor, O.M., F.R.S. on June 27, 1975, at the age of 89. For more than fifty years, he occupied a leading place in applied mathematics and continuum mechanics. It is appropriate that these proceedings on Modern Modeling of Continuum Phenomena be dedicated to his memory.

RICHARD C. DIPRIMA, EDITOR
Department of Mathematical Sciences
Rensselaer Polytechnic Institute

Troy, New York, September 1975

Lectures in Applied Mathematics
Volume 16, 1977

An Introduction to Continuum Theory[1]

Lee A. Segel

Introduction. Revolutions are a dime a dozen in some unstable areas; by contrast, a revolution that initiates a new form of activity which persists for two centuries certainly bears examination. Publication of the Declaration of Independence in 1776 marks one such revolution, and publication of Euler's laws of mechanics in the same year marks another. Commemoration of the American Bicentennial is well in hand; it is the story of the Eulerian continuum revolution and its aftermath that occupies us here.

Classical continuum mechanics has its triumphs (witness the modern supersonic jet) and its enduring challenges (e.g., the problems of turbulence, hurricane genesis, and earthquake prediction). Thus it must remain an important part of modern scientific teaching and research–for in common with other classical fields, continuum mechanics provides elegant, deep, and still-relevant distillations of a great period of creative effort. But there also is a "neoclassical" aspect of continuum studies that has emerged in the last decade or so. This involves extension of the earlier ideas into new domains such as chemistry, developmental biology, and ecology.

The present article is designed to serve as an introduction to both classical and neoclassical continuum theory. We have in mind an audience that is fairly sophisticated mathematically, but one that has not progressed in the physical sciences much further than a decent introductory college course.

In Part A we introduce the required basic definitions and theorems. Then we derive the differential equations that express the necessity for all physical continua to move in a way that conserves mass and that balances linear and angular momentum with the effects of appropriate forces and torques. The momentum balance laws are Euler's continuum extension of the Newtonian balance between force per unit mass and acceleration which is the keystone of *Principia*. That Part A is not an excessively lengthy scrutiny of a modest generalization is attested to by the fact that Euler's laws only appeared 89 years after *Principia* was published in 1687.

The first portion of Part B introduces the machinery of cartesian tensors. As we indicate, tensors not only form a compact language for the development of continuum mechanics, but in addition they are a natural vehicle for the expression of physical laws, which must be fundamentally the same in all coordinate systems.

AMS (*MOS*) *subject classifications* (1970). Primary 73B99; Secondary 92–02, 73–01, 76–01.

Excerpts are reprinted from *Mathematics applied to deterministic problems in the natural sciences* (1974) by C. C. Lin and L. A. Segel with permission of Macmillan Publishing Co.

[1] This work was partially supported by the Army Research Office and the National Science Foundation.

With tensors in hand, one can proceed swiftly to derive the particular constitutive equations that characterize a viscous fluid and a linearly elastic solid. To give a preliminary feeling for what the equations entail, we present three simple exact solutions of the viscous equations. To illustrate the generation of physical intuition we show how the important concept of vorticity permits one to understand the behavior demonstrated by such solutions as the result of an interaction among a small number of basic processes (in this case, boundary generation of vorticity plus its diffusion and convection). The section closes with a few steps into the domain of nonlinear elasticity.

Some of the neoclassical aspects of continuum theory are briefly discussed in Part C. (Others were mentioned in connection with the treatment of mass conservation in Part A.) "Subcontinuum" considerations of random walk lead to a deeper appreciation of the continuum equations, and also serve to introduce continuum models as currently used in microbiology. Pattern generation in biochemistry and ecology is the central topic treated in the concluding section.

Much of Part A is condensed (with the publisher's permission) from material found in Chapters 13 and 14 of Lin and Segel [**1974**] (denoted by **LS** below). A considerably expanded version of Part B forms a portion of a just published book by Segel [**1977**]. His conflict of interest is so apparent that the author will presumably be forgiven if he recommends the just-cited books for further development of the ideas presented here. Other standard references include the books of Serrin [**1959**], Rosenhead [**1963**], and Jaunzemis [**1967**].

A. FIELD EQUATIONS OF CONTINUUM MECHANICS

2. Kinematics of deformable media. We begin with the study of the motion of a continuum without regard to the forces that produce the motion. This is called *kinematics* as opposed to *dynamics*, the study of the influence of forces on motion.

As a continuum moves, it is natural to try to specify the position of each point as a function of the time t. To identify a given point, we specify its position $\mathbf{A}$ at a certain initial time t_0. If $\mathbf{x}$ is the position vector of this point at t, we describe the motion of the medium by the equations

$$\mathbf{x} = \mathbf{x}(\mathbf{A}, t), \qquad \mathbf{x}(\mathbf{A}, t_0) \equiv \mathbf{A}. \tag{1a, b}$$

In words, $\mathbf{x}$ is the position at time t of a point that was initially (at $t = t_0$) located at $\mathbf{A}$. The word *particle* is used to denote the point that starts at a fixed initial position $\mathbf{A}$ and moves according to (1). For fixed $\mathbf{A}$ the curve

$$\mathbf{x} = \mathbf{x}(\mathbf{A}, t), \qquad t \geqslant t_0,$$

is called a *particle path*.

Let $\mathbf{A}(\mathbf{x}, t)$ be the function of $\mathbf{x}$ and t obtained by solving (1) for $\mathbf{A}$. The result $\mathbf{A} = \mathbf{A}(\mathbf{x}, t)$ states that $\mathbf{A}$ is the initial position of the particle now (at time t) located at $\mathbf{x}$. Since $\mathbf{A}(\mathbf{x}, t)$ and $\mathbf{x}(\mathbf{A}, t)$ are a pair of inverse functions,

$$\mathbf{A}[\mathbf{x}(\mathbf{A}, t), t] \equiv \mathbf{A}, \qquad \mathbf{x}[\mathbf{A}(\mathbf{x}, t), t] \equiv \mathbf{x}. \tag{2a, b}$$

We shall postulate *impenetrability of matter*: $\mathbf{A}(\mathbf{x}, t)$ and $\mathbf{x}(\mathbf{A}, t)$ will be assumed to be single-valued invertible functions so that, at time t, for each $\mathbf{A}$ there exists a unique $\mathbf{x}$ and vice versa. We shall also postulate *smoothness of motion*: $\mathbf{A}(\mathbf{x}, t)$ and $\mathbf{x}(\mathbf{A}, t)$ will be assumed to have as many continuous partial derivatives as required for the various operations we shall perform on these functions.

As one example of the many dependent variables that will enter our discussion, consider the density. Let $\delta(\mathbf{A}, t)$ denote the mass per unit volume at time t of the particle initially located at $\mathbf{A}$. Let $\rho(\mathbf{x}, t)$ denote the mass per unit volume at time t, at the fixed point $\mathbf{x}$. *We assert that at time t the value of a dependent variable at the point* $\mathbf{x}$ *equals the value of the same dependent variable for the particle located at* $\mathbf{x}$ *at time t. This assertion could be called point-particle interchangeability*. It is really part of the definition of "particle".

In the case of density, point-particle interchangeability gives

$$\rho(\mathbf{x}, t) = \delta[\mathbf{A}(\mathbf{x}, t), t]. \tag{3}$$

To see this, observe that, by the definition of δ, the expression $\delta[\mathbf{A}(\mathbf{x}, t), t]$ describes the density now associated with a particle that was initially located at the initial location of the particle now at $\mathbf{x}$. ("Now" means "at time t".) This is the same as saying that $\delta[\mathbf{A}(\mathbf{x}, t), t]$ is the density associated with a particle now at $\mathbf{x}$.

Substituting $\mathbf{x} = \mathbf{x}(\mathbf{A}, t)$ into (3) and using (2a), we find

$$\rho[\mathbf{x}(\mathbf{A}, t), t] = \delta(\mathbf{A}, t). \tag{4}$$

Equations (3) and (4) illustrate the transformation between the *spatial description* and *material description*, where the terms are defined as follows.

Spatial description: $\mathbf{x}, t$ independent variables.
Material description: $\mathbf{A}, t$ independent variables.

As an example, let us consider the relation between material and spatial descriptions of velocity. In analogy with elementary point mechanics, we define the velocity of a particle by the rate of change of the particle's position with time:

$$\mathbf{V}(\mathbf{A}, t) \equiv \partial \mathbf{x}(\mathbf{A}, t)/\partial t. \tag{5}$$

To preserve point-particle interchangeability, we must define $\mathbf{v}(\mathbf{x}, t)$, the velocity in spatial coordinates, by

$$\mathbf{v}(\mathbf{x}, t) = \mathbf{V}[\mathbf{A}(\mathbf{x}, t), t]. \tag{6a}$$

Equations (6a) and (2a) imply that

$$\mathbf{v}[\mathbf{x}(\mathbf{A}, t), t] = \mathbf{V}(\mathbf{A}, t). \tag{6b}$$

3. Material derivative. Recall that in spatial and material variables, respectively, we denote the density by $\rho(\mathbf{x}, t)$ and $\delta(\mathbf{A}, t)$ so that

$$\rho(\mathbf{x}, t) = \delta[\mathbf{A}(\mathbf{x}, t), t], \qquad \delta(\mathbf{A}, t) = \rho[\mathbf{x}(\mathbf{A}, t), t], \tag{1}$$

as in (2.3) and (2.4). When we compute a partial derivative of δ with respect to t, we regard the initial position $\mathbf{A}$ as fixed. On the other hand, when we compute such a partial derivative of $\rho(\mathbf{x}, t)$, the density described in spatial variables, we regard the spatial position $\mathbf{x}$ as fixed. Since $\mathbf{A}$ and $\mathbf{x}$ are related, we can relate the two derivatives by the chain rule

$$\frac{\partial \delta(\mathbf{A}, t)}{\partial t} = \frac{\partial \rho}{\partial x_1}\bigg|_{\mathbf{x}=\mathbf{x}(\mathbf{A},t)} \frac{\partial x_1}{\partial t} + \frac{\partial \rho}{\partial x_2}\bigg|_{\mathbf{x}=\mathbf{x}(\mathbf{A},t)} \frac{\partial x_2}{\partial t} + \frac{\partial \rho}{\partial x_3}\bigg|_{\mathbf{x}=\mathbf{x}(\mathbf{A},t)} \frac{\partial x_3}{\partial t} + \frac{\partial \rho}{\partial t}\bigg|_{\mathbf{x}=\mathbf{x}(\mathbf{A},t)} \tag{2}$$

As in (2.5), $\partial x_i / \partial t$ evaluated at $(\mathbf{A}, t)$ is simply $V_i(\mathbf{A}, t)$, the ith velocity component expressed in material variables, $i = 1, 2, 3$. Our goal is an expression for the left-hand side of (2) in spatial coordinates. To obtain this, we must make the substitution $\mathbf{A} = \mathbf{A}(\mathbf{x}, t)$, which yields

$$\frac{\partial \delta(\mathbf{A}, t)}{\partial t}\bigg|_{\mathbf{A}=\mathbf{A}(\mathbf{x},t)} = \sum_{i=1}^{3} v_i(\mathbf{x}, t) \frac{\partial \rho(\mathbf{x}, t)}{\partial x_i} + \frac{\partial \rho(\mathbf{x}, t)}{\partial t}, \tag{3}$$

where we have used the relation

$$V_i(\mathbf{A}, t)|_{\mathbf{A}=\mathbf{A}(\mathbf{x},t)} = v_i(\mathbf{x}, t). \tag{4}$$

The notation D/Dt is frequently employed for the *material* or *substantial derivative* which appears on the left-hand side of (3). With this we have

$$\frac{\partial \delta(\mathbf{A}, t)}{\partial t}\bigg|_{\mathbf{A}=\mathbf{A}(\mathbf{x},t)} \equiv \frac{D\rho(\mathbf{x}, t)}{Dt}, \tag{5}$$

$$\frac{D\rho(\mathbf{x}, t)}{Dt} = \frac{\partial \rho(\mathbf{x}, t)}{\partial t} + \mathbf{v}(\mathbf{x}, t) \cdot \nabla \rho(\mathbf{x}, t). \tag{6}$$

We have used the density throughout our discussion only for concreteness. For any quantity, the material derivative tells the time rate of change of that quantity for a fixed particle. Thus, to give an important example, if $\mathbf{v}(\mathbf{x}, t)$ denotes the velocity of the particle at position $\mathbf{x}$ and time t, then $D\mathbf{v}(\mathbf{x}, t)/Dt$ denotes the acceleration *of the particle* located at position $\mathbf{x}$ at time t. That is,

$$\frac{\partial \mathbf{V}(\mathbf{A}, t)}{\partial t}\bigg|_{\mathbf{A}=\mathbf{A}(\mathbf{x},t)} = \frac{D\mathbf{v}(\mathbf{x}, t)}{Dt}; \qquad \frac{D\mathbf{v}}{Dt} = \frac{\partial \mathbf{v}}{\partial t} + (\mathbf{v} \cdot \nabla)\mathbf{v}. \tag{7}$$

The ith component of $D\mathbf{v}/Dt$ is given by the formula

$$\frac{Dv_i}{Dt} = \frac{\partial v_i}{\partial t} + \sum_{j=i}^{3} v_j \frac{\partial v_i}{\partial x_j}.$$

A formula such as (7) can be readily obtained by more physical arguments (which can easily be made rigorous). The acceleration of the particle at $(\mathbf{x}, t)$ is

$$\lim_{\Delta t \to 0} \frac{\mathbf{v}(\mathbf{x} + \Delta\mathbf{x}, t + \Delta t) - \mathbf{v}(\mathbf{x}, t)}{\Delta t} \tag{8}$$

where $\Delta\mathbf{x} \approx \mathbf{v}(\mathbf{x}, t)\Delta t$. But, to lowest order, the numerator of (8) is given by $\nabla \mathbf{v} \cdot \Delta\mathbf{x} + \mathbf{v}_t \Delta t = (\nabla \mathbf{v} \cdot \mathbf{v} + \mathbf{v}_t)\Delta t$, from which the result follows at once. The term $\mathbf{v}_t$ is called the *local acceleration*, in contrast to the *convective acceleration* $\mathbf{v} \cdot \nabla \mathbf{v}$. Count Rumford coined "convection" by combining Latin words for "with" and "wind". It should be clear that the convective contribution to the material derivative indeed arises because particles are carried along with the "wind", so that modifications in a particle property may be present even in a steady flow $(\partial/\partial t \equiv 0)$ owing to the changing spatial location of a given particle.

4. The Jacobian and its material derivative. Having introduced via $\mathbf{x} = \mathbf{x}(\mathbf{A}, t)$ the material description of a moving continuum, we can *make precise the notion that a specified portion of material occupies the region $R(t)$ at time t.* We specify the portion of material by giving its location at some instant of time, say $t = 0$. Denote by $R(0)$ the region occupied at $t = 0$ by the designated portion of material. Let the subsequent motion of any point $\mathbf{x}$ be described by $\mathbf{x} = \mathbf{x}(\mathbf{A}, t)$. Then the region $R(t)$ occupied by the designated material at time t is the image of $R(0)$ at time t under the mapping $\mathbf{x} = \mathbf{x}(\mathbf{A}, t)$.

Let $\mathcal{V}(0)$ denote the volume of $R(t)$. Then

$$\mathcal{V}(t) = \iiint_{R(0)} dA_1\, dA_2\, dA_3 \quad \text{and} \quad \mathcal{V}(t) = \iiint_{R(t)} dx_1\, dx_2\, dx_3. \tag{1a, b}$$

In (1b) we can change the variables of integration to A_1, A_2, and A_3, where $\mathbf{A} = \mathbf{A}(\mathbf{x}, t)$. With this,

$$\mathcal{V}(t) = \iiint_{R(0)} J(A_1, A_2, A_3, t)\, dA_1\, dA_2\, dA_3, \tag{2}$$

where $J(A_1, A_2, A_3, t) \equiv \partial(x_1, x_2, x_3)/\partial(A_1, A_2, A_3)$ is the Jacobian. No absolute value sign is needed on J, since the components of $\mathbf{x}$ and $\mathbf{A}$ are both determined in right-handed coordinate systems.

From (2) it follows that if $V(0)$ is the volume of a very small region centered about $\mathbf{A}$ then $V(t)/V(0) \approx J$. [See **LS** for a more precise formulation of this result.] One can say that the Jacobian at point $\mathbf{A}$ and time t represents the *dilatation* of an infinitesimal volume initially at $\mathbf{A}$, where the dilatation is the ratio of volume occupied by the infinitesimal material region at time t to its initial volume. The importance of the Jacobian is confirmed by our ability to describe it, as we just have, in a manner that does not depend on the use of a particular coordinate system.

In §7 we show that

$$\frac{\partial \mathbf{J}(A_1, A_2, A_3, t)}{\partial t} = \left(\frac{\partial v_1}{\partial x_1} + \frac{\partial v_2}{\partial x_2} + \frac{\partial v_3}{\partial x_3}\right)\Bigg|_{\mathbf{x}=\mathbf{x}(\mathbf{A},t)} J(A_1, A_2, A_3, t). \quad (3)$$

If we make the substitution $\mathbf{A} = \mathbf{A}(\mathbf{x}, t)$ [i.e., if we write (3) in spatial variables] and use the definition of $\nabla \cdot \mathbf{v}$, we have the *Euler expansion formula*,

$$\frac{DJ[\mathbf{A}(\mathbf{x}, t), t]}{Dt} = (\nabla \cdot \mathbf{v})J[\mathbf{A}(\mathbf{x}, t), t]. \quad (4)$$

Equations (3) and (4) look simpler when written as follows:

In material variables: $\partial J/\partial t = \left[(\nabla \cdot \mathbf{v})|_{\mathbf{x}=\mathbf{x}(\mathbf{A},t)}\right] J.$ (5)

In spatial variables: $DJ/Dt = (\nabla \cdot \mathbf{v})J.$ (6)

We emphasize that the derivatives required by the divergence $\nabla \cdot$ are with respect to spatial variables.

5. Conservation of mass. §4 gave a precise characterization of the idea that a specified portion of material occupies the region $R(t)$ at time t. *That the mass of a portion of material does not vary as time increases is one way to state what is meant by conservation of mass.* Thus we hypothesize that

$$\frac{d}{dt}\iiint_{R(t)} \rho(x_1, x_2, x_3, t)\, dx_1\, dx_2\, dx_3 = 0, \qquad R(t) \text{ arbitrary}, \quad (1)$$

where $\rho(\mathbf{x}, t)$ denotes the density at point $\mathbf{x}$, time t.

Since (1) holds over an arbitrary material region $R(t)$, we shall be able to transform it into the more tractable form of a differential equation by means of the *Dubois-Reymond lemma*, which follows.

DUBOIS-REYMOND LEMMA. *Suppose that*

$$\iiint_R f(\mathbf{x})\, dx_1\, dx_2\, dx_3 = 0 \quad (2)$$

for every region R contained in a domain D. If $f(\mathbf{x})$ is continuous for $\mathbf{x}$ in D, then $f(\mathbf{x}) \equiv 0$ for $\mathbf{x}$ in D.

To take the derivative inside the integral in (1), as required for (2), we shall transform to initial coordinates $\mathbf{A}$ via the material description $\mathbf{x} = \mathbf{x}(\mathbf{A}, t)$. The region of integration will thereby no longer depend on time. Thus, denoting the integral in (1) by $I(t)$, we have

$$\begin{aligned}\frac{dI}{dt} &\equiv \frac{d}{dt}\iiint_{R(t)} \rho(\mathbf{x}, t)\, dx_1\, dx_2\, dx_3 \\ &= \frac{d}{dt}\iiint_{R(0)} \rho[\mathbf{x}(\mathbf{A}, t), t]J(\mathbf{A}, t)\, dA_1\, dA_2\, dA_3.\end{aligned} \quad (3)$$

We can write (3) more simply as

$$\frac{dI}{dt} = \frac{d}{dt}\iiint_{R(0)} \delta(\mathbf{A}, t)J(\mathbf{A}, t)\, dA_1 dA_2 dA_3$$

by employing the material density $\delta(\mathbf{A}, t) \equiv \rho[\mathbf{x}(\mathbf{A}, t), t]$ of (3.1). We now take the time differential inside the integral and use partial derivative notation to indicate that $\mathbf{A}$ is held constant during the differentiations with respect to t:

$$\frac{dI}{dt} = \iiint_{R(0)} \left[\frac{\partial \delta(\mathbf{A}, t)}{\partial t} J(\mathbf{A}, t) + \delta(\mathbf{A}, t) \frac{\partial J(\mathbf{A}, t)}{\partial t} \right] dA_1\, dA_2\, dA_3. \tag{4}$$

Using the Euler expansion formula (4.5), we obtain

$$\frac{dI}{dt} = \iiint_{R(0)} \left\{ \frac{\partial \delta(\mathbf{A}, t)}{\partial t} + \delta(\mathbf{A}, t)(\nabla \cdot \mathbf{v})\big|_{\mathbf{x}=\mathbf{x}(\mathbf{A},t)} \right\} J(\mathbf{A}, t)\, dA_1\, dA_2\, dA_3.$$

Making the variable change $\mathbf{A} = \mathbf{A}(\mathbf{x}, t)$ and using (3.5), (3.1), and the relation between $\mathbf{x}$ and $\mathbf{A}$ given in (2.2b), we then obtain

$$\frac{dI}{dt} = \iiint_{R(t)} \left\{ \frac{D\rho}{Dt}(\mathbf{x}, t) + \rho(\mathbf{x}, t)\left[\nabla \cdot \mathbf{v}(\mathbf{x}, t) \right] \right\} dx_1\, dx_2\, dx_3. \tag{5}$$

But $dI/dt = 0$, by (1). From (5) we thereby conclude that

$$D\rho/Dt + \rho \nabla \cdot \mathbf{v} = 0, \tag{6}$$

where we have assumed that the integrand in (5) is continuous and therefore have used the Dubois-Reymond lemma. Equation (6), often called the *continuity equation*, is the desired differential equation form of *mass conservation*. Since $D\rho/Dt = \partial\rho/\partial t + \mathbf{v} \cdot \nabla \rho$, an alternative form of (6) is

$$\partial\rho/\partial t + \nabla \cdot (\rho \mathbf{v}) = 0. \tag{7}$$

A material form of mass conservation can be obtained starting from

$$\frac{d}{dt} \iiint_{R(t)} \rho(\mathbf{x}, t)\, d\tau = 0,$$

where $R(t)$ is a material region and $d\tau$ denotes a volume element. The above equation implies

$$\iiint_{R(t)} \rho(\mathbf{x}, t)\, d\tau = \text{constant} = \iiint_{R(0)} \delta(\mathbf{A}, 0)\, dA_1\, dA_2\, dA_3.$$

Thus, changing variables from $\mathbf{x}$ to $\mathbf{A}$ in the first integral, we obtain

$$\iiint_{R(0)} \left[\delta(\mathbf{A}, t) J(\mathbf{A}, t) - \delta(\mathbf{A}, 0) \right] dA_1\, dA_2\, dA_3 = 0.$$

If the integrand is continuous, we may conclude that

$$\delta(\mathbf{A}, t) J(\mathbf{A}, t) = \delta(\mathbf{A}, 0). \tag{8}$$

We have seen how the time derivative of an integral over the moving material region $R(t)$ can be calculated by the device of shifting from $R(t)$ to $R(0)$ and back. It is appropriate here to present two useful formulas that essentially record the results of using this device in commonly encountered contexts.

For a sufficiently smooth function F and material region $R(t)$, assuming conservation of mass, we have

$$\frac{d}{dt}\iiint_{R(t)} F\rho\, d\tau = \iiint_{R(t)} \frac{DF}{Dt}\rho\, d\tau. \tag{9}$$

To see this, note that the left side of (9) equals

$$\frac{d}{dt}\iiint_{R(0)} F[\mathbf{x}(\mathbf{A}, t)]\delta(\mathbf{A}, t)J(\mathbf{A}, t)\, dA_1\, dA_2\, dA_3$$

$$= \iiint_{R(0)} \frac{\partial F}{\partial t}[x(\mathbf{A}, t)]\delta(\mathbf{A}, t)J(\mathbf{A}, t)\, dA_1\, dA_2\, dA_3,$$

where we have used (8).

A result closely related to (9) is the following. For any sufficiently smooth function G and material region $R(t)$,

$$\frac{d}{dt}\iiint_{R(t)} G\, d\tau = \iiint_{R(t)} \frac{\partial G}{\partial t}\, d\tau + \iint_{\partial R(t)} G\mathbf{v}\cdot\mathbf{n}\, d\sigma. \tag{10}$$

Here $\mathbf{n}$ is the unit exterior normal to R, $d\tau$ is a volume element, $d\sigma$ is a surface element, and ∂R denotes the boundary of R. Both (9) and (10) are valid if F and G are replaced by vector fields $\mathbf{F}$ and $\mathbf{G}$. Equation (10) is called the *Reynolds transport theorem*.

To afford insight into the nontrivial matter of deriving basic continuum equations, **LS** provide four different derivations of the mass conservation equation in §14.1 of their book, with several more variants in various exercises. We provide one more derivation here.

We start with the statement that conservation of mass requires that *for a region fixed in space the rate of increase of the mass contained in this region must equal the net mass flux into the region.* Only flow of mass across the boundary can cause a change in the mass of the material contained in a fixed region, since mass is neither created nor destroyed. In mathematical terms, then, the italicized statement is

$$\frac{d}{dt}\iiint_{R} \rho(\mathbf{x}, t)\, d\tau = -\iint_{\partial R} \mathbf{n}(\mathbf{x}, t)\cdot\mathbf{v}(\mathbf{x}, t)\rho(\mathbf{x}, t)\, d\sigma. \tag{11}$$

Here R is an arbitrary region, fixed in space; ∂R is its boundary; and $\mathbf{n}$ is a unit exterior normal to ∂R. Using the divergence theorem, we can write (11) as

$$\iiint_{R} \left\{\frac{\partial\rho(\mathbf{x}, t)}{\partial t} + \nabla\cdot[\mathbf{v}(\mathbf{x}, t)\rho(\mathbf{x}, t)]\right\} d\tau = 0. \tag{12}$$

Since R is arbitrary, if the integrand is continuous then (7) follows by the Dubois-Reymond lemma.

The basic mass conservation equation has been extended so that it applies to a number of areas that are seemingly rather far removed from the original concept of a continuum. One example is provided by stellar dynamics where one studies the phase space distribution function $\Psi(x, y, z, u, v, w)$. By definition, $\Psi(x, y, z, u, v, w)dxdydzdudvdw$ gives the number of stars whose three spatial coordinates lie in the ranges $(x, x + dx)$, $(y, y + dy)$, $(z, z + dz)$ and whose three velocity components lie in the ranges $(u, u + du)$, $(v, v + dv)$, $(w, w + dw)$. By extending traditional ideas of mass conservation, one can derive the governing equation

$$\frac{\partial \Psi}{\partial t} + u\frac{\partial \Psi}{\partial x} + v\frac{\partial \Psi}{\partial y} + w\frac{\partial \Psi}{\partial z} + \frac{du}{dt}\frac{\partial \Psi}{\partial u} + \frac{dv}{dt}\frac{\partial \Psi}{\partial v} + \frac{dw}{dt}\frac{\partial \Psi}{\partial w} = 0. \quad (13)$$

We leave the derivation of (13) to an exercise provided at the end of this section. This exercise serves as a sample of a type of multi-part problem that the author has found useful in applied mathematics courses. One important technical feature of the exercise should be noted; if a student cannot do one part, he can take the answer for granted and proceed. Thus a small area of ignorance will not bring about a disaster. More important, homework assignments or examinations with such exercises test the instructor's success in achieving the major goal of the course, instilling understanding of the general applied mathematical approach. By contrast, examinations that emphasize manipulations promote the (hopefully false) notion that the instructor's main goal was to impart a sequence of lightly related "tricks" and methods.

Returning to stellar dynamics, the acceleration components in (13) remain unspecified. But these are proportional to the force, which in turn can be regarded as given by the gradient of the gravitational potential

$$V(x, y, z, t) = -G\iiint \frac{\rho(\xi, \eta, \zeta, t)\, d\xi\, d\eta\, d\zeta}{\left((x-\xi)^2 + (y-\eta)^2 + (z-\zeta)^2\right)^{1/2}}. \quad (14)$$

Here G is the universal gravitational constant, and ρ is the spatially averaged distribution of stellar matter.

In the last decade there has been a great deal of activity in stellar dynamics, particularly with regard to the complex and fascinating problem of understanding the spiral structure of galaxies. For a survey, see Bok (1972).

Mathematical biology is another area where a generalized mass balance equation has recently been used to good advantage. To give one example, Levin and Payne (1974) have proposed that natural ecological communities can profitably be characterized as a mosaic of interacting "islands", "holes", or "patches", where these are connected subsystems that differ from a relatively homogeneous reference background. Independent variables are taken to be time t, patch age a, and patch size ξ. Dependent variables are the patch density function n (precisely analogous to the mass density function Ψ), mean patch growth rate g (the average value of $d\xi/dt$), and patch extinction rate μ. At this point the reader should find no trouble in accepting the mass balance law

$$\frac{\partial n}{\partial t} + \frac{\partial n}{\partial a} + \frac{\partial}{\partial \xi}(gn) = -\mu n. \tag{15}$$

This equation will have a unique solution once one specifies the initial patch distribution $n(0, a, \xi)$ and the age-size specific birth rate of patches $n(t, 0, \xi)$. In what promises to be a series of significant papers, Levin and Payne (1974) provide some solutions of (15) using the method of characteristics. They describe observations and experiments that will be used to test the usefulness of their theory in a situation on the outer coast of Washington State where "patches are generated within stands of the competitively dominant mussels by the shearing force of waves, wave-driven logs, or perhaps even spontaneous decay of aged mussels".

Another biological example of the use of a continuum mass balance condition is provided by studies of cell population growth. In this connection, Rubinow (1973) discusses a special case of (15), namely

$$\frac{\partial n}{\partial t} + \frac{\partial}{\partial \xi}(gn) = -\mu n, \qquad \xi_0 < \xi \leqslant \xi_1.$$

Here $n = n(t, \xi)$ is the number of cells at time t characterized by "maturity" ξ, where ξ could for example be identified with cell volume or amount of DNA per cell. Now $g(\xi)$ has the interpretation of maturation velocity. Auxiliary conditions differ from those appropriate for the Levin-Payne model: The initial (maturity) distribution $n(0, \xi)$ is prescribed as before, but cell division at maturity is reflected in the boundary condition $n(t, \xi_0)g(\xi_0) = 2n(t, \xi_1)g(\xi_1)$. Rubinow (1973) compares predictions of maturity-time theory and age-time theory ($g \equiv 1$) with data on a growing population of the one-celled organism *Tetrahymena geleii*.

Applications of a generalized mass-balance condition to ecology are extensively discussed by Oster in this volume.

EXERCISE. The object of this exercise is to indicate how (13) is derived.

For simplicity we shall restrict consideration to a function that depends on a single spatial coordinate x, the corresponding velocity component u, and the time t. Thus $\Psi(x, u, t)$ is a smooth function with the property that to sufficient accuracy $\int_{u_1}^{u_2}\int_{x_1}^{x_2}\Psi(x, u, t)\,dx\,du$ gives the mass of material which is located between x_1 and x_2, and which has a velocity between u_1 and u_2.

(a) Consider the "box" in (x, u) space pictured in Figure 1. Explain fully what is meant by "the mass of the fluid occupying the box". That is, what can one say about the position and velocity of such fluid? This is an easy question.

(b) Write a double integral giving the mass of fluid in the box at time t.

(c) Show that the time rate of change of (b) is $\Psi_t(x^*, u^*, t)\Delta x \Delta u$, where $x_0 \leqslant x^* \leqslant x_0 + \Delta x$, $u_0 \leqslant u^* \leqslant u_0 + \Delta u$. Remember that u is an independent variable.

(d) What is meant by "the net amount of fluid mass passing inward through the side labeled S_1 in the time interval $(t, t + \Delta t)$"?

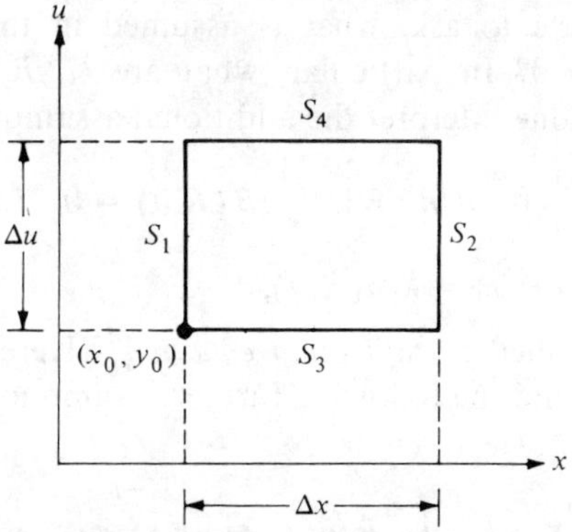

FIGURE 1. Imaginary "box" used to derive an equation for the distribution function $\Psi(x, u, t)$.

(e) Explain why (d) is given by

$$\int_{u_0}^{u_0+\Delta u}\left[\int_{x_0-u\Delta t}^{x_0}\Psi(x, u, t)\,dx\right]du.$$

(f) From (e) show that the rate of mass flow inward through side S_1 at time t is

$$\int_{u_0}^{u_0+\Delta u} u\Psi(x_0, u, t)\,du.$$

(g) Derive a similar expression valid for side S_2.
(h) Combine (f) and (g) to form an expression proportional to $\Delta x \Delta u$.
(i) Same as (d), only substitute S_3 for S_1.
(j) Explain why (i) is given by

$$\int_{x_0}^{x_0+\Delta x}\left[\int_{u_0-\int_t^{t+\Delta t}a(x,s)\,ds}^{u_0}\Psi(x, u, t)\,du\right]dx,$$

where $a(x, t)$ is the acceleration in the direction of x increasing. As only gravitational forces need be considered, the acceleration (force per unit mass) is independent of u.

(k) Use the methods of (f)–(h) to find an expression for the mass flux inward through S_3 and S_4.

(l) Since collisions between stars can be ignored, the final answer is obtained by equating (c) with (h) + (k) and taking the appropriate limit.

Additional note. Experience indicates that particularly successful examination questions can be derived from recent journal articles, for this gives students a gratifying feeling of learning up-to-date and relevant material. In this spirit, a suitable question could well be based on the paper of Cohen and Keener (1975). In the brief time allotted for examinations one cannot expect students to *derive* equations for new situations, but one can expect them to *interpret* them. Thus one might point out that Cohen and Keener deal with the formation of particles in a homogeneous supersaturated solution. Let $n(R, t)$ be the number of (assumed spherical) particles of radius R, at time t.

It would be legitimate to ask, what is assumed in the equation $\partial n/\partial t + G\partial n/\partial R - B + D = 0$? In particular, what are G, B, and D expected to denote? How should one interpret the additional assumptions

$$G = k_1 \Big/ \left[\int_0^\infty R^2 n \, dR \right], \qquad B(R, t) = 0 \quad \text{for } R \neq 0,$$
$$D(R, t) = k_2 n(R, t),$$

that lead to the so-called *cystallization equation*? (Here k_1 and k_2 are constants.) What lies behind the following further assumptions, that complete the formulation?

$$n(0, t) = k_3 \left[\int_0^\infty R^2 n(R, t) \, dR \right]^{-p}, \qquad p > 0;$$
$$n(R, 0) = F(R), \qquad F \text{ given.}$$

6. Balance of linear and angular momentum. In particle mechanics, a particle P that has mass m and moves with velocity $\mathbf{v}$ is said to possess (linear) momentum $m\mathbf{v}$. According to Newton's second law, the rate of change of the linear momentum of P is equal to the net force acting on P. To generalize this law to continuum mechanics, we define the linear momentum at time t possessed by material of density $\rho(\mathbf{x}, t)$ that occupies a region $R(t)$ as

$$\iiint_{R(t)} \rho(\mathbf{x}, t)\mathbf{v}(\mathbf{x}, t) \, d\tau, \tag{1}$$

where $\mathbf{v}$ is the velocity vector. Again, the time rate of change of the integral in (1) is hypothesized to be equal to the net forces acting on the material in R.

What kind of forces are expected? The force of gravity is the most familiar example of *body forces*, which in particle mechanics are proportional to the mass of a point particle. In continuum mechanics it is natural to pass from points to regions by means of an integral. We thus consider body forces on the material in R that are of the form

$$\iiint_{R(t)} \rho(\mathbf{x}, t)\mathbf{f}(\mathbf{x}, t) \, d\tau, \tag{2}$$

where $\mathbf{f}(\mathbf{x}, t)$ is the body force vector per unit mass. We also consider *surface forces*. These have the form

$$\iint_{\partial R(t)} \mathbf{t}(\mathbf{x}, t, \mathbf{n}) \, d\sigma, \tag{3}$$

where $\mathbf{x}$ is the position vector of the surface element, t is the time, and $\mathbf{n}$ is the unit exterior normal to ∂R at $\mathbf{x}$. The dependent variable $\mathbf{t}$ is called the *stress vector*.

In writing (3), we have implicitly assumed that force per unit area (*stress*) approaches a limit as area approaches zero. To see this, let S_k ($k = 1, 2, 3, \ldots$) denote a sequence of smooth similar surfaces of decreasing area, each a portion of ∂R and each containing the point $\mathbf{x}^{(0)}$. Let $\mathbf{n}^{(0)}$ be the unit exterior normal to ∂R at $\mathbf{x}^{(0)}$. Then if A_k is the area of S_k, the ith component of the force per unit area on S_k is

$$\frac{1}{A_k}\iint\limits_{S_k} t_i(\mathbf{x}, t, \mathbf{n})\, d\sigma = t_i(\tilde{\mathbf{x}}^{(k)}, t, \tilde{\mathbf{n}}^{(k)}), \qquad i = 1, 2, 3,$$

where we have used the integral mean value theorem. Here $\tilde{\mathbf{n}}^{(k)}$ is the unit exterior normal at $\tilde{\mathbf{x}}^{(k)}$, a point of S_k. If t_i is a continuous function of its variables,

$$\lim_{A_k\to 0;\mathbf{x}^{(0)} \text{ in } S_k} \frac{1}{A_k}\iint\limits_{S_k} t_i(\mathbf{x}, t, \mathbf{n})\, d\sigma = t_i(\mathbf{x}^{(0)}, t, \mathbf{n}^{(0)}), \tag{4}$$

since $\tilde{\mathbf{x}}$ must approach $\mathbf{x}^{(0)}$ as $A_k \to 0$. [Note that $\mathbf{n}$ is determined at every point $\mathbf{x}$ *on a given surface* so that in this situation $\mathbf{t}(\mathbf{x}, t, \mathbf{n})$ is actually a function only of $\mathbf{x}$ and t.]

Why ought one to assume the existence of a stress? Historically, this assumption emerged as a generalization of what were originally special assumptions concerning special materials. In stretching a metal bar, for example, it has been assumed since Hooke that the relative elongation is proportional to the applied force per unit area. The randomly colliding elastic-spheres model of a perfect gas leads to the introduction of a normal force per unit area (pressure), as the reader doubtless recalls from his introductory physics course.

The definitions of $\mathbf{f}$ and $\mathbf{t}$ are given fully by the assumptions that forces of the form (2) and (3) act on R. But it may be useful to provide more verbal descriptions of these important quantities.

Body force. At time t, a force per unit mass $\mathbf{f}(\mathbf{x}, t)$ is assumed to act at each point $\mathbf{x}$ of R.

Surface stress. At time t, consider a point $\mathbf{x}$ on the boundary ∂R of region R. Let $\mathbf{n}$ be the unit exterior normal to R at $\mathbf{x}$. Then $\mathbf{t}(\mathbf{x}, t, \mathbf{n})$ is the force per unit area exerted at the point $\mathbf{x}$ of ∂R *by* the material exterior to R *on* the material interior to R.

Combining our speculations thus far, we put forth the hypothesis that the rate of change of linear momentum in a material region $R(t)$ equals the contribution of the body forces plus the contribution of the surface forces:

$$\frac{d}{dt}\iiint\limits_{R(t)} \rho\mathbf{v}\, d\tau = \iiint\limits_{R(t)} \rho\mathbf{f}\, d\tau + \iint\limits_{\partial R(t)} \mathbf{t}\, d\sigma. \tag{5}$$

By the vector version of (5.9) we can take the time derivative in (5) inside the integral:

$$\frac{d}{dt}\iiint\limits_{R(t)} \rho\mathbf{v}\, d\tau = \iiint\limits_{R(t)} \rho\left(\frac{D\mathbf{v}}{Dt}\right) d\tau. \tag{6}$$

With this, (5) can be rewritten as

$$\iiint_{R(t)} \rho\left(\frac{D\mathbf{v}}{Dt}\right) d\tau - \iiint_{R(t)} \rho\mathbf{f}\, d\tau = \iint_{\partial R(t)} \mathbf{t}\, d\sigma. \tag{7}$$

Important consequences can be derived from (7) even without a clear idea of the forces involved. To do this, consider a family of similar regions R_L characterized by the length L. (An example is provided by the family of rectangular parallelepipeds with sides of lengths aL, bL, and cL.) Such regions have the same shape and their volume $\mathcal{V}_L$ and area $\mathcal{S}_L$ satisfy

$$\mathcal{V}_L = \lambda_R L^3, \qquad \mathcal{S}_L = \mu_R L^2,$$

for some constants λ_R and μ_R that depend only on the shape of the regions. Hence, for any continuous function G the integral mean value theorem implies that

$$\iiint_{R_L} G(\mathbf{x})\, d\tau = \lambda_R L^3 G(\tilde{\mathbf{x}}), \qquad \tilde{\mathbf{x}} \text{ in } R_L. \tag{8}$$

Let us apply (8) to the right side of (7), component by component. Since all volume integrals are proportional to L^3, by dividing by L^2 and then letting $L \to 0$, we find that

$$\lim_{L\to 0} \frac{1}{L^2} \iint_{\partial R_L(t)} \mathbf{t}\, d\sigma = \mathbf{0}. \tag{9}$$

Equation (9) is an important one. Insight into its meaning can be acquired by recognizing that if the material in the region R_L were subject to surface stresses and to no other forces, and if this material were in equilibrium, then necessarily the total force on the material would have to vanish. That is, $\iint_{\partial R_L} \mathbf{t}\, d\sigma = \mathbf{0}$. Actually, each region R_L is moving and is subject to body forces. But as $L \to 0$, the effect of the surface stress term becomes more and more dominant compared to the volume integrals $\iiint_{R_L} \rho D\mathbf{v}/Dt\, d\tau$ and $\iiint_{R_L} \rho\mathbf{f}\, d\tau$, which represent the effects of motion (inertia) and body forces. In the limit one obtains (9), which states, roughly speaking, that the surface force on an infinitesimal region, divided by its area $\mu_R L^2$, vanishes–just *as if* the region were in equilibrium. Equation (9) is thus known as the *principle of local stress equilibrium*.

We shall derive consequences of the principle of local stress equilibrium by applying it to various regions. We first consider a rectangular parallelepiped F_L, containing a fixed point $\mathbf{x}$ on one corner and having dimensions εL by L by L, $\varepsilon \ll 1$. Let $\mathbf{n}$, $-\mathbf{n}$, $\mathbf{n}_T$, $-\mathbf{n}_T$, $\mathbf{n}_R$, and $-\mathbf{n}_R$ be the unit exterior normals to the front, rear, top, bottom, right, and left faces of this "flake", respectively. (See Figure 2.) By the integral mean value theorem, the ith component of the surface integral in (9) can be written as

$$L^2\left[\, t_i(\mathbf{x}_F, t, \mathbf{n}) + t_i(\mathbf{x}_R, t, -\mathbf{n}) + \varepsilon Q_i\right], \qquad i = 1, 2, 3, \tag{10}$$

where

$$Q_i = t_i(\mathbf{x}_T, t, \mathbf{n}_T) + t_i(\mathbf{x}_B, t, -\mathbf{n}_T) + t_i(\mathbf{x}_{Ri}, t, \mathbf{n}_R) + \mathbf{t}_i(\mathbf{x}_L, t, -\mathbf{n}_R).$$

Here $\mathbf{x}_F$, $\mathbf{x}_R$, $\mathbf{x}_T$, $\mathbf{x}_B$, $\mathbf{x}_{Ri}$, and $\mathbf{x}_L$ are points in the front, rear, top, bottom, right, and left faces of the flake, respectively. Equation (10) comes from multiplying t_i at some point on each face by the area of that face. The area is L^2 for the front and rear faces, εL^2 for the remaining faces.

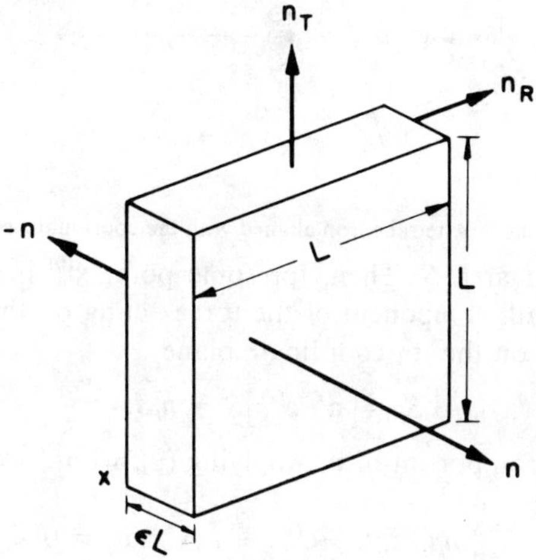

FIGURE 2. The "flake" F_L.

Combining (9) and (10), we obtain

$$t_i(\mathbf{x}, t, \mathbf{n}) + t_i(\mathbf{x}, t, -\mathbf{n}) + \varepsilon Q_i = 0, \qquad i = 1, 2, 3.$$

The first argument in each t_i is $\mathbf{x}$, since $\mathbf{x}_F$, $\mathbf{x}_B$, etc., all approach $\mathbf{x}$ as $L \to 0$, and since *we assume that* $\mathbf{t}$ *is continuous* in a domain containing $\mathbf{x}$. Using the continuity and hence the boundedness of Q_i, we observe that εQ_i can be made arbitrarily small so that

$$t_i(\mathbf{x}, t, \mathbf{n}) + t_i(\mathbf{x}, t, -\mathbf{n}) = 0, \qquad i = 1, 2, 3, \tag{11}$$

or, suppressing the dependence on $\mathbf{x}$ and t,

$$\mathbf{t}(-\mathbf{n}) = -\mathbf{t}(\mathbf{n}). \tag{12}$$

We have thus *deduced* the counterpart of Newton's third (action-reaction) law for continua.

We now show that $\mathbf{t}(\mathbf{n})$ can be expressed as a linear combination of $\mathbf{t}(\mathbf{e}^{(1)})$, $\mathbf{t}(\mathbf{e}^{(2)})$, and $\mathbf{t}(\mathbf{e}^{(3)})$. To do this, we apply (9) to the tetrahedron of Figure 3. The mutually perpendicular faces of area S_i are normal to the orthonormal basis vectors $\mathbf{e}^{(i)}$. These faces meet at $\mathbf{x}$. The "slanting" face has area S. We define L by $L \equiv S^{1/2}$.

The jth component of the force acting on the face of area S_i is, by the integral mean value theorem,

$$S_i t_j(\mathbf{x}^{(i)}, t, -\mathbf{e}^{(i)}), \qquad i, j = 1, 2, 3,$$

where $\mathbf{x}^{(i)}$ is a point in this face. Let us denote by $\mathbf{n}$ the unit exterior normal to

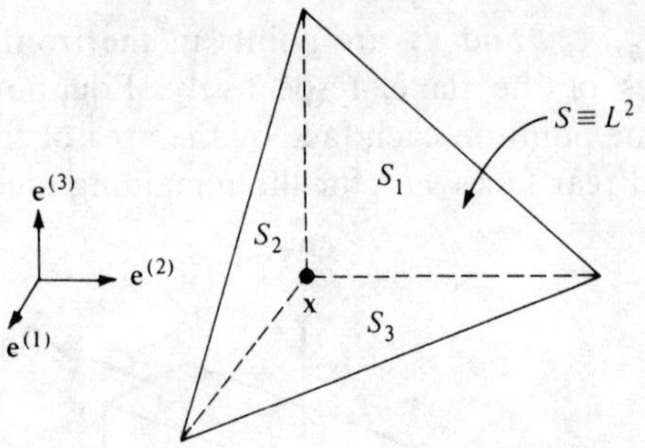

FIGURE 3. A tetrahedron aligned with the coordinate axes.

the slanting face of area S. Then, for some point $\mathbf{x}^{(0)}$ in the slanting face, $St_j(\mathbf{x}^{(0)}, t, \mathbf{n})$ is the jth component of the force acting on this face. Since S_i is the projection of S on the ith coordinate plane,

$$S_i = [\mathbf{n} \cdot \mathbf{e}^{(i)}]S = n_i S, \tag{13}$$

where n_i is the ith component of $\mathbf{n}$. Applying (9), with $L^2 \equiv S$, we obtain

$$\sum_{i=1}^{3} n_i t_j(\mathbf{x}, t, -\mathbf{e}^{(i)}) + t_j(\mathbf{x}, t, \mathbf{n}) = 0. \tag{14}$$

We have used the fact that all points $\mathbf{x}^{(q)}$ selected by the mean value theorem must approach $\mathbf{x}$, $q = 0, 1, 2, 3$. Using (12), we can rewrite (14) as

$$t_j(\mathbf{x}, t, \mathbf{n}) = \sum_{i=1}^{3} n_i t_j(\mathbf{x}, t, \mathbf{e}^{(i)}), \qquad j = 1, 2, 3. \tag{15}$$

We define T_{ij} by

$$T_{ij}(\mathbf{x}, t) \equiv t_j(\mathbf{x}, t, \mathbf{e}^{(i)}), \qquad i, j = 1, 2, 3. \tag{16}$$

Now we rewrite (15) finally as

$$t_j(\mathbf{x}, t, \mathbf{n}) = \sum_{i=1}^{3} n_i T_{ij}(\mathbf{x}, t), \qquad j = 1, 2, 3. \tag{17}$$

Using what can be regarded as a generalization of dot product notation, we sometimes write (17), in "direct notation", as

$$\mathbf{t}(\mathbf{x}, t, \mathbf{n}) = \mathbf{n} \cdot \boldsymbol{T}(\mathbf{x}, t).$$

The *very important relation* (17) shows that the dependence of the stress vector $\mathbf{t}$ on the exterior unit normal vector $\mathbf{n}$ is necessarily of a special linear form. This dependence is a consequence of the principle of local stress equilibrium. The nine quantities T_{ij}, $i, j = 1, 2, 3$, are called *components of the stress tensor* $\boldsymbol{T}$. The significance of the term *tensor* will be discussed in detail below. It will, however, be worth remembering that the *first* component of T_{ij} refers to the unit vector that is normal to the *face* on which the *stress*, denoted by the *second* component of T_{ij}, is acting. [Refer to definition (16).] A good mnemonic is

F$^{\text{irst}}_{\text{ace}}$ S$^{\text{econd}}_{\text{tress}}$

We now observe that the jth component of the surface term in the momentum balance equation (7) can be written

$$\iint_{\partial R(t)} t_j \, d\sigma = \iint_{\partial R(t)} \sum_{i=1}^{3} n_i T_{ij} \, d\sigma = \iiint_{R(t)} \sum_{i=1}^{3} \frac{\partial}{\partial x_i} (T_{ij}) \, d\tau, \tag{18}$$

where in the last equation we have used the divergence theorem. With this (7) can be written in component form as

$$\iiint_{R(t)} \left[\rho \frac{Dv_j}{Dt} - \rho f_j - \sum_{i=1}^{3} \frac{\partial}{\partial x_i} (T_{ij}) \right] d\tau = 0, \qquad j = 1, 2, 3.$$

Since $R(t)$ is arbitrary, if we assume that the integrand is continuous we can write

$$\rho \frac{Dv_j}{Dt} = \rho f_j + \sum_{i=1}^{3} \frac{\partial}{\partial x_i} (T_{ij}), \qquad j = 1, 2, 3. \tag{19}$$

In direct notation, (19) is written

$$\rho \frac{D\mathbf{v}}{Dt} = \rho \mathbf{f} + \nabla \cdot \boldsymbol{T}. \tag{20}$$

Relations (19) or (20) are *Cauchy's differential equations expressing for any continuum the balance of linear momentum.* The only hypotheses used in obtaining these equations were sufficient smoothness of the functions involved and mass conservation [employed to obtain (6)].

Let us now turn to the question of angular momentum. If a mass point at $\mathbf{x}$ possesses momentum $\mathbf{P}$, its *moment of momentum* or *angular momentum* about $\mathbf{0}$ is defined to be $\mathbf{x} \wedge \mathbf{P}$. Suppose that a system of mass points is subject to mutual *central* forces (like gravity) which are directed along the lines joining the interacting particles. It is then a *theorem* of point mechanics that the rate of change of each particle's moment of momentum is equal to the sum of the torques acting on the particle.

In continuum mechanics we deal not only with centrally acting body forces like gravity, but also with contact forces that differ greatly from material to material. Now we must make the *assumption* that in any "nonmoving" or inertial coordinate system, the rate of change of moment of momentum is equal to the torques exerted by body forces and surface stresses. To formulate this assumption more precisely, we must generalize the concepts of moment of momentum and torque used in point mechanics, but this is straightforward. We thus postulate

$$\frac{d}{dt} \iiint_{R(t)} (\mathbf{x} \wedge \rho \mathbf{v}) \, d\tau = \iiint_{R(t)} (\mathbf{x} \wedge \rho \mathbf{f}) \, d\tau + \iint_{\partial R(t)} (\mathbf{x} \wedge \mathbf{t}) \, d\sigma. \tag{21}$$

From this it follows that

$$T_{ij}(\mathbf{x}, t) = T_{ji}(\mathbf{x}, t), \tag{22}$$

as will be domonstrated at the end of §9.

An energy balance relation and an entropy inequality are needed to complete the field equations, but we can omit these matters here.

The general field equations must be supplemented by constitutive equations, which describe the nature of a particular material. Development of constitutive equations is easier if the machinery of cartesian tensors is employed, and this is the path that we shall take, starting in the next section. The reader may prefer to skip to the first paragraphs of §10, which immediately open up a wide vista of inviscid fluid theory.

B. CONSTITUTIVE EQUATIONS AND CLASSICAL EXAMPLES

7. Vectors, summation convention, permutation symbol. We turn now to the development of vector and tensor formalisms that will prove useful in our further study of continuum mechanics.

Consider two sets of cartesian axes, unprimed and primed, with common origin 0. Denote unit vectors in these two systems by $\mathbf{e}^{(i)}$ and $\mathbf{e}^{(j)\prime}$ respectively. Let the components of a vector $\mathbf{x}$ be denoted by x_i and x_j', so that

$$\mathbf{x} = \sum_{i=1}^{3} x_i \mathbf{e}^{(i)} = \sum_{j=1}^{3} x_j' \mathbf{e}^{(j)\prime}. \tag{1}$$

We wish to find the relation between the two sets of components. To this end, we take the scalar product of (1) with $\mathbf{e}^{(k)}$. We obtain

$$\sum_{i=1}^{3} x_i (\mathbf{e}^{(i)} \cdot \mathbf{e}^{(k)}) = \sum_{j=1}^{3} x_j' (\mathbf{e}^{(j)\prime} \cdot \mathbf{e}^{(k)}).$$

Noting that $\mathbf{e}^{(i)} \cdot \mathbf{e}^{(k)} = \delta_{ik}$, where δ_{ik} is the Kronecker symbol, and making the definition

$$l_{kj} \equiv \mathbf{e}^{(k)} \cdot \mathbf{e}^{(j)\prime}, \tag{2}$$

we obtain

$$x_k = \sum_{j=1}^{3} l_{kj} x_j', \qquad k = 1, 2, 3. \tag{3}$$

Similarly

$$x_j' = \sum_{k=1}^{3} l_{kj} x_k, \qquad j = 1, 2, 3. \tag{4}$$

It is convenient to introduce various notational conventions. In a given equation, an index (subscript or superscript) is called a *dummy* if it is used in a summation. Other indices are called *free*. In (3) k is a free subscript and j is a dummy subscript.

Range convention. A free index takes on the values 1, 2, 3. The same free indices must appear in each term of an equation.

Summation convention. No index will be written more than twice in a single term. Suppose an index, say i, appears twice in a single term. Then the symbol $\Sigma_{i=1}^{3}$ will be understood to precede that term.

In terms of these summation conventions, the important transformation laws (3) and (4) can be written simply as

$$x_k = l_{kj}x'_j, \qquad x'_j = l_{kj}x_k. \tag{5a,b}$$

Note that according to its definition (2), l_{kj} gives the jth component of the vector $\mathbf{e}^{(k)}$ in the primed system. This leads to the following result

$$\mathbf{e}^{(i)} \cdot \mathbf{e}^{(j)} = l_{ip}\mathbf{e}^{(p)'}l_{jq}\mathbf{e}^{(q)'} = l_{ip}l_{jq}\delta_{pq} = l_{ip}l_{jp}.$$

That is

$$\delta_{ij} = l_{ip}l_{jp}. \tag{6}$$

Similarly

$$\delta_{ij} = l_{pi}l_{pj}. \tag{7}$$

Note that $l_{ji} \neq l_{ij}$, in general. To remember definition (2) memorize "*the second subscript goes with the prime*". The same phrase makes it easy to remember the transformation laws (5a) and (5b), and to avoid errors in future formulas.

DEFINITION. The *permutation symbol* or *alternator* ε_{ijk}, $i, j, k = 1, 2, 3$, has 27 values, one for each of the 3^3 different sets of subscripts. These are defined as follows:

$$\begin{aligned} \varepsilon_{ijk} &= 0 \quad \text{if two of the integers } i, j, k \text{ are equal,} \\ &\qquad \text{or if all three are equal;} \\ \varepsilon_{ijk} &= 1 \quad \text{if } (ijk) \text{ is an even permutation of } (123); \\ \varepsilon_{ijk} &= -1 \quad \text{if } (ijk) \text{ is an odd permutation of } (123). \end{aligned} \tag{8}$$

DEFINITION. The determinant $|A|$ of a three-by-three matrix $A \equiv (A_{ij})$ is given by

$$|A| = \varepsilon_{ijk}A_{1i}A_{2j}A_{3k}. \tag{9a}$$

From this definition there follow various well-known properties of determinants, for example the fact that the determinant of the transpose of a matrix is equal to the determinant of the original matrix, i.e.

$$|A| = \varepsilon_{ijk}A_{i1}A_{j2}A_{k3}. \tag{9b}$$

In addition, we shall find it useful to record the following less well-known properties.

THEOREM.

$$\varepsilon_{rst}|A| = \varepsilon_{ijk}A_{ir}A_{js}A_{kt} = \varepsilon_{ijk}A_{ri}A_{sj}A_{tk}. \tag{10a,b}$$

PROOF. To prove (10a), we first note that unless r, s, and t are all different, the right side is the determinant of a matrix with at least two identical columns, and so equals zero. The left side is zero in this case by the definition of the alternating symbol. If $(rst) = (123)$, then (10a) is just (9b). If (rst) is an even (odd) permutation of (123) the right side is the determinant of a matrix which differs from A by an even (odd) number of row interchanges and so equals (minus) $|A|$. So does the left side. □

COROLLARY. $|AB| = |A|\,|B|$.

THEOREM.

$$\begin{vmatrix} A_{ip} & A_{iq} & A_{ir} \\ A_{jp} & A_{jq} & A_{jr} \\ A_{kp} & A_{kq} & A_{kr} \end{vmatrix} = \varepsilon_{ijk}\varepsilon_{pqr}|A|. \tag{11}$$

PROOF. Similar to that of the preceding theorem. □

THEOREM (THE "ED" RULE).

$$\varepsilon_{ijk}\varepsilon_{pqk} = \delta_{ip}\delta_{jq} - \delta_{iq}\delta_{jp}. \tag{12}$$

PROOF. In (11), let $A_{ij} = \delta_{ij}$ so $|A| = 1$, put $r = k$, and evaluate the resulting determinant. □

EXAMPLE. Let position $\mathbf{x}$ depend on initial position $\mathbf{A}$ at time t. We shall find a compact formula for the time-derivative of the Jacobian

$$J \equiv \partial(x_1, x_2, x_3)/\partial(A_1, A_2, A_3).$$

We begin by observing that

$$\dot{J} = \varepsilon_{ijk}\frac{\partial \dot{x}_1}{\partial A_i}\frac{\partial x_2}{\partial A_j}\frac{\partial x_3}{\partial A_k} + \ldots, \tag{13}$$

where we omit two further terms whose form is obvious by symmetry. Introducing the velocity $V_i \equiv \dot{x}_i$ we have, by the chain rule,

$$\dot{J} = \varepsilon_{ijk}\frac{\partial v_1}{\partial x_p}\bigg|\frac{\partial x_p}{\partial A_i}\frac{\partial x_2}{\partial A_j}\frac{\partial x_3}{\partial A_k} + \ldots = \frac{\partial v_1}{\partial x_p}\bigg|\delta_{1p}J + \ldots.$$

The vertical line indicates that the substitution $\mathbf{x} = \mathbf{x}(\mathbf{A}, t)$ must be made after the partial differentiation. Thus

$$\dot{J} = \left(\frac{\partial v_i}{\partial x_i}\right)\bigg|J \equiv (\nabla\cdot\mathbf{v})|J. \tag{14}$$

THEOREM. *Let $|L|$ denote the determinant of the transformation matrix L (with components l_{ij}). Then*

$$\begin{aligned} |L| &= 1 \quad \textit{for proper rotations,} \\ |L| &= -1 \quad \textit{for improper rotations.} \end{aligned} \tag{15}$$

PROOF. It follows from (6) that $|L|^2 = 1$. But $|L|$ is a continuous function of the elements of the transformation matrix L. And $|L| = 1$ for the identity transformation (wherein $l_{ij} = \delta_{ij}$) by direct computation, and therefore for all proper rotations by continuity. Improper rotations have $|L| = -1$, for they can be obtained by continuous transformation after the particular reflection wherein

$$L = \begin{bmatrix} 1 & 0 & 0 \\ 0 & 1 & 0 \\ 0 & 0 & -1 \end{bmatrix} \quad \text{and} \quad |L| = -1. \quad \square$$

8. Consistency–a motivation for tensors. We attempt here to provide motivation for the formal definition of a tensor, which will be given in the next section.

Versions of the same scientific law in different coordinate systems must be consistent. As an example of what we mean by this, consider Newton's second law. In a certain cartesian coordinate system let F_i and a_i denote the components of the force acting on a particle and the components of its acceleration respectively. Let F_i' and a_i' denote corresponding quantities in another cartesian coordinate system. In the first system, Newton's second law can be written

$$F_i = ma_i. \tag{1}$$

In the second system the law must have exactly the same form, namely

$$F_i' = ma_i'. \tag{2}$$

One reason for this similarity can be found in the definitions of the components of vectors like $\mathbf{F}$:

The components F_i are the projections of $\mathbf{F}$ on the fixed mutually perpendicular unit vectors $\mathbf{e}^{(i)}$.

The components F_i' are the projections of $\mathbf{F}$ on the fixed mutually perpendicular unit vectors $\mathbf{e}^{(i)'}$.

These definitions show that components in the unprimed and primed systems are conceptually identical and only notationally different. If (1) is true, therefore, (2) must be true, because the correct notational change of adding primes to vector components has been made.

Let us check directly that (2) holds if and only if (1) holds. Since the primed and unprimed components of a vector are linked by (7.5b), (2) can be written

$$l_{pi}F_p = ml_{pi}a_p \quad \text{or} \quad l_{pi}(F_p - ma_p) = 0. \tag{3a,b}$$

Equations (3b) can be regarded as three homogeneous equations for the three unknowns $F_p - ma_p$. By (7.15) the determinant of the coefficients is either $+1$ or -1, so that (3b) has only the trivial solution $F_p - ma_p = 0$, which is the same as (1). By reversing the above reasoning, one can show that (1) implies (2).

We distinguish between the general statement of a law and its particular numerical version. The *general statement* can be expressed independently of any coordinate system [$\mathbf{F} = m\mathbf{a}$] while the *particular version* relates numbers obtained using a particular coordinate system [$F_i = ma_i$]. In general, different particular versions of a given law, that is relations between different sets of numbers, are obtained if different coordinate systems are used [$F_i = ma_i$, $F_i' = ma_i'$]. These different versions are found from the general statement of a law by applying the coordinate rules appropriate to different coordinate systems. *Consistency demands that the same result be obtained if a given version of the law is obtained directly from the general statement* [$F_p = ma_p$] *or is obtained indirectly by transformation of another version* [$F_p = ma_p$ follows from $F'_i = ma'_i$ and the appropriate transformation law (7.5b)].

In the previous paragraph, as an example of a general coordinate-independent statement of a physical law we cited $\mathbf{F} = m\mathbf{a}$. This example used vectors, entities that have a coordinate-free interpretation which involves length, direction, and addition by means of the parallelogram law. Combinations of vectors, sometimes involving derivative operators, also have coordinate-free interpretations and therefore can also be used in the general statement of a physical law. To mention the simplest example, the scalar product can be described in a coordinate-free manner as the product of the lengths of the two vectors involved times the cosine of the angle between them.

But we cannot restrict ourselves to combinations of vectors in formulating physical laws. More complicated entities called tensors are necessary. We turn now to a discussion that will motivate the detailed definition of a tensor.

Recall from (6.17) that the stress vector $\mathbf{t}(\mathbf{x}, t, \mathbf{n})$ is a special function of the unit normal vector $\mathbf{n}$ given (using summation notation) by

$$t_j(\mathbf{x}, t, \mathbf{n}) = n_i T_{ij}(\mathbf{x}, t). \tag{4}$$

Here T_{ij} is by definition the jth component of the stress vector acting on the surface element whose exterior normal points in the direction of the ith cartesian base vector $\mathbf{e}^{(i)}$. We shall examine (4) in the light of the consistency requirement. This will lead naturally to the concept of a second order tensor.

Consider a new cartesian coordinate system with mutually perpendicular unit base vectors $\mathbf{e}^{(i)\prime}$. Using the same reasoning which we used in proceeding from (1) to (2), we see that in the new system (4) must become

$$t'_j = n'_i T'_{ij}. \tag{5}$$

Here T'_{ij} is the jth component (the projection on $\mathbf{e}^{(j)\prime}$) of the stress vector acting on the surface element whose exterior normal points in the direction of $\mathbf{e}^{(i)\prime}$. We know the relations between the components of the vectors $\mathbf{t}$ and $\mathbf{n}$ in the two systems. Assuming that (4) and (5) are consistent, let us determine what the relations must be between the quantities T_{ij} and the quantities T'_{ij}.

Using appropriately modified versions of (7.5b) we substitute for t'_j and n'_i in (5):

$$l_{pj} t_p = l_{qi} n_q T'_{ij}. \tag{6}$$

We substitute for t_p by using p instead of j as a free subscript in (4):

$$l_{pj} n_i T_{ip} = l_{qi} n_q T'_{ij}. \tag{7}$$

Employing a frequently used trick, we multiply both sides of (7) by l_{rj}:

$$l_{rj} l_{pj} n_i T_{ip} = l_{rj} l_{qi} n_q T'_{ij}. \tag{8}$$

But by (7.6)

$$l_{rj} l_{pj} = \delta_{rp}, \tag{9}$$

with which (8) becomes

$$n_i T_{ir} = l_{rj} l_{qi} T'_{ij}. \tag{10}$$

We rewrite (10), changing the dummy subscript on the left and reordering the factors on the right:

$$n_i T_{ir} = l_{rj} l_{qi} n_q T'_{ij}. \tag{11}$$

Now (4) is true for any unit vector **n**. In particular, it is certainly true if **n** has the components (1, 0, 0). For this **n**, (11) implies that

$$T_{1r} = l_{1i} l_{rj} T'_{ij}. \tag{12}$$

Equation (12) plus similar ones corresponding to vectors **n** with components (0, 1, 0) and (0, 0, 1) can all be written compactly as

$$T_{qr} = l_{qi} l_{rj} T'_{ij}. \tag{13}$$

We remind the reader that because of the summation convention, (13) symbolizes nine equations, corresponding to $q, r = 1, 2, 3$.

We have established that *if* (4) *and* (5) *are consistent then* (13) *must hold*. But (13) can be verified directly. A straightforward calculation, essentially the reverse of the one in the previous paragraph, then demonstrates the consistency of (4) and (5).

It is important to note that our derivation of (13) did not use any special properties of the quantities involved in (4) and (5). We needed only these facts:

(a) **t** and **n** satisfy the vector transformation law (7.5);

(b) we can determine the nine quantities T_{ij} in any cartesian coordinate system;

(c) **n** is arbitrary.

It turns out that there are many physical laws of the form (4) whose constituents satisfy (a), (b), and (c) and which are consistent with respect to cartesian coordinate systems. Sets of nine quantities satisfying the transformation law (13) are therefore common, and consequently they have been given the special name *second order cartesian tensor*. (Hence the name *stress tensor* for $\boldsymbol{T}$.)

In essence, our discussion of (4) indicates that if a two-subscripted entity appears in a physical law then that entity must represent components of a second order tensor.

Comparing the transformation law for the 3^1 components of a vector (or *first order tensor*)

$$t_i = l_{ip} t'_p \tag{14}$$

with that for the 3^2 components of a second order tensor

$$T_{ij} = l_{ip} l_{jq} T'_{pq}, \tag{15}$$

there is no difficulty in generalizing to the 3^n components of an nth order tensor. This is no empty generalization, for higher order tensors are of frequent occurrence, as we shall see.

9. Introduction to cartesian tensors. This rather formal section presents a number of definitions and theorems concerning cartesian tensors. The adjective "cartesian" may be left out, but "tensor" and "coordinate system" always refer to cartesian tensors and to cartesian coordinate systems in ordinary three-dimensional (Euclidean) space. Straightforward proofs will be omitted.

DEFINITION. A *zeroth order tensor* or *scalar* s is such that (a) in any cartesian coordinate system there is a rule for associating s with a real number; (b) this real number is the same no matter what cartesian coordinate system is employed.

An nth *order tensor* $\boldsymbol{T}$, $n = 1, 2, 3, \ldots,$ is such that (a) in any cartesian coordinate system there is a rule for associating the tensor $\boldsymbol{T}$ with a unique ordered set of 3^n quantities $T_{i_1 \cdots i_n}$ (called the *components* of $\boldsymbol{T}$); (b) if $T_{i_1 \cdots i_n}$ and $T'_{j_1 \cdots j_n}$ are the components of $\boldsymbol{T}$ in two different cartesian coordinate systems then

$$T_{i_1 \cdots i_n} = l_{i_1 j_1} \cdots l_{i_n j_n} T'_{j_1 \cdots j_n}. \tag{1}$$

Note. For a second order tensor, (1) can be written $T_{pq} = l_{pm} l_{qn} T'_{mn}$.

Notation. By $[\boldsymbol{T}]_{ij} = T_{ij}$ we mean that in a certain cartesian coordinate system T_{ij} is the (ij)th component of $\boldsymbol{T}$.

EXAMPLE. Let $\boldsymbol{I}$ be an entity with nine components given by $[\boldsymbol{I}]_{ij} = \delta_{ij}$. Then $\boldsymbol{I}$ is a second order tensor (called the *unit tensor*), for, using (7.6),

$$l_{im} l_{jn} \delta_{mn} = l_{in} l_{jn} = \delta_{ij},$$

which is the required transformation law.

EXAMPLE. Let $\boldsymbol{E}$ be an entity with 27 components given by $[\boldsymbol{E}]_{ijk} = \varepsilon_{ijk}$. Then $\boldsymbol{E}$ is a third order tensor (called the *alternating tensor*) *providing* that transformations between coordinate systems are restricted to *proper rotations*. To see this we must demonstrate the transformation law

$$\varepsilon_{ijk} = l_{ip} l_{jq} l_{kr} \varepsilon_{pqr}.$$

By (7.10b) the right side of this equation is $\varepsilon_{ijk} L$ and the result follows since L, the determinant of the transformation matrix, is unity for proper rotations. [See (7.15).]

Note that if reflections were not prohibited there would be transformations for which $L = -1$, and $\boldsymbol{E}$ would not be a tensor. When we speak of "the tensor $\boldsymbol{E}$" and when we derive consequences from the tensorial character of $\boldsymbol{E}$, we shall be implicitly assuming that reflections are prohibited.

DEFINITION. Let $\boldsymbol{A}$ and $\boldsymbol{B}$ be second order tensors with components A_{ij} and B_{ij}. Let c be a scalar. The *sum*, *difference*, and *scalar multiple* are defined by

$$[\boldsymbol{A} + \boldsymbol{B}]_{ij} = A_{ij} + B_{ij}; \ [\boldsymbol{A} - \boldsymbol{B}]_{ij} = A_{ij} - B_{ij}; \ [c\boldsymbol{A}]_{ij} = cA_{ij}. \tag{2}$$

Similar definitions are employed for nth order tensors.

DEFINITION. The *zero tensor* $\mathbf{0}$ is an entity whose components in any cartesian coordinate system are all zero. Two tensors are *equal* if their difference is the zero tensor; that is

$$A = B \Leftrightarrow A - B = 0. \tag{3}$$

(It follows that to prove that tensors are equal one needs only show that they have the same components in a single coordinate system.)

THEOREM. *Tensors of the same order form a vector space under addition and scalar multiplication.*

DEFINITION. A *contraction* of an nth order tensor, $n \geqslant 2$, is formed by setting two of its indices equal and performing the resulting sum.

An example of how contraction leads to a new tensor whose order is two less than the original tensor is provided by the following theorem.

THEOREM. *Let $\boldsymbol{T}$ be a fourth order tensor. Define $\boldsymbol{U}$ by*

$$U_{in} \equiv [\boldsymbol{U}]_{in} = T_{ijjn}. \tag{4}$$

Then $\boldsymbol{U}$ is a second order tensor.

PROOF. The rule for obtaining the components of $\boldsymbol{U}$ is given by (4). Since $\boldsymbol{T}$ is a fourth order tensor $T_{ijmn} = l_{ip}l_{jq}l_{mr}l_{ns}T'_{pqrs}$, so, setting $j = m$,

$$T_{ijjn} = l_{ip}l_{jq}l_{jr}l_{ns}T'_{pqrs}.$$

But $l_{jq}l_{jr} = \delta_{qr}$ as in (7.7), so $T_{ijjn} = l_{ip}l_{ns}T'_{prrs}$. By (4), the above equation is equivalent to $U_{in} = l_{ip}l_{ns}U'_{ps}$, which is the required transformation law. □

DEFINITION. Let $\boldsymbol{A}$ be an nth order tensor and $\boldsymbol{B}$ be an mth order tensor. Then the *tensor product* $\boldsymbol{AB}$ is defined by

$$[\boldsymbol{AB}]_{i_1 \cdots i_{m+n}} = A_{i_1 \cdots i_n} B_{i_{n+1} \cdots i_{m+n}}. \tag{5}$$

THEOREM. *Let $\boldsymbol{W} = \boldsymbol{AB}$, where $\boldsymbol{A}$ is an nth order tensor and $\boldsymbol{B}$ is an mth order tensor. Then $\boldsymbol{W}$ is an $(n + m)$th order tensor.*

DEFINITION. Let $\boldsymbol{A}$ and $\boldsymbol{B}$ be second order tensors with components A_{ij} and B_{pq}. Then the *contraction product* $\boldsymbol{A} \cdot \boldsymbol{B}$ is defined by

$$[\boldsymbol{A} \cdot \boldsymbol{B}]_{iq} = A_{ij}B_{jq}.$$

The contraction product of tensors of arbitrary order is defined similarly, by summing over adjacent indices.

THEOREM. *Let $\boldsymbol{A}$ be an mth order tensor and $\boldsymbol{B}$ be an nth order tensor. Then $\boldsymbol{A} \cdot \boldsymbol{B}$ is a tensor of order $m + n - 2$.*

PROOF. $\boldsymbol{A} \cdot \boldsymbol{B}$ is a contraction of the $(m + n)$th order tensor $\boldsymbol{AB}$. □

THEOREM ("*quotient rule*"). *Suppose that in any cartesian coordinate system there is a rule associating a unique ordered set of nine quantities with $\boldsymbol{T}$. If, for an arbitrary vector* **a**, $\mathbf{a} \cdot \boldsymbol{T}$ *is a vector then $\boldsymbol{T}$ is a second order tensor.*

PROOF. Essentially given in §8.

The following is another quotient rule of a rather general character.

THEOREM. *Suppose that in any coordinate system there is a rule associating a unique set of* 3^n *quantities with* $\boldsymbol{T}$. *Also suppose that, for an arbitrary mth order tensor* $\boldsymbol{A}$, $\boldsymbol{TA}$ *is a tensor of order* $m + n$. *Then* $\boldsymbol{T}$ *is an nth order tensor.*

Special results for second order tensors.

DEFINITION. Let $\boldsymbol{A}$ be a second order tensor with components A_{ij}. Then the *transpose* of $\boldsymbol{A}$ is denoted by $\boldsymbol{A}^{\mathrm{Tr}}$ and is defined by $[\boldsymbol{A}^{\mathrm{Tr}}]_{ij} = A_{ji}$. If $\boldsymbol{A} = \boldsymbol{A}^{\mathrm{Tr}}$ then $\boldsymbol{A}$ is said to be *symmetric*. If $\boldsymbol{A} = -\boldsymbol{A}^{\mathrm{Tr}}$ then $\boldsymbol{A}$ is said to be *antisymmetric* (or *skew-symmetric*).

THEOREM. *Let* $\boldsymbol{T}$ *be a second order tensor. Then one can write, uniquely,*

$$\boldsymbol{T} = \boldsymbol{S} + \boldsymbol{A} \tag{7}$$

where $\boldsymbol{S}$ *is a symmetric order tensor and* $\boldsymbol{A}$ *is an antisymmetric second order tensor. In fact*

$$\boldsymbol{S} \equiv \tfrac{1}{2}(\boldsymbol{T} + \boldsymbol{T}^{\mathrm{Tr}}), \qquad \boldsymbol{A} = \tfrac{1}{2}(\boldsymbol{T} - \boldsymbol{T}^{\mathrm{Tr}}). \tag{8}$$

The following easily verified representation theorem for antisymmetric tensors will prove useful.

THEOREM. *Corresponding to any antisymmetric tensor* $\boldsymbol{A}$ *there exists a vector* $\boldsymbol{\omega}$, *given by*

$$\omega_i = \tfrac{1}{2}\varepsilon_{ipq}A_{pq}, \tag{9a}$$

such that

$$\boldsymbol{A} = \boldsymbol{E}\cdot\boldsymbol{\omega}, \qquad \textit{i.e.}, A_{pq} = \varepsilon_{pqr}\omega_r. \tag{9b}$$

Matrices are an aid in performing certain manipulations with the components of second order tensors. For example, in a given coordinate system $[\boldsymbol{A}]_{ij}$ can be written in the ith row and jth column of a 3×3 matrix. We shall denote this matrix by $(\boldsymbol{A})$. For second order tensors $\boldsymbol{A}$ and $\boldsymbol{B}$, $(\boldsymbol{A}\cdot\boldsymbol{B}) = (\boldsymbol{A})(\boldsymbol{B})$, i.e., the matrix of the contraction product of $\boldsymbol{A}$ and $\boldsymbol{B}$ is the product of the $\boldsymbol{A}$ and $\boldsymbol{B}$ matrices.

Matrix eigenvalue theory goes through for second order tensors, with only slight changes. Thus, the eigenvalues λ and right eigenvectors $\mathbf{v}$ of a tensor $\boldsymbol{T}$ satisfy $\boldsymbol{T}\cdot\mathbf{v} = \lambda\mathbf{v}(\mathbf{v} \neq \mathbf{0})$. A real symmetric tensor $\boldsymbol{T}$ has real eigenvalues and a set of mutually orthonormal eigenvectors. If these vectors are chosen as the basis of a cartesian coordinate system (*principal axes*), then the corresponding components of $\boldsymbol{T}$ form a diagonal matrix, with the eigenvalues along the diagonal.

Isotropic tensors. There are certain materials like water that appear to have no internal preferred direction. Their constitutive equations must be expressed in terms of special tensors whose components are always the same. We shall now study this class of "isotropic" tensors.

DEFINITION. Tensors, like $\boldsymbol{I}$ and $\boldsymbol{E}$, whose components are given by the same set of numbers in all coordinate systems are called *isotropic*. It follows immediately from the definition of zero order tensors that they are isotropic.

THEOREM. *There are no nontrivial isotropic tensors of order one.*

PROOF. Suppose $\mathbf{v}$ is an isotropic vector, and suppose that its components are v_i relative to a certain basis $\mathbf{e}^{(i)}$. Consider another set of basis vectors satisfying

$$\mathbf{e}^{(1)'} = \mathbf{e}^{(2)}, \qquad \mathbf{e}^{(2)'} = \mathbf{e}^{(3)}, \qquad \mathbf{e}^{(3)'} = \mathbf{e}^{(1)},$$

so that

$$v_1' = v_2, \qquad v_2' = v_3, \qquad v_3' = v_1. \tag{10}$$

Since $\mathbf{v}$ is isotropic

$$v_1' = v_1, \qquad v_2' = v_2, \qquad v_3' = v_3. \tag{11}$$

Combining (10) and (11) we find

$$v_1' = v_2 = v_2' = v_3 = v_3' = v_1. \tag{12}$$

Consider a third set of basis vectors $\mathbf{e}^{(i)''}$ satisfying

$$\mathbf{e}^{(1)''} = \mathbf{e}^{(2)}, \qquad \mathbf{e}^{(2)''} = -\mathbf{e}^{(1)}, \qquad \mathbf{e}^{(3)''} = \mathbf{e}^{(3)}.$$

By the accompanying transformation law $v_2'' = -v_1$, but by isotropy $v_2'' = v_2$, so $v_2 = -v_1$; this is consistent with (12) if and only if $v_1 = v_2 = v_3 = 0$. The only isotropic vector is thus the zero vector. □

THEOREM. *Any isotropic tensor of order* 2 *can be written* $\lambda \boldsymbol{I}$ *for some scalar* λ. *Any isotropic tensor of order* 3 *can be written* $\lambda \boldsymbol{E}$ *for some scalar* λ. (*If reflections are allowed* $\boldsymbol{E}$ *is not a tensor so the only isotropic tensor of order* 3 *is* $\boldsymbol{0}$.) *If* $\boldsymbol{T}$ *is an isotropic tensor of order* 4 *then*

$$\begin{aligned}[\boldsymbol{T}]_{pqrs} = \lambda\delta_{pq}\delta_{rs} &+ \mu(\delta_{pr}\delta_{qs} + \delta_{ps}\delta_{qr}) \\ &+ \kappa(\delta_{pr}\delta_{qs} - \delta_{ps}\delta_{qr})\end{aligned} \tag{13}$$

for some scalars λ, μ, *and* κ.

PROOF. The ideas of the previous proof suffice. □

The calculus of tensor functions. Until now we have considered *tensor constants*. This was a necessary preliminary to a discussion of *tensor functions* whose components are functions of one or more variables.

Note. When a tensor is a function of position $\mathbf{x}$ we sometimes speak of a *tensor field*.

The transformation laws for second order tensor fields are

$$T_{ij}[\mathbf{x}] = l_{im}l_{jn}T'_{mn}[\mathbf{x}'(\mathbf{x})] \tag{14a}$$

and

$$T'_{mn}[\mathbf{x}'] = l_{im}l_{jn}T_{ij}[\mathbf{x}(\mathbf{x}')] \tag{14b}$$

where $\mathbf{x} = \mathbf{x}(\mathbf{x}')$ and $\mathbf{x}' = \mathbf{x}'(\mathbf{x})$ denote the transformation laws between vector components $x_p = l_{pq}x_q'$, $x_q' = x_p l_{pq}$.

The concepts of limit, continuity, and derivative are defined for tensor functions in exact analogy with their definition for scalar functions. For example if we denote the partial derivative with respect to time by ∂_t then

$$\partial_t \boldsymbol{T}(\mathbf{x}, t) = \lim_{\Delta t \to 0} \left[\boldsymbol{T}(\mathbf{x}, t + \Delta t) - \boldsymbol{T}(\mathbf{x}, t) \right] / \Delta t.$$

Since the l_{ij} are constants, differentiation of the transformation law shows that the time derivative of a tensor (if it exists) is a tensor of the same order.

There are a number of notations connected with the process of differentiating a tensor field with respect to a position coordinate x_i. Two frequently used notations for $\partial T_{ij} / \partial x_p$ are $\partial_p T_{ij}$ and $T_{ij,p}$. Direct notation is achieved by means of the "del" operator ∇ defined by

$$[\nabla]_p \equiv \partial_p \equiv \partial / \partial x_p.$$

Thus, by definition, if $\boldsymbol{T}$ is a second order tensor, $[\nabla \boldsymbol{T}]_{pij} \equiv \partial_p T_{ij} \equiv T_{ij,p}$. In words, $\nabla \boldsymbol{T}$ is often called "the gradient of $\boldsymbol{T}$" and is sometimes written "grad $\boldsymbol{T}$". Divergence and curl are also defined by generalizations of their familiar definitions for vectors:

$$[\operatorname{div} \boldsymbol{T}]_j \equiv [\nabla \cdot \boldsymbol{T}]_j \equiv \partial_i T_{ij} = T_{ij,i},$$

$$[\operatorname{curl} \boldsymbol{T}]_{iq} \equiv [\nabla \wedge \boldsymbol{T}]_{iq} = \varepsilon_{ijp} \partial_j T_{pq} = \varepsilon_{ijp} T_{pq,j}.$$

THEOREM. *If* $\boldsymbol{T}$ *is a tensor field of order* m *then* $\nabla \boldsymbol{T}$ *is a tensor field of order* $m + 1$.

THEOREM. *If* $\boldsymbol{T}$ *is a tensor field of order* m *then* $\nabla \cdot \boldsymbol{T}$ *is a tensor field of order* $m - 1$.

THEOREM. *If* $\boldsymbol{T}$ *is a tensor field of order* m *then* $\nabla \wedge \boldsymbol{T}$ *is also a tensor field of order* m.

The subscript notation and the ed rule make it easy to derive useful identities. Here is an example, involving the convective acceleration term $(\mathbf{v} \cdot \nabla)\mathbf{v}$.

THEOREM.

$$\mathbf{v} \wedge (\nabla \wedge \mathbf{v}) = \tfrac{1}{2} \nabla (\mathbf{v} \cdot \mathbf{v}) - (\mathbf{v} \cdot \nabla)\mathbf{v}. \tag{15}$$

PROOF.

$$\begin{aligned} [\mathbf{v} \wedge (\nabla \wedge \mathbf{v})]_m &= \varepsilon_{mni} v_n \varepsilon_{ijk} \partial_j v_k = (\delta_{mj}\delta_{nk} - \delta_{mk}\delta_{jn}) v_n \partial_j v_k \\ &= v_n \partial_m v_n - v_n \partial_n v_m = \tfrac{1}{2} \partial_m (v_n v_n) - v_n \partial_n v_m \\ &= \left[\tfrac{1}{2} \nabla (\mathbf{v} \cdot \mathbf{v}) - (\mathbf{v} \cdot \nabla)\mathbf{v} \right]_m. \quad \square \end{aligned}$$

We have space here to give only one of the numerous applications to continuum mechanics of the formalism we have developed. We choose to show how the symmetry of the stress tensor can be deduced from the integral form of angular momentum balance (6.21):

$$\frac{d}{dt}\iiint\limits_{R(t)} (\mathbf{x} \wedge \rho\mathbf{v})\, d\tau = \iiint\limits_{R(t)} (\mathbf{x} \wedge \rho\mathbf{f})\, d\tau + \iint\limits_{\partial R(t)} (\mathbf{x} \wedge \mathbf{t})\, d\sigma. \tag{16}$$

Using (5.9) and the fact that $D\mathbf{x}/Dt = \mathbf{v}$, we see that the left-hand side of (16) can be written

$$\iiint\limits_{R(t)} \frac{D}{Dt}(\mathbf{x} \wedge \mathbf{v})\rho\, d\tau = \iiint\limits_{R(t)} \left(\mathbf{x} \wedge \frac{D\mathbf{v}}{Dt}\right)\rho\, d\tau. \tag{17}$$

We manipulate the second term on the right side of (16) by passing to subscript notation and using the relation (6.17) linking the stress vector with the stress tensor, plus the divergence theorem:

$$\left[\iint\limits_{\partial R(t)} (\mathbf{x} \wedge \mathbf{t})\, d\sigma\right]_i = \iint\limits_{\partial R(t)} \varepsilon_{ijk} x_j t_k\, d\sigma = \iint\limits_{\partial R(t)} \varepsilon_{ijk} x_j n_p T_{pk}\, d\sigma$$

$$= \iiint\limits_{R(t)} \varepsilon_{ijk}\partial_p (x_j T_{pk})\, d\sigma = \iiint\limits_{R(t)} \varepsilon_{ijk}(\delta_{pj} T_{pk} + x_j \partial_p T_{pk})\, d\sigma.$$

Consequently

$$\iiint\limits_{R(t)} \mathbf{x} \wedge \left[\rho \frac{D\mathbf{v}}{Dt} - \rho\mathbf{f} - \nabla \cdot \boldsymbol{T}\right] d\tau = \iiint\limits_{R(t)} \boldsymbol{T}_s\, dT \tag{18}$$

where $[\boldsymbol{T}_s]_i \equiv \varepsilon_{ijk} T_{jk}$. The left side of (18) vanishes by virtue of the momentum balance law (6.20). Since R is arbitrary, it must be that $\boldsymbol{T}_s = 0$. Thus

$$(T_{23} - T_{32}, T_{31} - T_{13}, T_{12} - T_{21}) = 0,$$

and the symmetry of $\boldsymbol{T}$ is now apparent. □

10. Equations of viscous flow and Hookean elasticity. In this section we shall formulate the basic constitutive equations for classical fluid and solid mechanics. We begin with physical observations that lead, rather simply, to equations that respectively characterize motionless and inviscid fluids.

We shall find it useful to distinguish between *shear stresses* that act on an element in a direction perpendicular to its normal **n** and *normal stresses* that act along **n**. It is noteworthy that shear stresses are particularly apparent in liquids like cold molasses, which we call "very viscous". As an illustration, think of how difficult it is to slice through cold molasses with a knife. The edge of the knife blade is so thin that this resistance must be due to shear stresses acting on the flat part of the blade. The faster you try to slice, the harder it is. If you turn the knife sideways and try to "spread" the molasses, the faster you want to make one layer of molasses flow over another, the harder you have to push. These observations could be summed up in the qualitative graph of Figure 4. Note that small relative speeds are associated with small shear stresses. One expects that the shear stresses will vanish for a motionless fluid, wherein relative velocities of adjacent material are always zero.

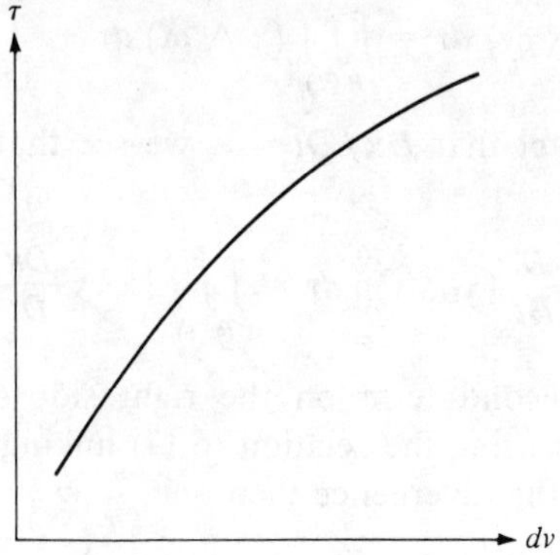

FIGURE 4. Qualitative plot of the shear stress τ versus dv, the relative speed of adjacent layers of fluid.

We are led to assume that *for a motionless fluid, stresses are purely normal.* That is, $\mathbf{t}$ must be in the direction $\mathbf{n}$, so

$$\mathbf{t}(\mathbf{x}, t, \mathbf{n}) = -p(\mathbf{x}, t, \mathbf{n})\mathbf{n} \tag{1}$$

for a scalar function p. At the moment we cannot rule out a dependence of p on $\mathbf{n}$. But from the relation (6.17) between the stress vector and the stress tensor, we find using (1) that

$$-p(\mathbf{x}, t, \mathbf{n})n_j = \sum_{i=1}^{3} n_i T_{ij}(\mathbf{x}, t),$$

which implies that

$$T_{ij}(\mathbf{x}, t) = -p(\mathbf{x}, t, \mathbf{n})\delta_{ij}. \tag{2}$$

Since the left side of (2) is independent of $\mathbf{n}$, it must be that $p(\mathbf{x}, t, \mathbf{n})$ is actually independent of $\mathbf{n}$, i.e., the pressure p on an element of area located at point $\mathbf{x}$ must be independent of the normal vector $\mathbf{n}$. The constitutive equation for a motionless fluid thus takes the final form

$$T_{ij}(\mathbf{x}, t) = -p(\mathbf{x}, t)\delta_{ij}. \tag{3}$$

Our discussion of slicing cold molasses associated large shear stresses with what are commonly called "fluids of high viscosity". On the other hand, it is easy to slice a knife through water or air. It therefore appears reasonable to assume, as a first approximation at any rate, that shear stresses are always zero in these important fluids. Thus we define a class of *inviscid fluids*, for which, even when they are in motion, the stress tensor satisfies the constitutive equation (3). The equation for balance of linear momentum (6.20) now becomes the *Euler momentum equation for inviscid flow*

$$\rho D\mathbf{v}/Dt = -\nabla p + \rho\mathbf{f}. \tag{4}$$

Often it is appropriate to assume that the fluid is of uniform density, in which case the mass conservation equation (5.6) becomes simply

$$\nabla \cdot \mathbf{v} = 0. \tag{5}$$

Equations (4) and (5) provide four equations for the three components of velocity and the pressure. Appropriate boundary conditions stem from the requirement that no fluid penetrate the boundaries.

The way is now open for the development of rich and useful theories, for example those concerned with surface waves on water, and internal waves in stratified fluids. Admission of compressibility effects often only requires the additional equation of state $p = A\rho^\gamma$ (where A and γ are constants), but a wealth of striking new phenomena become accessible to mathematical analysis. A large number of correspondences between theory and experiment leaves no doubt that the inviscid fluid flow equations provide an excellent basis for the solution of broad classes of fluid mechanics problems. That these equations are surprisingly deficient in some circumstances is shown by D'Alembert's paradox: uniform flow of an inviscid fluid past a bounded obstacle exerts no drag. Let us then turn to the formulation of constitutive equations that will include viscous effects, i.e., resistance to shear.

Viscous fluids. Since shear stresses depend on relative velocities, it will prove useful to look more closely at the relative velocities of two nearby points **x** and **y**. We regard **x** as a fixed reference point.

Omitting explicit indication of time dependence, we have

$$v_i(\mathbf{y}) - v_i(\mathbf{x}) = v_{i,j}(\mathbf{x})[y_j - x_j] + \cdots. \tag{6}$$

Here $\cdots$ denotes $c_{ij}(y_j - x_j)$, where $\lim_{\mathbf{y}\to\mathbf{x}} c_{ij} = 0$ (assuming that **v** has continuous second partial derivatives in a domain containing **x**). In the language of differentials, if $\mathbf{dx} \equiv \mathbf{y} - \mathbf{x}$ then

$$dv_i = v_{i,j}(\mathbf{x})dx_j, \qquad \text{i.e., } \mathbf{dv} = \mathbf{dx} \cdot \nabla\,\mathbf{v}. \tag{7}$$

We now split up $\nabla\,\mathbf{v}$ into its symmetric and antisymmetric parts, which we denote by $\boldsymbol{D}$ and $\boldsymbol{W}$. Thus

$$\nabla\,\mathbf{v} = \boldsymbol{D} + \boldsymbol{W} \text{ where } \boldsymbol{D} = \boldsymbol{D}^T \text{ and } \boldsymbol{W} = -\boldsymbol{W}^T. \tag{8}$$

By (9.8),

$$D_{ij} = \tfrac{1}{2}(v_{j,i} + v_{i,j}), \qquad W_{ij} = \tfrac{1}{2}(v_{j,i} - v_{i,j}). \tag{9a,b}$$

From (7) and (8),

$$\mathbf{dv} = \mathbf{dx} \cdot \boldsymbol{D} + \mathbf{dx} \cdot \boldsymbol{W} \tag{10}$$

or

$$\mathbf{v}(\mathbf{y}) = \mathbf{v}(\mathbf{x}) + (\mathbf{y} - \mathbf{x}) \cdot \boldsymbol{W}(\mathbf{x}) + (\mathbf{y} - \mathbf{x}) \cdot \boldsymbol{D}(\mathbf{x}) + \cdots. \tag{11}$$

Since higher order terms are neglected in (11) one often says that (11) describes the velocity field in an infinitesimal neighborhood of the fixed reference point **x**.

Because of the symmetry of $\boldsymbol{D}$ and the antisymmetry of $\boldsymbol{W}$, $\mathbf{v}(\mathbf{y})$ in (11) can be regarded as the superposition of a uniform translation with velocity $\mathbf{v}(\mathbf{x})$, a solid body rotation with angular velocity vector $\frac{1}{2}$ curl $\mathbf{v}(\mathbf{x})$, and pure elongations along the three mutually perpendicular principal axes of $\boldsymbol{D}(\mathbf{x})$ at a rate per unit length equal to the respective eigenvalues of $\boldsymbol{D}(\mathbf{x})$. (The name *deformation tensor* or *rate of strain tensor* is given to $\boldsymbol{D}$.) The role of $\boldsymbol{D}$ is easily seen by writing $(\mathbf{y}-\mathbf{x})\cdot\boldsymbol{D}$ in principal axes. The motion associated with $(\mathbf{y}-\mathbf{x})\cdot\boldsymbol{W}$ can be determined with the aid of the representation theorem for antisymmetric tensors given in §9. From (9.9b) we see that [with the abbreviation $\mathbf{r}\equiv\mathbf{y}-\mathbf{x}$]

$$(\mathbf{r}\cdot\boldsymbol{W})_q = r_p W_{pq} = r_p\varepsilon_{pqs}\omega_s = \varepsilon_{qsp}\omega_s r_p = (\omega\wedge\mathbf{r})_q,$$

the last expression being the well-known velocity field for solid body rotation with angular velocity ω. Furthermore, from (9.9a) and a little manipulation it follows that

$$\omega_i = \tfrac{1}{2}\varepsilon_{ipq}W_{pq} = \tfrac{1}{2}\varepsilon_{ipq}v_{q,p} = \tfrac{1}{2}(\nabla\wedge\mathbf{v})_i.$$

With the foregoing discussion as background, we turn to formulation of the simplest constitutive equation that takes into account the presence of viscous shear stresses. We shall assume that *the stress tensor* $\boldsymbol{T}$ *at a point depends only on the velocity gradient tensor at that point*. Due to a change in its make-up, the response of the fluid could be different at different locations in space and time; we rule out consideration of such cases by *assuming homogeneity in space and time*. We must bear in mind the fact that for a given distribution of stress, *the response of a fluid generally varies with its thermodynamic state*. (An example is the fact that molasses becomes less viscous as the temperature increases.) Moreover, *our constitutive equation must be such that it reduces to the hydrostatic relation* (3) when the fluid is motionless. Thus, we assume that

$$T_{ij} = -p\delta_{ij} + f_{ij}(v_{r,s}, \rho, \theta), \qquad f_{ij}(0,\rho,\theta) = 0, \tag{12}$$

where we have selected the density ρ and the temperature θ to represent the thermodynamic state of the fluid.

We now make the simplifying *assumption that the dependence of* $\boldsymbol{T}$ *on* $\boldsymbol{D}$ *is linear*, with which (12) becomes

$$T_{ij} = -p\delta_{ij} + C_{ijrs}v_{r,s}. \tag{13}$$

The coefficients C_{ijrs} will depend on ρ and θ. By the quotient rule, these coefficients are the components of a fourth order tensor.

Since T_{ij} is symmetric, it must be that

$$C_{ijrs} = C_{jirs}. \tag{14}$$

To see this, we interchange i and j in (13) and then use the symmetry of the stress tensor $\boldsymbol{T}$ and the Kronecker delta to obtain

$$T_{ij} = -p\delta_{ij} + C_{jirs}v_{r,s}. \tag{15}$$

Upon subtraction of (15) from (13) we find that

$$B_{rs}v_{r,s} = 0 \quad \text{where } B_{rs} = C_{ijrs} - C_{jirs}. \tag{16a,b}$$

Since (16a) holds for arbitrary $v_{r,s}$ we conclude that $B_{rs}=0$, i.e., that (14) holds.

As a final simplifying assumption, *we restrict our consideration to fluids that are isotropic*, by which we mean that there is no preferred direction in the fluid. Consequently the quantities C_{ijrs} must be the components of an *isotropic* tensor, so that [by (9.13)]

$$C_{ijrs} = \lambda\delta_{ij}\delta_{rs} + \mu(\delta_{ir}\delta_{js} + \delta_{is}\delta_{jr}) + \kappa(\delta_{ir}\delta_{js} - \delta_{is}\delta_{jr}) \tag{17}$$

for some scalars λ, μ, and κ. It is easy to see that the symmetry requirement (14) holds if and only if $\kappa = 0$. Now, using the decomposition of $v_{r,s}$ into its symmetric and antisymmetric parts, we can write

$$C_{ijrs}v_{r,s} = C_{ijrs}D_{rs} + C_{ijrs}W_{rs}. \tag{18}$$

But with $\kappa = 0$, (17) shows that C_{ijrs} is symmetric in its last two indices and we know that W_{rs} is antisymmetric–so the second term in the above equation is zero. Therefore, in its final form the constitutive equation shows that the stress tensor is determined by just the deformation portion $\boldsymbol{D}$ of the velocity gradient tensor. This final form is

$$T_{ij} = (-p + \lambda D_{rr})\delta_{ij} + 2\mu D_{ij}. \tag{19}$$

The *viscosity coefficients* λ and μ depend on the density ρ and temperature θ (or any other convenient pair of thermodynamic variables). In what follows we shall be concerned only with the special case in which density is uniform. Then the mass conservation equation (5.6) gives

$$\nabla \cdot \mathbf{v} = 0. \tag{20}$$

Since $D_{ii} = v_{i,i} = \nabla \cdot v = 0$, the constitutive equation (19) reduces to

$$T_{ij} = -p\delta_{ij} + 2\mu D_{ij} \quad \text{or} \quad \boldsymbol{T} = -p\boldsymbol{I} + 2\mu\boldsymbol{D}, \tag{21}$$

where μ is the sole remaining *viscosity coefficient*. Using (21) and dividing by ρ, we find that the momentum balance condition (6.20) gives the *Navier-Stokes equation*

$$D\mathbf{v}/Dt = \mathbf{f} - \rho^{-1}\nabla p + \nu\nabla^2\mathbf{v}. \tag{22}$$

Here $\nu = \mu/\rho$ is called the *kinematic viscosity*. This can be regarded as a constant, so that (20) and (22) comprise four equations for the four unknown quantities: the pressure and the three velocity components. (At a pressure of one atmosphere and at a temperature of 20°C, for air $\nu \approx 0.15\text{cm}^2/\text{sec}$ and for water $\nu \approx 0.01\ \text{cm}^2/\text{sec}$.)

Equation (22) differs from the inviscid Euler equation (4) due to the presence in the former of the term $\nu\nabla^2\mathbf{v}$. Addition of this term has the important mathematical consequence of increasing from one to two the highest order of spatial derivatives present. A new boundary condition is called for (beside the requirement that fluid does not penetrate the boundary), namely that a viscous fluid cannot slip along a solid boundary.

Elastic solids. The development of the counterpart of the Navier-Stokes equations in elasticity uses many of the ideas we have already discussed, so quick progress can be made. We begin by noting that the simplest constitutive equations of elasticity are in essence just a tensorial version of Hooke's law, that the stress required to stretch a bar is proportional to the amount of stretch. "Amount of stretch" is quantified by comparing the distance between points before and after deformation. In doing this it is convenient to work in terms of the *displacement vector* **U**, which is defined to be the difference between a particle's present position and its initial position. That is

$$U_i(\mathbf{A}, t) = x_i(\mathbf{A}, t) - \mathbf{A}. \tag{23}$$

Consider the position of nearby points after displacement. We have

$$x_i(\mathbf{B}, t) - x_i(\mathbf{A}, t) = B_i - A_i + U_{i,j}(\mathbf{A}, t)\cdot(B_j - A_j) + \dots . \tag{24}$$

Let us decompose the *displacement gradient tensor* $U_{i,j}$ into the sum of a symmetric tensor η_{ij} and an asymmetric tensor ω_{ij}:

$$\eta_{ij} = \tfrac{1}{2}(U_{i,j} + U_{j,i}), \qquad \omega_{ij} = \tfrac{1}{2}(U_{i,j} - U_{j,i}). \tag{25a,b}$$

Using exactly the same tools as in the fluid mechanics situation we can assert that an "infinitesimal" vector **B** − **A** is distorted by a stretching along the three mutually perpendicular principal axes of the *linearized material strian tensor* $\boldsymbol{\eta}(\mathbf{A}, t)$ and is also subject to a rigid rotation specified by the *linearized material rotation tensor* $\boldsymbol{\omega}(\mathbf{A}, t)$.

An important simplifying feature of a large class of problems in elasticity is that displacements are very small. This means that the distinction between material and spatial coordinates can be ignored to first approximation. For example, we can ignore the distinction between $\boldsymbol{\eta}$ and the linearized spatial strain tensor

$$\varepsilon_{i,j}(\mathbf{x}, t) \equiv \tfrac{1}{2}\left[u_{i,j}(\mathbf{x}, t) + u_{j,i}(\mathbf{x}, t)\right]. \tag{25c}$$

In fluid mechanics one assumes that the stress tensor T_{ij} is a linear function of the *velocity* gradient tensor $v_{r,s}$. Here we assume that T_{ij} is a linear function of the *displacement* gradient tensor $u_{r,s}$. There is no counterpart of the hydrostatic pressure that is present in the absence of fluid motion; in the absence of elastic distortion we assume that the stress is everywhere zero. Thus instead of (13) we postulate

$$T_{ij} = C_{ijrs}u_{r,s}. \tag{26}$$

Using the same arguments as in the fluids case, we find that an isotropy assumption now leads to the *generalized Hooke's law*

$$T_{ij} = 2\mu\varepsilon_{ij} + \lambda\varepsilon_{kk}\delta_{ij}. \tag{27}$$

Note that

$$T_{kk} = (2\mu + 3\lambda)\varepsilon_{kk}, \tag{28}$$

so that (27) can be inverted to give

$$\varepsilon_{ij} = \frac{T_{ij}}{2\mu} - \frac{\lambda T_{kk}\delta_{ij}}{2\mu(2\mu + 3\lambda)}. \tag{29}$$

Connection with the elementary version of Hooke's law can be established by considering the stretching by a stress N of a cylinder aligned along the x_1 axis. Here we expect that

$$T_{11} = N, \qquad T_{ij} = 0 \quad \text{otherwise.} \tag{30}$$

In the *tension test* under consideration, the ratio of axial stress N to axial strain ε_{11} is called *Young's modulus* E. The ratio of lateral contraction to longitudinal extension is designated by *Poisson's ratio* ν, i.e., $-\varepsilon_{22}/\varepsilon_{11} = -\varepsilon_{33}/\varepsilon_{11} = \nu$. One finds that

$$E = \frac{\mu(2\mu + 3\lambda)}{\lambda + \mu}, \qquad \nu = \frac{\lambda}{2(\lambda + \mu)},$$
$$\lambda = \frac{E\nu}{(1 + \nu)(1 - 2\nu)}, \qquad \mu = \frac{E}{2(1 + \nu)}. \tag{31}$$

For steel, $E \approx 3 \times 10^7$ pounds per square inch $\approx 2 \times 10^{12}$ dynes/cm^2 and $\nu \approx 0.3$.

Final formulation of the governing equations for linear elasticity can be achieved by combining the constitutive equation (27) with the appropriate field equations. It can be shown that the mass conservation equation need not be employed; it gives a seldom-used correction to the assumption that the density at a particular spatial point $\mathbf{x}$ is the (given) density $\rho(\mathbf{x})$ of the particle that was initially located at $\mathbf{x}$. This assumption is of course based on the fact that our considerations are limited to instances where displacements are small. Another welcome consequence of small displacements is that the acceleration of the particle at $\mathbf{x}$ is given simply by $\partial^2 \mathbf{u}(\mathbf{x}, t)/\partial t^2$. Thus the momentum equation becomes

$$\rho \partial^2 u_i/\partial t^2 = T_{ij,j} + \rho f_i, \tag{32}$$

where f_i is the body force per unit mass. Often boundary conditions involve only the displacement vector $\mathbf{u}$ in which case one combines (27), (25a) and (32) to obtain the *Navier equations*

$$\rho \partial^2 u_i/\partial t^2 = \mu u_{i,jj} + (\mu + \lambda) u_{j,ji} + \rho f_i \tag{33a}$$

or

$$\rho \partial^2 \mathbf{u}/\partial t^2 = \mu \nabla^2 \mathbf{u} + (\mu + \lambda) \nabla (\nabla \cdot \mathbf{u}) + \rho \mathbf{f}. \tag{33b}$$

There is a close analogy between the Navier equations above and the Navier-Stokes equations (22). In the former case the displacement $\mathbf{u}$ is prescribed on the boundary, in the later the velocity $\mathbf{v}$ is typically prescribed.

Another type of boundary condition is common in elasticity, wherein the stress is prescribed on the boundary. In such a situation one naturally seeks to formulate the problem entirely in terms of stresses. The momentum balance law (32) only provides three equations for the six independent components of the stress tensor. Additional conditions come from the requirement that if (27) is used to relate stress to strain and (25c) to relate strain to displacement, then the displacement must be single-valued. Are there three (single-valued) displacement functions u_1, u_2, and u_3 that satisfy the six distinct equations provided by (25c) where ε_{ij} is a given symmetric tensor? An equivalent question is, given a strain tensor ε_{ij} and an antisymmetric rotation tensor ω_{ij}, under what circumstances can we guarantee that there is a displacement vector u_i satisfying $u_{i,j} = \varepsilon_{ij} + \omega_{ij}$? The equivalence is not hard to demonstrate mathematically; physically the situation is clear, for a rigid rotation can be introduced at will without affecting the basic situation.

From (9.9b) we have the representation $\omega_{ij} = \varepsilon_{ijk}\Omega_k$, and a straightforward calculation using (9.9a) and (25b) shows that $\Omega_{k,k} = \frac{1}{2}\varepsilon_{ijk}\omega_{ij,k} = 0$.

Our problem now takes the following form: given the symmetric tensor ε_{ij} and the vector Ω_k, with $\Omega_{k,k} = 0$, find necessary and sufficient conditions for the existence of a vector u_i such that

$$u_{i,j} = \varepsilon_{ij} + \varepsilon_{ijk}\Omega_k. \tag{34}$$

For fixed i, the question is "can we write a given vector field as the gradient of a scalar field"? In this form the answer is well known: assuming that the derivatives are continuous and that the region in question is simply connected, a necessary and sufficient condition is that the curl vanishes. We therefore require

$$\varepsilon_{spj}(\varepsilon_{ij} + \varepsilon_{ijk}\Omega_k)_{,p} = 0,$$

or, using the "ed rule" (7.12), that $-\Omega_{s,i} = \varepsilon_{spj}\varepsilon_{ij,p}$.

The auxiliary quantity $\boldsymbol{\Omega}$ disappears if the vanishing curl condition is applied once again to each component of Ω_s, yielding the *compatibility conditions*

$$\varepsilon_{rmi}\varepsilon_{spj}\varepsilon_{ij,pm} = 0. \tag{35}$$

Using (7.11), we can write these equations in the form

$$\varepsilon_{rs,ii} + \varepsilon_{ii,rs} - \varepsilon_{si,ri} - \varepsilon_{ri,si} = 0. \tag{36}$$

Applying the strain-stress law (27) we find the stress compatibility conditions

$$\nabla^2 T_{rs} + T_{kk,rs} - T_{ri,si} - T_{si,ri} = \frac{\nu}{1+\nu}\left[\delta_{rs}\nabla^2 T_{kk} + T_{kk,rs}\right], \tag{37}$$

which can be used to supplement the momentum equation (32) when it is desirable to formulate the governing equations entirely in terms of the stresses.

11. Exact solutions to the Navier-Stokes equations. In this section we present three simple exact solutions of the Navier-Stokes equations. We shall briefly discuss the use of such solutions in determining the value of the viscosity μ.

Solution 1. *Plane Couette flow*. Consider viscous incompressible fluid contained between the infinite parallel plates $y = 0$ and $y = d$. Suppose that the bottom plate is stationary and that the top plate moves parallel to itself with uniform speed U. Body forces can be ignored (for they will not affect the velocity field and will only add a hydrostatic term to the pressure).

Since the plates are infinite it seems reasonable to look for a solution in which the only nonzero velocity component is u, the component parallel to the plates. It is also reasonable to guess that u is a function only of the vertical component y. With the assumptions $v = w = 0$ and $u = u(y)$, mass conservation is automatically satisfied. In the assumed absence of body force, the momentum equation (10.22) implies

$$-\rho^{-1}p_x + \nu u_{yy} = 0, \quad p_y = 0, \quad p_z = 0. \tag{1 a,b,c}$$

From (1b) and (1c) it follows that the pressure can depend only on x and t. Since u is assumed to depend only on y, we deduce from (1a) that p_x and u_{yy} must be constants. We consider a flow due entirely to the effects of tangential viscous stresses and therefore look for a solution with $p_x = 0$. Then $u_{yy} = 0$. From the adherence boundary conditions

$$u = 0 \text{ at } y = 0, \qquad u = U \text{ at } y = d,$$

we obtain the final exact solution

$$u = Uy/d, \qquad v = w = 0, \qquad p = \text{constant}.$$

The tangential stress exerted by the moving fluid on the lower plate is given by

$$T_{21} = 2\mu D_{21} = \mu(\partial v_1/\partial x_2 + \partial v_2/\partial x_1).$$

In our current notation, the above equation is written

$$T_{yx} = \mu(\partial u/\partial y + \partial v/\partial x) = \mu U/d. \tag{2}$$

The stress is directly proportional to U and inversely proportional to d. If we could move a plate parallel to another plate at constant speed, then we presumably could measure μ by finding the proportionality constant. The plates would have to be large enough so that end and side effects would be negligible.

Solution 2. *Plane Poiseuille flow*. To see how viscosity impedes flow forced by pressure gradients, let us consider horizontal flow between motionless parallel plates at $y = \pm d$ forced by a constant horizontal pressure gradient $p_x = -C$, C a constant. With the assumptions $u = u(y)$, $v = w = 0$, $p = -Cx + \text{constant}$, all equations are satisfied providing $C + \mu u_{yy} = 0$. Adherence requires $u = 0$ at $y = \pm d$, so

$$u = (C/2\mu)(d^2 - y^2). \tag{3}$$

Of chief physical interest here is the *mass flux*–the mass of fluid per unit time, per unit length in the z direction, that passes a given plane $x = \text{constant}$. This is given by

$$\int_{-d}^{d} \rho u \, dy = \frac{2}{3} C\nu^{-1}d^3. \tag{4}$$

The pressure can be measured, so that comparison of a measured mass flux with (4) could in principle provide an alternative way of measuring viscosity.

The solution just presented is called *plane Poiseuille flow*. A more useful solution is that for (ordinary) Poiseuille flow in which fluid flows along a circular pipe because of an axial pressure gradient. Instead of (4), mass efflux E depends on an imposed pressure gradient of magnitude D according to

$$E = (\pi/8)D\nu^{-1}R^4. \tag{5}$$

A practical method of measuring viscosity consists of determining the pressure gradient D and velocity flux E and solving for ν from (5).

Solution 3. *Rayleigh impulsive flow*. We turn to a problem that illustrates how viscous effects develop with time. Consider fluid in the domain $y > 0$ which for $t \leqslant 0$ is motionless and is bounded below by a motionless horizontal plate. Suppose that at time $t = 0$ the plate instantaneously begins to move, in its own plane, with speed U. Presumably, viscous effects will bring the entire body of fluid into motion. How does this happen?

We assume $u = u(y, t)$, $v = w = 0$. As above, this requires $p_y = p_z = 0$, p_x = constant. We take $p_x = 0$ so that the motion is due entirely to viscous effects. The governing equations then require

$$u_t = \nu u_{yy}, \qquad t > 0, \tag{6}$$

while, by assumption, $u \equiv 0$ for $t \leqslant 0$. The adherence boundary condition implies

$$\text{for } t > 0: \qquad u(0, t) = U. \tag{7}$$

A person with considerable insight might realize that in this simple problem the speed U plays only a limited role. All that the solutions will tell us is how u/U passes from its initial value of zero to its ultimate value of unity, at a given spatial position and for a fluid of a given viscosity. That is, it is expected that

$$u/U = f(y, t, \nu), \tag{8}$$

the point being that U does not appear on the right side of (8). Mathematics can be used to verify our physical insight. With the change of variable $u' = u/U$, (6) and (7) become

$$u'_t = \nu u'_{yy}, \quad u'(0, t) = 1 \qquad (t > 0). \tag{9a,b}$$

Since these equations do not involve U, u' cannot depend on U, from which (8) follows.

The ratio u/U is dimensionless. It must therefore depend only on a dimensionless combination of y, t, and ν. Since the dimensions of ν are length^{2}time^{-1}, y, t, and ν can only occur in the combination y^2t^{-1}/ν. More precisely, $u/U = g(y^2/\nu t)$ for some function g. It is convenient to write this last equation in the equivalent form:

$$u' = u/U = F(\eta) \quad \text{where } \eta = \tfrac{1}{2}\, y(\nu t)^{-1/2}, \tag{10}$$

for then the final answer will be expressible in terms of a certain tabulated function.

To test the conclusion (10) [obtained by physical reasoning, but also easily accessible to similarity methods], we substitute into the governing equation (9a). We obtain

$$F'' + 2\eta F' = 0. \tag{11}$$

For any fixed t, $y \to 0$ implies $\eta \to 0$ so (9b) implies

$$F(0) = 1. \tag{12}$$

Note that for fixed y, $\eta \to 0$ as $t \to \infty$ so (12) requires that at any fixed location $u/U \to 1$ as $t \to \infty$. This is a reasonable result.

Equation (11) is of second order, so there must be a boundary condition in addition to (12). We require at any fixed time that the velocity far from the plate approaches zero:

$$\lim_{y\to\infty} u(y, t) = 0, \qquad t > 0. \tag{13}$$

Trouble threatens, however, when further thought indicates that it is just as reasonable to require at any fixed position that the velocity approaches zero as $t\downarrow 0$:

$$\lim_{t\downarrow 0} u(y, t) = 0, \qquad y > 0. \tag{14}$$

It is reassuring that both (13) and (14) amount to the same condition on $F(\eta)$, namely

$$\lim_{\eta\to\infty} F(\eta) = 0. \tag{15}$$

We have chosen to arrive at our final problem in a manner that emphasizes the interaction between the physical background of a problem and its ultimate formulation. However arrived at, the problem posed by (11), (12) and (15) is not at all difficult. Its solution is $u/U = F(\eta)$ where

$$F(\eta) = 1 - \frac{2}{\pi^{1/2}} \int_0^\eta e^{-\xi^2}\, d\xi \equiv 1 - \operatorname{erf}(\eta). \tag{16}$$

Here $\operatorname{erf}(\eta)$ is the tabulated *error function*.

We can ascertain the main features of our solution without resorting to tables. Merely from the fact that the solution is a function of $\eta \equiv \frac{1}{2}y(\nu t)^{-1/2}$ we see that the place where $u/U =$ constant is at a certain value of $y(\nu t)^{-1/2}$. Thus the portion of fluid that moves with less than a given percent of its final velocity U is farther from the plate than a plane whose position is proportional to $(\nu t)^{1/2}$. From a table we ascertain that $F(\eta) = 0.005$ when $\eta = 2$, so that when $\eta > 2$ or $y > 4(\nu t)^{1/2}$ the fluid speed has not reached even half of one percent of its final value.

From the highly simplified situation typical of an exact solution we wish only to draw the most general type of conclusion. Such a conclusion in the present case is that the *purely diffusive cause of setting fluid into motion takes effect in a region whose thickness is of order* $(\nu t)^{1/2}$. This spreading at a rate proportional to $t^{1/2}$ is the single most important characteristic of diffusion.

12. Vorticity changes in viscous fluid motion. In the introduction to his *Lectures in Physics*, Feynman quotes Dirac's statement "I understand what an equation means if I have a way to figure out the characteristics of its solution without solving it." In this section we shall indicate how such an understanding of the Navier-Stokes equations can be achieved, to some extent at any rate, by means of the vorticity concept. First we shall show, in a somewhat informal manner,[2] how vorticity changes in a viscous fluid. Then we shall demonstrate that the form of the exact solutions found in the previous section can be readily understood in terms of vorticity.

Recall that if a fluid moves with velocity $\mathbf{v}$ the *vorticity* is defined by $\omega \equiv \text{curl } \mathbf{v}$. From the results stated below (10.11) we see that $\boldsymbol{\omega}$ may be interpreted as twice the angular velocity vector of the solid body rotation which, together with translation and elongation, makes up the local fluid motion.

In an inviscid fluid of uniform density, an equation for the evolution of vorticity can be obtained by taking the curl of the Navier-Stokes equations (10.22). Employing (9.15) one obtains

$$\partial\omega/\partial t + \mathbf{v}\cdot\nabla\,\omega = \omega\cdot\nabla\,\mathbf{v} + \nu\,\nabla^2\omega. \tag{1}$$

Let us first consider how vorticity changes in the absence of viscosity. With $\nu = 0$, (1) may be written

$$D\omega/Dt = \omega\cdot\nabla\,\mathbf{v}. \tag{2}$$

As discussed at the end of §3, the $\mathbf{v}\cdot\nabla\,\omega$ contribution to the left side of (2) represents a change in vorticity due to convection. As a start in understanding the $\omega\cdot\nabla\,\mathbf{v}$ term, consider two material points P and Q. Suppose that, at time t, P is located at $\mathbf{x}$ and Q at $\mathbf{x} + \varepsilon\mathbf{n}$, a distance ε along some unit vector $\mathbf{n}$. Neglecting $O(\Delta t)^2$ terms, we see that the original vector $\varepsilon\mathbf{n}$ that joins P to Q will have been transformed after time Δt into a new vector given by

$$\varepsilon\mathbf{n} + [\mathbf{v}(\mathbf{x} + \varepsilon\mathbf{n}) - \mathbf{v}(\mathbf{x})]\Delta t = \varepsilon\mathbf{n} + (\mathbf{n}\cdot\nabla\,\mathbf{v})\varepsilon\Delta t + O(\varepsilon^2). \tag{3}$$

Subtracting the original vector $\varepsilon\mathbf{n}$ and dividing by Δt, we conclude that to lowest order in ε the quantity $\varepsilon\mathbf{n}\cdot\nabla\,\mathbf{v}$ is the rate of change per unit time of a (short) material vector $\varepsilon\mathbf{n}$ emanating from $\mathbf{x}$. We can write

$$D(\varepsilon\mathbf{n})/Dt = \varepsilon\mathbf{n}\cdot\nabla\,\mathbf{v} \qquad (\text{as } \varepsilon\to 0). \tag{4}$$

Using (4) and (2) we see that if $\omega - \varepsilon\mathbf{n}$ is initially zero then it remains zero, for

$$D(\omega - \varepsilon\mathbf{n})/Dt = 0. \tag{5}$$

(The result holds if the differential equation has a unique solution, for $\omega - \varepsilon\mathbf{n} \equiv \mathbf{0}$ is certainly *one* solution of (5) that satisfies the initial conditions.)

What we have shown is this. Units can be chosen so that ω can initially be identified with a material vector $\varepsilon\mathbf{n}$ of small length ε. The smaller ε is, the

[2] A rigorous treatment can be found in Serrin [**1959**].

more nearly it is true that ω continues to be identified with, and to move like, the material vector $\varepsilon\mathbf{n}$. Thus (2) can be interpreted as describing evolution of the vorticity vector equivalent to that which would occur if that vector were an (infinitesimally long) material vector. Relative motion of the fluid at the two ends of the vorticity vector cause it to undergo a rigid rotation (relative velocity normal to ω) and a contraction or elongation (relative velocity along ω). The former motion alters the orientation of the vorticity; the latter changes vorticity by (positive or negative) *vortex stretching*. (A closely related production of spin occurs if a twirling ice skater pulls in his extended arms.) One can think of *vortex lines* throughout the fluid passing from one vorticity vector to the "adjacent" one, and vorticity being altered by the tilting and stretching of these lines.

We have demonstrated that vorticity in an inviscid fluid changes by convection and stretching. These mechanisms can only magnify or relocate existing vorticity; they cannot create vorticity where none existed. An initially irrotational invisicid fluid therefore remains irrotational. But if $\nabla \wedge \mathbf{v} = \mathbf{0}$ then there exists a velocity potential ϕ such that $\mathbf{v} = \nabla \phi$. If the further assumption of incompressibility is made then $\nabla \cdot \mathbf{v} = 0$ so that $\nabla^2 \phi = 0$. Thus the mathematics of inviscid incompressible fluid flow is a branch of potential theory.

The introduction of viscosity brings into the picture two new ways for vorticity to change. First, the $\nu \nabla^2 \omega$ term in (1) is responsible for vorticity diffusion; local spin at one location will, through shear stresses, engender spin at an adjacent location, etc. Secondly, the necessity for the fluid to adhere at the boundary brings about a shear. Thus, in a viscous fluid the boundary is a source (or sink) of vorticity so that an initially irrotational *viscous* fluid will not remain irrotational.

To obtain formal confirmation of the boundary's role as a vorticity source, note from the vorticity equation (1) that diffusion is associated with a (tensor) vorticity flux $-\nu \nabla \omega$. Consider a flat plate along $z = 0$; here $u = v = w = 0$ so

$$\frac{\partial u}{\partial x} = 0, \quad \frac{\partial v}{\partial y} = 0, \quad \frac{\partial w}{\partial z} = -\left(\frac{\partial u}{\partial x} + \frac{\partial v}{\partial y}\right) = 0,$$

where we have used the mass conservation condition $\nabla \cdot \mathbf{v} = 0$. The x-vorticity generated by the plate is given by computing the following expression at $z = 0$:

$$-\nu \frac{\partial \omega_x}{\partial z} = \nu \frac{\partial^2 v}{\partial z^2} = \nu \nabla^2 v = \frac{1}{\rho} \frac{\partial p}{\partial y}. \tag{6}$$

We see that vorticity is generated even in the limit $\nu \to 0$.

It can be shown that there is a unique velocity field $\mathbf{v}(\mathbf{x}, t)$ that corresponds to a given vorticity field $\omega(\mathbf{x}, t)$, subject only to the restriction that the *normal* velocity component of the fluid equals that of an adjacent boundary (with suitable conditions at infinity). In a short time interval $(t, t + \Delta t)$ the vorticity

$\omega(\mathbf{x}, t)$ is convected, stretched, and diffused according to (1), but additional vorticity appears *near the boundary* (from the boundary source). This provides the unique $\omega(\mathbf{x}, t + \Delta t)$ that gives identical normal and tangential fluid and boundary velocities, as required by the "no slip" boundary condition for viscous fluids.

Pressure perturbations propagate with the speed of sound (infinitely fast in an incompressible fluid). Vorticity changes are far slower; this is a fundamental reason why study of vorticity is so revealing. (For further information, start with M. J. Lighthill's article in Rosenhead [**1963**].)

We now restrict consideration to two-dimensional flows wherein $\mathbf{v} = \mathbf{v}(x, y, t)$ and the vorticity vector is $\boldsymbol{\omega} = (0, 0, \omega)$, $\omega \equiv \partial u/\partial y - \partial v/\partial x$. The stretching term vanishes; vorticity in the interior changes only from convection and diffusion, according to the following simplified version of (1):

$$\partial\omega/\partial t + \mathbf{v}\cdot\nabla\,\omega = \nu\,\nabla^2\omega. \tag{7}$$

All the exact solutions of §11 were two-dimensional; let us examine their vorticity distributions. We begin with the plane Poiseuille solution (11.3) where the vorticity $\omega = \partial u/\partial y = -(C/\mu)y$ is maximum at the wall $y = \pm d$ and decreases to zero at the center. Convection does not play a role here, as there is no variation along streamlines. There is continual upward flow of vorticity (into the fluid from the lower boundary, throughout the fluid, and out at the upper boundary) at the constant rate

$$-\nu\partial\omega/\partial y = -(\mu\rho^{-1})\partial\omega/\partial y = C\rho^{-1}. \tag{8}$$

The pressure gradient forces the fluid past the boundary, but the fluid must adhere at the wall, so a spin is imparted to the fluid layer adjacent to the wall. This spin sets fluid in the "next" layer spinning, etc. (vorticity diffuses). Exactly the same process proceeds from the boundary at $y = d$, but the spin is in the opposite direction. At the center of the fluid, opposite spins meet and annihilate each other, giving rise to the absence of vorticity noted above.

When the horizontal plate of Solution 3, §11 is snatched into motion at speed U, there instantaneously develops a discontinuity in the horizontal velocity component u. The horizontal speed is U at the plane and zero just above. At the initial instant, the vorticity ω can be regarded as infinite at this discontinuity and zero elsewhere.

A plane of velocity discontinuities is called a *vortex sheet*. Here the instantaneous start of the motion from rest creates a vortex sheet, and diffusion of vorticity from this sheet accounts for the entire supply of vorticity in the flow thereafter. This picture is confirmed by the exact solution (11.16), which yields

$$\omega = \partial u/\partial y = -U(\pi\nu t)^{-1/2}\exp(-\eta^2), \qquad t > 0,\ \eta \equiv y/(4\nu t)^{1/2}, \tag{9}$$

$$\partial\omega/\partial y = yU(4\pi\nu^3t^3)^{-1/2}\exp(-\eta^2), \qquad t > 0. \tag{10}$$

Upon substituting $y = 0$ into (10) we obtain zero. Thus there is no flux of vorticity through the boundary after the motion starts. But (9) shows that vorticity is relatively appreciable when and only when η is of order of magnitude unity or less, that is when y is not greater in order of magnitude than $(\nu t)^{1/2}$. As we have already seen, the spreading of an effect like $(\nu t)^{1/2}$ is characteristic of diffusion with "diffusivity" ν. Here we see in addition an overall decrease in intensity like $t^{-1/2}$, as the fixed amount of vorticity is spread over an ever-widening area.

Suppose that a second plate were below the plate which instantaneously begins to move with speed U. Diffusion would rearrange the fixed amount of vorticity that is brought into being until ultimately the vorticity was evenly spread between the two plates. This is precisely what we see in Solution 1 of §11, the ultimate steady flow with constant vorticity throughout.

13. A glimpse of nonlinear elasticity. In this section we provide a brief introduction into deformations that (as previously postulated) can be characterized by position- or strain-gradient tensors, but we shall no longer linearize all equations by assuming that displacements are small. We shall show that local deformation can still be regarded as a combination of rotation and stretching along mutually perpendicular axes. Analysis of one example, that of simple shear, will reveal some interesting differences between the linear and nonlinear cases. The book of Jaunzemis [**1967**] is recommended as a good further reference and a bridge to modern research in finite deformation theory.

$$dx_i = x_{i,j}(\mathbf{A})dA_j, \tag{1}$$

which we write in the forms

$$dx_i = F_{ij}(\mathbf{A})dA_j \quad \text{or} \quad \mathbf{dx} = \boldsymbol{F} \cdot \mathbf{dA}. \tag{2}$$

Here $\boldsymbol{F}$ is a common notation for the position-gradient tensor $\nabla\,\mathbf{x}$. We assume that $\boldsymbol{F}$ is nonsingular, so that each initial line element $\mathbf{dA}$ is transformed into a unique line element $\mathbf{dx}$ and conversely. Under these circumstances the transformation is characterized by the following theorem.

Polar Decomposition Theorem. [For a proof, see for example Jaunzemis [**1967**], p. 77.] *If $\boldsymbol{F}$ is nonsingular then it has a unique right decomposition of the form*

$$\boldsymbol{F} = \boldsymbol{R} \cdot \boldsymbol{U}. \tag{3}$$

Here R is orthogonal,

$$\boldsymbol{R}^T = \boldsymbol{R}^{-1}; \tag{4}$$

and $\boldsymbol{U}$ is symmetric and positive. (A symmetric tensor is termed positive if all its eigenvalues are positive.)

There is a unique left decomposition of the form

$$\boldsymbol{F} = \boldsymbol{V} \cdot \boldsymbol{R} \tag{5}$$

where $\boldsymbol{R}$ is the same matrix that appears in (3) *and $\boldsymbol{V}$ is positive and symmetric.*

A transformation associated with an orthogonal matrix[3] such as R of course preserves length, for

$$y_i = R_{ij}x_j \quad \text{implies} \quad y_i y_i = R_{ij}R_{ik}x_j x_k = x_k x_k \tag{6}$$

since $R_{ij}R_{ik} = (R^T R)_{jk} = \delta_{jk}$. Thus the transformation induced by $\boldsymbol{R}$ is a rotation and $\boldsymbol{R}$ is called a *rotation tensor*. Moreover, $\boldsymbol{U}$ and $\boldsymbol{V}$ induce stretching (or contraction) along their three mutually perpendicular eigenvectors and are termed the *right* and *left stretch* tensors respectively.

We now show that use of the polar decomposition theorem leads to a characterization of strain that, as it should, reduces to the results in §10 when strains are small. Comparing (2) and (10.24) in the form

$$dx_i = dA_i + U_{i,j}dA_j = (\delta_{ij} + U_{i,j})dA_j \tag{7}$$

we see the position gradient tensor $\boldsymbol{F}$ and the displacement gradient tensor $\nabla \mathbf{U}$ are related by

$$F_{ij} = \delta_{ij} + U_{i,j} \quad \text{or} \quad \boldsymbol{F} = \boldsymbol{I} + \nabla \mathbf{U}. \tag{8}$$

We write the tensors of the polar decomposition $\boldsymbol{F} = \boldsymbol{R} \cdot \boldsymbol{U}$ in the following way.

$$\boldsymbol{R} = \boldsymbol{I} + \tilde{\boldsymbol{R}}, \qquad \boldsymbol{U} = \boldsymbol{I} + \tilde{\boldsymbol{U}}.^4 \tag{9}$$

Let us assume that the rotation described by $\boldsymbol{R}$ and the stretching described by $\boldsymbol{U}$ are "small". We interpret this to mean that $\boldsymbol{R}$ and $\boldsymbol{U}$ are "nearly" unit tensors, so that products of $\tilde{\boldsymbol{R}}$ and $\tilde{\boldsymbol{U}}$ can be neglected in calculations such as the following:

$$\boldsymbol{I} = \boldsymbol{R}^T \cdot \boldsymbol{R} = (\boldsymbol{I} + \tilde{\boldsymbol{R}}^T)\cdot(\boldsymbol{I} + \tilde{\boldsymbol{R}}) = \boldsymbol{I} + \tilde{\boldsymbol{R}}^T + \tilde{\boldsymbol{R}} + \dots . \tag{10}$$

From (10) we already see that in a linearized theory when corrections are entirely neglected

$$\tilde{\boldsymbol{R}}^T + \tilde{\boldsymbol{R}} = \boldsymbol{0}, \tag{11}$$

so that $\tilde{\boldsymbol{R}}$ is antisymmetric. Moreover the symmetry of $\tilde{\boldsymbol{U}}$ follows from the symmetry of $\boldsymbol{U}$. Thus since

$$\boldsymbol{F} = \boldsymbol{R} \cdot \boldsymbol{U} = (\boldsymbol{I} + \tilde{\boldsymbol{R}})\cdot(\boldsymbol{I} + \tilde{\boldsymbol{U}}) = \boldsymbol{I} + \tilde{\boldsymbol{R}} + \tilde{\boldsymbol{U}} + \dots, \tag{12}$$

comparison with (8) shows that in linearized theory

$$\nabla \mathbf{U} = \tilde{\boldsymbol{R}} + \tilde{\boldsymbol{U}}. \tag{13}$$

Decomposition of a tensor into the sum of a symmetric and asymmetric tensor is unique so it must be that

$$\tilde{R}_{ij} = \tfrac{1}{2}(U_{i,j} - U_{j,i}), \qquad \tilde{U}_{ij} = \tfrac{1}{2}(U_{i,j} + U_{j,i}). \tag{14a,b}$$

---

[3]As here we switch freely between tensors and the matrices that represent them in some cartesian coordinate system. Thus R is a matrix that gives components of $\boldsymbol{R}$, and $(R^T R)_{jk}$ is the jkth component of the matrix product formed by the transpose of R and R itself.

[4]Do not confuse the displacement vector $\mathbf{U}$ and the stretch tensor $\boldsymbol{U}$.

In the limit of small deformations, the two approaches thus yield exactly the same result.

We remark that the fundamental way to decompose a transformation is to regard it as equivalent to a *succession* of simpler transformations. This is reflected in our discovery (for example) that $\boldsymbol{F}$ can be written as the *product* of the simpler tensors $\boldsymbol{R}$ and $\boldsymbol{U}$. In the linear theory of elasticity, however, it is sufficient to write a deformation as the *sum* of simpler deformations. This is because one can analyze each deformation separately after which (because the governing equations are linear) one can add the results to obtain an analysis of the full deformation.

In nonlinear elasticity the material strain tensor $\boldsymbol{\eta}$ is defined as follows. For some particle **A** consider dx_i as defined in (7) and another similarly defined line element dx'_i. From the definition

$$dx_i dx'_i - dA_i dA'_i \equiv 2\eta_{jk} dA_j dA'_k \tag{15}$$

there follows

$$\eta_{jk} = \tfrac{1}{2}\left(U_{j,k} + U_{k,j} + U_{i,j}U_{i,k}\right). \tag{16}$$

Note that when only linear terms are retained, (16) reduces to the earlier version of $\boldsymbol{\eta}$ given in (10.25a).

The stretch tensors $\boldsymbol{U}$ and $\boldsymbol{V}$ are not simply related to $\boldsymbol{F}$. On the other hand, if we define the *right and left Cauchy-Green deformation tensors* $\boldsymbol{C}$ and $\boldsymbol{B}$ by

$$\boldsymbol{C} \equiv \boldsymbol{F}^T \cdot \boldsymbol{F} \quad \text{and} \quad \boldsymbol{B} = \boldsymbol{F} \cdot \boldsymbol{F}^T \tag{17}$$

we see from (3) and (4) that

$$\boldsymbol{C} = \boldsymbol{U}^T \cdot \boldsymbol{R}^T \cdot \boldsymbol{R} \cdot \boldsymbol{U} = \boldsymbol{U}^T \cdot \boldsymbol{U} = \boldsymbol{U} \cdot \boldsymbol{U} \quad \text{and} \quad \boldsymbol{B} = \boldsymbol{V} \cdot \boldsymbol{V}. \tag{18}$$

The material strain tensor $\boldsymbol{\eta}$ is related to $\boldsymbol{C}$ by

$$\boldsymbol{\eta} = \tfrac{1}{2}(\boldsymbol{C} - \boldsymbol{I}), \tag{19}$$

and a similar connection can be established between $\boldsymbol{B}$ and the nonlinear version of the spatial strain tensor of (10.25c).

We wish now to write down a fairly general constitutive equation for an isotropic material. Not surprisingly, for such materials it can be shown that the rotational part of the deformation cannot effect the stretch. Of the various possibilities, with essentially no loss of generality we select $\boldsymbol{C}$ to characterize the stretching; as we have seen, it is one of the stretch measures that is relatively easy to compute.

Making some restriction in generality, but still retaining wide scope, we limit further consideration to polynomial equations of the form

$$\boldsymbol{T} = \sum_{n=0}^{N} a_n \boldsymbol{C}^n. \tag{20}$$

One expects that for an isotropic material the scalar coefficients a_i can only be functions of the principal invariants C_1, C_2, and C_3 of $\boldsymbol{C}$. This can be demonstrated (Jaunzemis [**1967**], pp. 430–432). Considerable further simplification can be obtained by invoking the Cayley-Hamilton theorem to write

$$\boldsymbol{T} = \alpha \boldsymbol{C}^{-1} + \beta \boldsymbol{I} + \gamma \boldsymbol{C}. \tag{21}$$

To get some inkling of the possible effects of nonlinearity, let us consider the case of simple shear. In an inverse approach to the problem, let us assume a standard shearing displacement of the form

$$x_1 = A_1 + 2s, \quad x_2 = A_2, \quad x_3 = A_3, \quad s \text{ a constant.} \tag{22}$$

It remains to show that the governing equations are satisfied by (22).

In terms of the component matrix T that represents $\boldsymbol{T}$ in the given coordinate system, the constitutive equation (21) for $C_{ij} \equiv x_{p,i}x_{p,j}$ becomes

$$T = \alpha \begin{bmatrix} 1+4s^2 & -2s & 0 \\ -2s & 1 & 0 \\ 0 & 0 & 1 \end{bmatrix} + \beta \begin{bmatrix} 1 & 0 & 0 \\ 0 & 1 & 0 \\ 0 & 0 & 1 \end{bmatrix} + \gamma \begin{bmatrix} 1 & 2s & 0 \\ 2s & 1+4s^2 & 0 \\ 0 & 0 & 1 \end{bmatrix}. \tag{23}$$

From (23) we see that the coefficients, being functions only of the matrix invariants, must be functions of s^2. We also observe that the equilibrium equations in the absence of a body force are satisfied ($T_{ij,j} = 0$), since T_{ij} is constant. In addition we note from (19) that the components of the material strain tensor $\boldsymbol{\eta}$ are given by

$$\boldsymbol{\eta} = \begin{bmatrix} 0 & s & 0 \\ s & 2s^2 & 0 \\ 0 & 0 & 0 \end{bmatrix}. \tag{24}$$

In linear elasticity, the fundamental result concerning simple shear is that the shear stress T_{12} is proportional to the shear strain $\eta_{12} \approx \varepsilon_{12}$ through a constant elastic modulus μ. Here we have

$$T_{12} = 2\hat{\mu}(s^2)\eta_{12}, \qquad \hat{\mu}(s^2) \equiv \gamma(s^2) - \alpha(s^2). \tag{25}$$

Since $\hat{\mu}(s^2) = \hat{\mu}(0) + O(s^2)$, the fact that the nonlinear shear modulus $\hat{\mu}$ is not strictly constant is not expected to be noticeable until rather large shear is imposed. This expectation is in accord with observations.

In contrast with linear elasticity, more than shear stresses are required to sustain simple shear. In particular, whatever the response functions, $T_{22} - T_{11} = 2sT_{12}$. Thus, the normal stresses T_{11} and T_{22} must be present (if $s \neq 0$), and they cannot be equal. Furthermore, it is easily shown that normal stresses must be imposed on the boundary to obtain a pure shear deformation. If these stresses are not supplied, one expects some (positive or negative) dilatation of the material to result from attempts to shear it. Such *Kelvin effects* are indeed observed.

For fixed thermodynamic variables, linear elasticity requires specification of only two constant moduli λ and μ. One of these can be found in principle from a single comparison between shear stress and shear deformation. By contrast, our constitutive equation of nonlinear elasticity (21) requires determination of three functions (α, β, and γ) of three variables (C_1, C_2, and C_3). Because $C_1 = C_2$ and $C_3 = 1$ in simple shear, these functions can be determined only to a limited degree by a study of this problem.

C. NEOCLASSICAL CONTINUUM PROBLEMS

14. A variety of models for bacterial chemotaxis. A gross phenomenological model, frequently of a deterministic continuum character, is often the best first step toward analyzing a new set of observations. Understanding may be enhanced if one then formulates stochastic "molecular" or "particle" models from which the continuum models can be derived in suitable limits. (A stochastic framework is appropriate since details of particle behavior are highly sensitive to changes in initial conditions, and in any case are generally not of interest.) Moreover, the particle approach can reveal instances where inadequacies in the original models can be remedied, at some expense in complication but without abandoning the profound simplification of a deterministic continuum framework. These points will be illustrated here in connection with some biological problems involving the collective motions of bacteria.

Organisms are said to be *chemotactic* when they tend to move toward or away from relatively high concentrations of certain chemicals. Chemotaxis in flagellated[5] bacteria is now under intensive investigation by biologists, primarily to elucidate the molecular details of the process in these simple "model" organisms. For the reception of chemical signals by the bacteria and its transduction into flagellar action exhibits characteristics of nervous system activity in higher creatures.

J. Adler[6] initiated modern studies of bacterial chemotaxis in experiments which exhibited "travelling bands" of bacteria that moved steadily down a capillary tube. The bacteria were placed in one end of the tube, and they consumed (in a typical experiment) the local oxygen supply. Preferring relatively high oxygen concentrations, most of the bacteria moved off down the tube, creating a favorable oxygen gradient as they moved. The bacteria that remained behind would form a band if they were introduced into a fresh tube.

Equations for chemotaxis in bacterial populations. Keller and Segel [**1971**] employed a continuum model in a first analysis of Adler's experiments. In particular, since the distance between bacteria was typically very small compared to the "bandwidth" these authors felt that it would be reasonable to represent bacterial density by a smooth function b. Let us now write a general balance law for b, assuming it to be a function of a single spatial variable x and the time t. Such a law states that the rate of change of the net amount of bacteria in a given fixed region equals the net rate at which b flows

[5] "Flagellated" is not used in the sense of "whipped" but rather "possessing whip-like appendages that induce motion".

[6] References to Adler's work, and to the work of other biologists not specifically cited here, can be found in theoretical papers that we shall mention specifically below.

across the boundary plus the net creation rate (birth rate minus death rate). This law is said to be "general", for it must apply to any substance. Let us define the *flux density* $J(x_1, t)$ as the net rate per unit length at which b crosses the line $x = x_1$, where J is to be positive (negative) if there are more bacteria crossing in the direction x increasing (decreasing). The term $Q(x, t)$ is the net rate per unit area at which bacteria are being created. With these definitions we can write

$$\frac{\partial}{\partial t}\int_x^{x+\Delta x} b(s, t)\, ds = J(x, t) - J(x + \Delta x, t) + \int_x^{x+\Delta x} Q(s, t)\, ds. \quad (1)$$

Using the integral mean value theorem, dividing by Δx, and then taking the limit as $\Delta x \to 0$, we obtain the *general balance law in differential equation form*,

$$\frac{\partial a}{\partial t}(x, t) = -\frac{\partial J}{\partial x}(x, t) + Q(x, t). \quad (2)$$

In the present case it turns out that both births and deaths can be ignored so

$$Q \equiv 0. \quad (3)$$

To obtain an expression for the flux J, let us first consider a situation when attractant gradients are not present and bacteria move about randomly. Under these conditions there appears to be a random flux J_r from regions of high bacterial concentration to regions of low concentration. The flux magnitude depends on the concentration difference between x and nearby points. The simplest description of this situation appears to be

$$J_r(x, t) = F\left[\frac{\partial b(x, t)}{\partial x}\right] \quad (4)$$

for some function F. If bacterial concentration is uniform there will be as many bacteria moving to the left as to the right, so we assume that

$$F(0) = 0. \quad (5)$$

For sufficiently small values of s, we can approximate the graph of F by a straight line. Designating the slope of this line by $-\mu$, we obtain

$$J_r(x, t) = -\mu \partial b(x, t)/\partial x. \quad (6)$$

If $J = J_r$ then (2), (3) and (6) imply the diffusion equation

$$\frac{\partial a}{\partial t} = \frac{\partial}{\partial x}\left(\mu \frac{\partial a}{\partial x}\right).$$

To account for chemotaxis to an attractant of density $s(x, t)$, we add to J_r an additional contribution J_c. By reasoning analogous to that which led us to suppose that $J_r \sim \partial b/\partial x$ we hypothesize that J_c is proportional to the attractant gradient, at least when it is small. For a given gradient, the flux should be proportional to bacterial density. Thus we assume that

$$J_c = \chi b \partial s / \partial x, \tag{7}$$

where the proportionality factor χ is the chemotactic sensitivity. In (6) and (7) the coefficients μ and χ are both expected to be positive, since amoebae tend to move away from amoeba concentrations and toward attractant concentrations. Both μ and χ in general will be functions of the attractant concentra tion s. Assuming that $J = J_r + J_c$, i.e., that

$$J = \mu \partial b / \partial x - \chi b \partial s / \partial x, \tag{8}$$

we obtain the final equation for the change in bacterial density

$$\frac{\partial b}{\partial t} = \frac{\partial}{\partial x}\left(\mu \frac{\partial b}{\partial x} - \chi b \frac{\partial s}{\partial x}\right). \tag{9}$$

Let the attractant s be consumed at a rate k per bacterium, and diffuse with diffusion constant D, so that

$$\partial s / \partial t = -kb + D \partial^2 s / \partial x^2. \tag{10}$$

Equations (9) and (10) form a pair of nonlinear partial differential equations for b and s.

On balance laws. Before proceeding, let us emphasize the ubiquity of the general balance equation

$$\partial a / \partial t + \partial J / \partial x = 0 \tag{11a}$$

or its three-dimensional generalization

$$\partial a / \partial t + \nabla \cdot \mathbf{J} = 0. \tag{11b}$$

(Here we have a flux density *vector* $\mathbf{J}$ whose ith component is the rate of flow of the given substance in the direction of the ith coordinate vector.) For example (10) is of this form, with a negative "creation" term $-kb$ and a constitutive equation that posits a flux proportional to the substance gradient. The mass conservation equation (5.7) can be written in the form (11b) with $\mathbf{J} = \rho \mathbf{v}$. This is a convective flux term, with mass being carried across a unit area perpendicular to the x_i axis at rate ρv_i (since v_i volume units cross this unit area per unit time, and ρ is the mass per unit volume). We speak of mass *conservation* rather than balance since the source term Q is zero.

Writing the momentum equation (6.19) in the form

$$\rho \partial v_i / \partial t = -\rho v_k v_{i,k} + \rho f_i + T_{ki,k}, \tag{12}$$

and taking into account the mass conservation requirement $\partial \rho / \partial t = -(\rho v_k)_{,k}$, one readily obtains for the rate of change of (linear) momentum

$$\frac{\partial}{\partial t}(\rho v_i) = \rho f_i + J_{ik,k}, \qquad J_{ik} \equiv \rho v_i v_k + T_{ki}. \tag{13}$$

Here is another general balance equation, but the momentum flux is represented by the *tensor* J_{ik}, since momentum itself is a vector. In this formulation the body force $\rho \mathbf{f}$ plays the role of a momentum source. As comprehensive discussions of mechanics show, the concept of a momentum flux tensor, with both a stress and a "velocity correlation" component, plays an important role both in devising certain calculational procedures and in the interpretation of otherwise counterintuitive results.

Analyses of bacterial chemotaxis. Returning to our main theme, we note that it is not appropriate here to provide a full review of work on the travelling band problem for (9) and (10). Suffice it to say that Keller and Segel [**1971**] found travelling wave solutions to these equations under certain simplifying but reasonable assumptions. Scribner, Segel, and Rogers [**1974**] performed a numerical analysis of the initial value problem and showed the process of band formation; indeed, not all the simulated bacteria joined the band. It should be mentioned that (in accord with observation) the model considered by Scribner et al. was slightly altered, in a biologically reasonable way, from a system of equations which is expected to produce exact travelling waves, and that bacteria slowly leaked from the trailing edge of the wave, as has been observed in experiment. (A. Novick and Segel (unpublished) have recently formulated an analytic description of this type of slowly deforming wave.) The class of wave solutions originally found by Keller and Segel [**1971**] has been broadened by Odell and Keller [**1976**], while Keller and Odell [**1976**] have also performed a full analysis of the functional forms of $\mu(s)$ and $\chi(s)$ that permit exact travelling wave solutions. Rosen [**1975**] has provided an interesting analytical approximation to the problem treated numerically by Scribner et al.

Segel and Jackson [**1973**] have emphasized, among other things, that properties such as chemotactic sensitivity cannot be expected to have the same value for every bacterium in the population. Only if the distributions of such parameters are sharply peaked around their mean values can one expect analyses of the type presented to provide an accurate picture of average behavior. Variability is so prevelant in biology that one must always be on the lookout for inaccuracies in analyses restricted to mean properties.

Papers representative of theoretical work by other authors are those of Nossal and Weiss [**1973**], and Lapidus and Schiller [**1974**].

Nonlinear diffusion waves in biology. We remark that wave motion is not normally associated with diffusion, but theoretical biology is replete with examples, like travelling bacterial bands, of wave propagation governed by nonlinear diffusion equations. (Some other examples are mentioned in the next section.) Rigorous work has been done in this area; the paper of Aronson and Weinberger [**1975**] provides a recent example. To be relevant to problems of chemotaxis, such work must be extended to cover systems of equations and equations wherein the nonlinearity is not merely added to a basic diffusion equation.

Certainly the most famous biological example of nonlinear diffusive travelling waves are those found by Hodgkin and Huxley to satisfy the phenomenological equations they derived for the propagation of nerve impulses. Recently there has been a number of papers that deal analytically with the Hodgkin-Huxley equations, or with various "stripped-down" versions of them. Typically these equations have more than one travelling wave solution, so questions of stability naturally arise. Rinzel [**1975**] provides a recent and interesting contribution to these questions, and a convenient starting point for a literature search.

Chemotaxis at the bacteria level–a random walk model. Biologists have made considerable progress in their quest to understand the detailed mechanisms involved in bacterial chemotaxis. They have shown that the bacteria normally swim for an exponentially distributed length of time in a reasonably straight line ("run"), then tumble or "twiddle", choosing a new direction more or less at random after a short period of noncoordinated swimming. Chemotaxis results from an increase in run length when the bacteria are moving toward an attractant. Attractant gradients are sensed by a comparison of present with earlier chemical concentrations. Comparison is mediated by several thousand receptor molecules, of one or perhaps a few types, located on the bacterial surface.

To probe deeper into the problem, it is useful to consider a random walk model of bacterial chemotaxis. To this end, imagine a one-dimensional situation in which bacteria move to the left or right along a line. Let x denote position along this line and let the superscripts "+" and "−" distinguish right- and left-moving bacteria respectively. At time t, let $n^+(x, t)$ [$n^-(x, t)$] denote the number of bacteria that arrive at point x from the left [right]. Let $p^+(x, t)$ be the probability that a bacterium persists after arriving at x from the left at time t, i.e., the probability that the particle continues from x in the positive direction. Let $q^+(x, t)$ be the corresponding reversal probability, so that

$$p^+(x, t) + q^+(x, t) = 1. \tag{14}$$

Correspondingly, persistence and reversal probabilities for left-moving bacteria will be denoted by p^- and q^-.

Suppose that the bacteria carry out a random walk, subject to the above probabilities, with a fixed step length Δ and with a fixed step duration τ. Then the following equations govern the phenomenon:

$$n^-(x - \Delta, t + \tau) = n^+(x, t)q^+(x, t) + n^-(x, t)p^-(x, t), \tag{15a}$$

$$n^+(x + \Delta, t + \tau) = n^+(x, t)p^+(x, t) + n^-(x, t)q^-(x, t). \tag{15b}$$

Equation (15a) is based on the fact that the bacteria which arrive at $x - \Delta$ from the right, at time $t + \tau$, must have been located at point x at time t. The first term on the right side of (15a) gives the number of bacteria that reached $x - \Delta$ by arriving at x from the left and reversing. The second term gives the number that arrived at x from the right and persisted. Equation (15b) can be obtained similarly.

Limiting differential equations. Approximation of (15) in the case of small Δ and τ will yield partial differential equations that we can compare with the equations which we obtained earlier from a strictly phenomenological approach to population motions. The issues involved in this approximation have been discussed in some detail by Segel [**1977**a]. It turns out that the appropriate "step" is a small fraction of the distance between tumbles.

Thus decreasing the step size Δ should leave the speed $v \equiv \Delta/\tau$ unaltered. But as Δ is decreased, there is a proportionate increase in the number of steps between tumbles. As a consequence, the appropriate limit in the present case is

$$\Delta \to 0, \quad \tau \to 0, \quad \Delta/\tau \equiv v \text{ fixed}, \quad q^{\pm} = O(\Delta). \tag{16}$$

We thus write

$$n^{\pm}(x, t; \Delta) = n_0^{\pm}(x, t) + \Delta n_1^{\pm}(x, t) + \dots \tag{17}$$

and

$$q^{\pm}(x, t; \Delta) = \Delta q_1^{\pm}(x, t) + O(\Delta^2),$$
$$p^{\pm}(x, t; \Delta) = 1 - \Delta q_1^{\pm}(x, t) + O(\Delta^2). \tag{18}$$

Inserting these expansions into (15a) and (15b) we find at lowest order that

$$\frac{\partial n_0^+}{\partial x} + \frac{1}{v}\frac{\partial n_0^+}{\partial t} = -n_0^+ q_1^+ + n_0^- q_1^-, \tag{19a}$$

$$-\frac{\partial n_0^-}{\partial x} + \frac{1}{v}\frac{\partial n_0^-}{\partial t} = n_0^+ q_1^+ - n_0^- q_1^-. \tag{19b}$$

To interpret these equations, and for later use, we introduce reversal probabilities per unit time $\sigma^{\pm}$ and bacterial densities $b^{\pm}$ for right- and left-moving bacteria:

$$\sigma^{\pm} \equiv (1/\tau) q^{\pm} = v q_1^{\pm} + O(\Delta^2), \quad n_0^{\pm}/\Delta \equiv b^{\pm}. \tag{20}$$

By making the further definitions

$$J^{\pm} = \pm v b^{\pm}, \qquad Q^{\pm} = \pm(b^- \sigma^- - b^+ \sigma^+), \tag{21}$$

we can write (19a) and (19b) in the form of balance laws

$$\partial b^{\pm}/\partial t = -\partial J^{\pm}/\partial x + Q^{\pm}. \tag{22}$$

Indeed, a moment of contemplation shows that J^+ and J^- give the flux of the two species of particles, while Q^+ and Q^- are "creation terms" as they should be. If we add the two forms of (22) and employ the definitions

$$b \equiv b^+ + b^-, \qquad J \equiv J^+ + J^-, \tag{23}$$

we obtain the proper bacterial conservation law

$$\partial b/\partial t = -\partial J/\partial x. \tag{24}$$

Subtraction of the two versions of (22) yields

$$\frac{\partial J}{\partial t} + (\sigma^+ + \sigma^-) J = v^2 \frac{\partial b}{\partial x} + v b(\sigma^- - \sigma^+). \tag{25}$$

It turns out that the $\partial J/\partial t$ term can be neglected when (as is usually the case) the velocity at which the population "drifts" is small compared to v. Then

$$J = -\left[\frac{v^2}{\sigma^+ - \sigma^-}\right]\frac{\partial b}{\partial x} + \left[\frac{v(\sigma^- - \sigma^+)}{\sigma^- + \sigma^+}\right]b, \tag{26}$$

and we recover our earlier phenomenological expression (8) for the flux, but now with expressions for the effective diffusivity or *motility* μ and the chemotactic sensitivity χ in terms of the reversal probabilities σ^+ and σ^- and the swimming speed v. The population parameters have been related to parameters that characterize the behavior of individual organisms.

As is discussed further in the more detailed presentation of this random walk model in Segel [**1977**a], retention of the $\partial J/\partial t$ gives rise to an inertia or history effect. That such effects are not generally found in elementary phenomenological theories reminds us that although basic balance equations are expected to have very wide validity, constitutive equations relate to a particular type of material and/or various plausible simplifying assumptions–so these latter equations must be viewed with an extra degree of skepticism.

A model for chemotaxis that incorporates receptor action. Contact with the forefront of biological experimentation requires that information concerning receptor action be incorporated into the model–i.e., that information on a level below that of the individual organism be taken into account. A first step in this direction has been taken by Segel [**1976**]. A one-dimensional model is employed in this analysis, with the assumption that the reversal probabilities σ^+ and σ^- are entirely determined at time t by the state of a set of receptor molecules at that time. The state is described by such quantities as the percentage of receptor sites "filled" by attractor. Although $\sigma^+(x, t)$ and $\sigma^-(x, t)$ both refer to reversal probabilities at the same point x, the time scale for chemical reactions is assumed to be such that a short term "memory" results in different average receptor states in left- and right-moving cells respectively. The character of the appropriate basic assumptions is suggested by our random walk model. The equations are similar to (21) and (22), but in addition to keeping track of the numbers of left- and right-moving cells, one must also monitor the number of various types of chemicals on left- and right-moving bacteria, and how they change by chemical reaction as well as by the reversal and convection mechanisms that appear in (21). A rather large system of nonlinear partial differential equations results, but useful information can be obtained without undue labor under the assumption that the concentration of the attractant is a slowly varying function of the spatial coordinate x. A detailed development of this approach can be found in Segel [**1976**] and [**1977**c].

15. The interaction-diffusion equation in chemistry and ecology. Classical continuum mechanics concerns itself with the behavior of solids and fluids. Problems typically require exploring the properties of partial differential equations with respect to such matters as stability of equilibrium, nature of wave propagation, and extent of diffusion. In recent years attention has increasingly been paid to problems of similar character that arise in such fields as physical chemistry, biochemistry, developmental biology and ecology. This section is devoted to a selective overview of a class of such problems, with emphasis on those areas to which the author has contributed

Turing [**1952**] initiated a major area of research with a suggestion that biological patterns could be engendered by the joint working of chemical reaction and diffusion. In support of this suggestion Turing studied the development of spatial inhomogeneity in certain informative examples of what later came to be known as reaction-diffusion equations. In two variables, such equations take the form

$$\partial C_i / \partial t = R_i(C_1, C_2) + D_i \nabla^2 C_i, \qquad i = 1, 2, \tag{1}$$

where the R_i, typically low order polynomials, give the effect of chemical reaction and the D_i are diffusion constants. Suppose that the equations (1) admit a solution $C_1 \equiv C_1^{(0)}$, $C_2 \equiv C_2^{(0)}$, that is uniform in space and time. Turing showed that under certain conditions such solutions can become unstable, and that the "most dangerous" perturbations will be of a well-defined and finite chemical wavelength. Such instabilities can be regarded as "caused" by diffusion. The reason is that in linear stability theory the omission of diffusion is equivalent to restricting consideration to disturbances of infinite wavelength–and, as we have just mentioned, it is finite wavelength disturbances that will destabilize the system.

How does diffusion, normally regarded as a mechanism that damps fluctuations, act to destabilize a steady uniform state? This question was considered by Segel and Jackson [**1972**] for a two-component system. The essence of their explanation is as follows. For a two-component system to permit diffusive instabilities of the type under investigation, one of the components must be a *stabilizer* whose perturbations not only decay but which also, because of the coupling between reactions, bring about the decay of the autocatalytic *destabilizer*. Diffusion indeed damps all perturbations, but relatively rapid diffusion can make the stabilizer disappear so fast that it has no time to damp a destabilizer fluctuation.

The polynomial terms in (1) are of the same nature as terms used in ecology to model the interaction between species. Random migration is the ecological equivalent of diffusion. The question then naturally arises, can reasonable ecological models be shown to yield the possibility of diffusive instability? Segel and Jackson [**1972**] showed that this question has an affirmative answer by analyzing a predator-prey interaction subject to random migration. Segel and Levin [**1976**] applied the methods of nonlinear stability theory to this problem, and thereby exhibited circumstances in which a steady-state nonuniform environment can arise from the instability of a spatially homogeneous situation. Their analysis used a particularly straightforward form of the nonlinear stability theory that was first developed for hydrodynamical problems. To see the essence of the matter, consider (1) after introduction of the vector of perturbations $\mathbf{V} = (C_1 - C_1^{(0)}, C_2 - C_2^{(0)})$. We can write

$$-(\partial \mathbf{V}/\partial t) + \mathbf{M}_\theta \cdot \mathbf{V} = \mathbf{N}(\mathbf{V}) \tag{2}$$

where $\mathbf{N}$ denotes the nonlinear terms. Dependence of the linear matrix operator $\mathbf{M}_\theta$ on the principal parameter of the problem is indicated explicitly by θ; in this case, θ is the ratio of diffusivities.

The zero perturbation corresponds to the exact uniform solution, so $\mathbf{N}(\mathbf{0}) = \mathbf{0}$. Linear stability theory considers solutions of the form

$$\mathbf{V}(x, t) = \mathbf{K} \exp(\sigma t) \cos qx. \tag{3}$$

(Here x and t denote spatial and temporal variables, while $\mathbf{K}$, σ, and q are constants.) This requires analysis of an eigenvalue problem of the form

$$\mathbf{M}_{\theta,q} \cdot \mathbf{K} = \sigma_{\theta,q}\mathbf{K}. \tag{4}$$

As θ increases, a positive value of $\sigma_{\theta,q}$ (instability) is first encountered when $\theta = \theta_c$, for a perturbation of the *critical wavenumber* q_c. Let the *critical eigenvector* $\mathbf{K}_c$ be defined by

$$\mathbf{M}_{\theta_c,q_c} \cdot \mathbf{K}_c = \sigma^{(1)}_{\theta_c,q_c}\mathbf{K}_c, \tag{5}$$

where the superscript "one" denotes the eigenvalue of largest real part.

We wish to analyze the effect of nonlinear terms upon the growth of a disturbance having the critical wavenumber. Segel and Levin [**1975**] accomplish this by means of the following successive approximations calculation.

$$\mathbf{V}^{(1)} = \varepsilon A(t_0, t_1, t_2, \ldots)\mathbf{K}_c \cos q_c x, \qquad t_i \equiv \varepsilon^i t,$$

$$-\partial \mathbf{V}^{(n)}/\partial t + \mathbf{M}_{\theta,q_c} \cdot \mathbf{V}^{(n)} = \mathbf{N}(\mathbf{V}^{(n-1)}) + O(\varepsilon^{n+1}), \qquad n = 2, 3, \ldots. \tag{6}$$

Here the $O(\varepsilon^{n+1})$ terms are not explicitly computed, for they cannot be accurately determined at the given stage. A multiple time-scale dependence of the amplitude is postulated, for otherwise a uniformly valid solution is not obtained by successive approximations.

Imposition of the requirement $A = O(1)$ uniformly in t turns out to require that the perturbation amplitude scale ε be related to the amount by which θ exceeds its critical value for instability by

$$\theta - \theta_c = O(\varepsilon^2), \quad \text{so } \sigma^{(1)}_{\theta,q_c} = \bar{\sigma}\varepsilon^2 + O(\varepsilon^4). \tag{7}$$

Here $\bar{\sigma}$ is an $O(1)$ constant. Moreover,

$$\frac{\partial A}{\partial t_0} = 0, \quad \frac{\partial A}{\partial t_1} = 0, \quad \frac{\partial A}{\partial t_2} = \bar{\sigma}A - \gamma A^3, \tag{8}$$

where the so-called *Landau constant* γ is positive and $O(1)$. Thus (in contrast to linear stability theory, which predicts continual exponential growth) the amplitude A approaches a constant $(\bar{\sigma}/\gamma)^{1/2}$. In addition, spatial harmonics are generated and there is a change in the mean. Details, and a parallel treatment of a discrete "patchy" environment, can be found in the original paper. Also see Levin and Segel [**1976**] for a discussion of the possibility that diffusive instability could be relevant to the observed patchiness of plankton distribution in the ocean.

The spatial patterns mentioned so far have been essentially static. Dynamic structure can occur, most dramatically in the Zhabotinskiĭ-Belousov reaction (Winfree [**1975**]). This reaction provides beautifully colored and clearly visible shifting patterns that can easily be generated in a flat dish by adding appropriate reagents. It has thus been selected as a prototype problem by a number of investigators; a recent survey that mathematicians will find appealing is that of Kopell and Howard [**1974**].

As has most recently been reported in a paper by Hardt, Naparstek, Segel, and Caplan [**1976**] another prototypical instance of spatio-temporal structures has been provided by the study of the digestive enzyme papain bound (i.e., constrained) to a thin membrane. In contrast to the Zhabotinskiĭ-Belousov system, the chemical kinetics here are fully understood and simple. Papain (concentration E) essentially catalyzes the breaking off of a hydrogen ion (concentration H) from a certain substrate (concentration S) on which the enzyme acts. The major feature of the system is that the catalytic "efficiency" of the enzyme has a maximum at a certain value of H. Also important is the fact that there is almost instantaneous equilibrium between the combination of hydrogen and hydroxyl ions (concentration OH) to form water, and the reverse ionization of the water.

Assuming a single value D_H for hydrogen and hydroxyl diffusion coefficients, one is led to equations of the following form for a membrane bounded by the planes $y = \pm a$.

$$\frac{\partial S}{\partial t} = -R(H, S) + D_S \nabla^2 S, \qquad \frac{\partial F}{\partial t} = R(H, S) + D_H \nabla^2 F,$$

$$F \equiv H - OH, \qquad (H)(OH) = K_W,$$

$$R = \frac{k_2 k_3 E}{k_2 + k_3 + k_3 K_S S^{-1}}, \quad k_2 = \frac{c_1}{1 + c_2 H + c_3 H^{-1}}, \quad k_3 = \frac{c_4}{1 + c_5 H}. \tag{9}$$

Here D_S, D_H, K_W, K_S and the c_i are constants. Outside the membrane one postulates the existence of an *unstirred layer* in which there occurs diffusion but no reaction. Concentrations and their normal derivatives are assumed continuous at the layer-membrane boundary. At the edge of the layer $S = S_0$, $H = H_0$ (so $F = F_0$), where S_0 and H_0 are given concentrations in a well-mixed medium into which the membrane is placed. (The unstirred layer is a representation of the boundary layer in which mixing efficiency is low.) At the center of the membrane, symmetry requires vanishing of the normal derivatives $\partial S/\partial y$ and $\partial F/\partial y$.

The full problem is obviously somewhat complicated. Progress has been made by means of intuition acquired from a "two-compartment model" in which the membrane is regarded a homogeneous domain that communicates by passive diffusion with another such domain, which represents the exterior. Analysis of the resulting ordinary differential equations shows that generally there are two stable steady states, with slow and fast reactions respectively. In light of these results, numerical exploration of the S_0, H_0 parameter plane revealed the possibility of an oscillating front that moves back and forth periodically, normally to the membrane boundary (Caplan, Naparstek, and Zabusky [**1973**])–and the oscillations were in fact found experimentally. Recently Hardt et al. [**1976**] have shown that there are, in addition, parameter ranges permitting propagation of signals along the membrane.

We conclude with a brief and selective citation of some other papers concerned with the interaction-diffusion equation, to provide some feeling for the scope of present activity.

Kernevez and Thomas [**1975**] have given a rather extensive mathematical discussion of problems arising from analysis of reactions catalyzed by enzymes bound in membranes. Various existence and uniqueness theorems are proved, multiplicity of solutions is demonstrated in certain instances, numerical results are presented, and some relevant optimal control problems are discussed.

The work of Mitchell and Murray [**1973**] provides a good example of mathematical clarification of an important biological problem. These authors investigate the facilitated diffusion of oxygen by hemoglobin. Here the flux of the oxygen-hemoglobin complex across a layer provides a mechanism for oxygen transfer that supplements simple diffusion of the oxygen itself. A singular perturbation approach to the reaction-diffusion equations provides easily calculated results that are in good accord with experiment for a broad range of parameters.

Another example of reaction-diffusion theory that is closely guided by experimental results is work on the development of patterns in hydra recently summarized by Meinhardt and Gierer [**1974**]. Here the interaction of a short-range activating chemical and a long-range inhibitor is stressed. The "range" requirements on activation and inhibition are the same as the relatively fast diffusivity of the stabilizer found in Turing's [**1952**] models and explained (as we have noted) by Segel and Jackson [**1972**]. Among other contributions, Meinhardt and Gierer have introduced a slight but noteworthy change in emphasis: instead of looking for symmetry-breaking instabilities they search for situations (apparently more in accord with biological reality) in which small pre-existing inhomogeneities are amplified into marked patterns. In the language of elasticity theory, the computer simulations of Meinhardt and Gierer are concerned with imperfection sensitive situations.

Representative of considerable work by the "Brussells school" is Goldbeter's [**1973**] report of simulations that show periodic formation, collision, and annihilation of sharp wavefronts in a system wherein the product of a reaction exerts positive feedback through a conformational change of the catalyzing enzyme.

At the moment, together with Hanna Parnas of the Hebrew University, Jerusalem, I am working on a computer simulation of slime mold aggregation. Here the behavior of the chemotactic slime mold amoebae is represented by an algorithm containing rules for movement and chemical secretion that take into account such factors as delays and refractory periods whose existence is indicated by experiment. Destruction of a secreted attractant by an enzyme is an important facet of the problem. The formulation that was mentioned in the previous section can be regarded as a rough average, perhaps too rough, of the model considered in our simulation.

One feature of the system that caused us some difficulty is a "stiffness" owing to the fact that the diffusion constants for the chemicals are much larger than the effective diffusion constant that describes the random component of the amoeba motion. It seems possible to surmount this difficulty by letting the chemical "diffuse" for several time increments before computing the net amoeba motion for the elapsed period. Preliminary results indicate that we can anticipate at least partial progress toward our goal of seeing which factors are dominant in producing the various types of organized behavior that is observed in slime mold aggregation.[7]

Stiffness is a factor common to the biological problem just mentioned and an important practical application of reaction-diffusion equations, the analysis of chemical reaction and transport in air pollution (Reynolds, Roth and Seinfeld [**1973**]). Most of the work in this area has been numerical. The degree of stiffness that seems to have been overcome (at least in some cases) is indicated by the fact that relevant time constants can differ by a factor of 10^{14} (Chang, Hindmarsh, and Madsen [**1973**]).

Chemical and biochemical pattern formation, spatial inhomogeneities in species distribution and in morphogenesis, spatio-temporal patterns of pollutants–these applications assure that studies of the interaction-diffusion equations will play a prominent part in neoclassical continuum mechanics.

References

D. G. Aronson and H. F. Weinberger (1975), *Nonlinear diffusion in population genetics, combustion, and nerve propagation*, Proc. Tulane Program in Partial Differential Equations, Lecture Notes in Math., Springer-Verlag, New York.

B. J. Bok (1972), *Updating galactic spiral structure*, Amer. Sci. **60**, 708–722.

S. R. Caplan, A. Naparstek and N. J. Zabusky (1973), *Chemical oscillations in a membrane*, Nature **245**, 364–367.

J. S. Chang, A. C. Hindmarsh and N. K. Madsen (1973), *Simulation of chemical kinetics transport in the stratosphere*, Stiff Differential Systems, Plenum Press, New York, pp. 51–65. MR **49** #8359.

D. S. Cohen and J. P. Keener (1975), *Oscillatory processes in the theory of particulate formation in supersaturated chemical solutions*, SIAM J. Appl. Math. **28**, 307–318.

A. Goldbeter (1973), *Patterns of spatiotemporal organization in an allosteric enzyme model*, Proc. Nat. Acad. Sci. U.S.A. **70**, 3255–3259.

S. Hardt, A. Naparstek, L. A. Segel and S. R. Caplan (1976), *Spatio-temporal structure formation and signal propagation in a homogeneous enzymatic membrane*, Proc. Internat. Sympos. on Analysis and Control of Immobilized Enzyme Systems, North-Holland, Amsterdam.

W. Jaunzemis (1967), *Continuum mechanics*, Macmillan, New York.

E. F. Keller and G. M. Odell (1976), *Necessary and sufficient conditions for traveling bands of chemotactic bacteria*, Math. Biosci. **27**, 309–317.

[7]NOTE ADDED IN PROOF. First results of the amoeba simulations can be found in Parnas and Segel [**1977**].

E. F. Keller and L. A. Segel (1970), *Initiation of slime mold aggregation viewed as an instability*, J. Theoret. Biology **26**, 399–415.

______ (1971), *Travelling bands of chemotactic bacteria: a theoretical analysis*, J. Theoret. Biology **30**, 235–248.

J. P. Kernevez and D. Thomas (1975), *Numerical analysis and control of some biochemical systems*, Appl. Math. Optimization **1**, 222–285.

N. Kopell and L. N. Howard (1974), *Pattern formation in the Belousov reaction*, Lectures on Math. in the Life Sciences, vol. 7, Amer. Math. Soc., Providence, R.I., pp. 201–216.

I. R. Lapidus and R. Schiller (1975), *Bacterial chemotaxis in a fixed nutrient gradient*, J. Theoret. Biology **53**, 215–222.

S. A. Levin and R. T. Paine (1974), *Disturbance, patch formation, and community structure*, Proc. Nat. Acad. Sci. U. S. A. **71**, 2744–2747.

C. C. Lin and L. A. Segel (1974), *Mathematics applied to deterministic problems in the natural sciences*, Macmillan, New York.

H. Meinhardt and A. Gierer (1974), *Applications of a theory of biological pattern formation based on lateral inhibition*, J. Cell. Sci. **15**, 321–346.

P. J. Mitchell and J. D. Murray (1973), *Facilitated diffusion: the problem of boundary conditions*, Biophysik **9**, 177–190.

R. Nossal and G. Weiss (1973), *Analysis of a densitometry assay for bacterial chemotaxis*, J. Theoret. Biology **41**, 143–147.

G. M. Odell and E. F. Keller (1976), *Traveling bands of chemotactic bacteria revisited*, J. Theoret. Biology **56**, 243–248.

H. Parnas and L. A. Segel (1977), *Computer evidence concerning the chemotactic signal in aggregating Dictyostelium discoideum*, J. Cell Sci. (in press).

S. R. Reynolds, P. M. Roth and J. H. Seinfeld (1973), *Mathematical modeling of photochemical air pollution*. I, Atmospheric Environment **7**, 1033–1061; II, ibid. **8**, 97–130, 1974); III, ibid. **8**, 563–596, 1974.

J. Rinzel (1975), *Neutrally stable traveling wave solutions of nerve conduction equations*, J. Math. Biol. **2**, 205–217.

G. Rosen (1975), *Analytical solution to the initialvalue problem for traveling bands of chemotactic bacteria*, J. Theoret. Biology **49**, 311–321.

L. Rosenhead (editor) (1963), *Laminar boundary layers*, Clarendon Press, Oxford. MR **27** #5433.

S. I. Rubinow (1973), *Mathematical problems in the biological sciences*, Regional Conference Series in Appl. Math., no. 10, SIAM, Philadelphia, Pa.

T. L. Scribner, L. A. Segel and E. H. Rogers (1974), *A numerical study of the formation and propagation of travelling bands of chemotactic bacteria*, J. Theoret. Biology **46**, 189–219.

J. B. Serrin, Jr. (1959), *Mathematical principles of classical fluid mechanics*, Encyclopedia of Physics, Handbuch der Physik, Band 8/1, Strömungsmechanik I, Springer-Verlag, Berlin, pp. 125–263. MR **21** #6836b.

L. A. Segel (1976), *Incorporation of receptor kinetics into a model for bacterial chemotaxis*, J. Theoret. Biol. **57**, 23–42.

______ (1977a), *Mathematical models for cellular behavior*, A Study in Mathematical Biology (S. Levin, ed.), Mathematical Association of America (to appear).

______ (1977b), *Mathematics applied to continuum mechanics*, Macmillan, New York.

______ (1977c), *A theoretical study of receptor mechanisms in bacterial chemotaxis*, SIAM J. Appl. Math. (in press).

L. A. Segel and J. L. Jackson (1973), *Theoretical analysis of chemotactic movement in bacteria*, J. Mechanochem. and Cell Motility **2**, 25–34.

______ (1972), *Dissipative structure: an explanation and an ecological example*, J. Theoret. Biology **37**, 545–559.

L. A. Segel and S. A. Levin (1976), *Application of nonlinear stability theory to the study of the effects of diffusion on predator-prey interactions*, Topics in Statistical Mechanics and Biophysics, R. Piccirelli, ed., American Institute of Physics AIP Conference Proceedings **27**, 123–152.

L. A. Segel and B. Stoeckley (1972), *Instability of a layer of chemotactic cells, attractant, and degrading enzyme*, J. Theoret. Biology **37**, 561–585.

A. Turing (1952), *The chemical basis for morphogenesis*, Philos. Trans. Roy. Soc. London B **237**, 37–72.

A. Winfree (1975), *Rotating solutions to reaction/diffusion equations in simply-connected media*, Mathematical Aspects of Chemical and Biochemical Problems and Quantum Chemistry, SIAM-AMS Proc., vol. 8, Amer. Math. Soc., Providence, R. I., pp. 17–31.

DEPARTMENT OF APPLIED MATHEMATICS, THE WEIZMANN INSTITUTE OF SCIENCE, REHOVOT, ISRAEL (Current address)

DEPARTMENT OF MATHEMATICAL SCIENCES, RENSSELAER POLYTECHNIC INSTITUTE, TROY, NEW YORK 12181

Lectures in Applied Mathematics
Volume 16, 1977

Perturbation Theory

Donald S. Cohen[1]

Preface. These notes formed the basis for five seventy-five minute lectures presented at the Summer Seminar on Continuum Modeling held at Rensselaer Polytechnic Institute, Troy, New York in July, 1975. In such a short amount of time it is impossible to cover very much except a quick introduction to a few of the currently useful techniques. Indeed, an entire academic year's graduate course is devoted to perturbation and asymptotic methods at the California Institute of Technology, and prerequisite for the course is a knowledge of standard asymptotic methods for integrals and linear ordinary differential equations.

Our attitude will very much be that we will use all means at our disposal in an attempt to understand qualitatively and/or quantitatively the problem at hand. We shall see that a technique or method (such as singular perturbation or multi-scaling) is simply a name for a large class of methods applicable in widely disparate situations, but there are common conceptual properties. Thus, rather than attempt any general definitions (which would be of questionable value anyway) we shall present the ideas via several examples.

To understand the properties of a difficult problem almost always requires more than a knowledge of the properties of the various differential and integral operators involved. It is just not enough to think formally about the class of problems of which the one of interest is a special case, and the ability to go further is, in fact, almost an art which can best be learned by a study of how previous concrete problems have been successfully analyzed.

The basic properties of many of the nonlinear problems we shall study are of such a nature that they are not contained in linear systems, and thus linearization a priori precludes the phenomena we are seeking. Furthermore, akin to concluding periodicity of the cosine from the first three terms of its Taylor's series, power series expansions of solutions in terms of a small parameter may be of little use. We shall see that techniques such as singular perturbation and multi-scaling provide genuine extensions to the classical methods of linearization and regular power series expansions in terms of a small parameter, and when coupled with the classical techniques they provide powerful tools for studying some difficult problems.

AMS (*MOS*) *subject classifications* (1970). Primary 41–02, 41A60, 35E15, 35B25; Secondary 80A30.

[1]This work was partially supported by the U. S. Army Research Office under Contract DAHC-04-68-C-0006 and the National Science Foundation under Grant GP-32157X2.

I. Introduction to singular perturbation theory. There commonly occur many types of differential equations containing a small parameter ε for which the classical asymptotic methods yielding expansions of Poincaré type (i.e., essentially expressions analytic in ε in the entire domain of interest) are not appropriate. We shall study one class of such problems.

Consider

$$\varepsilon y'' + a(x)y' + b(x)y = 0, \qquad 0 < x < 1, 0 < \varepsilon \ll 1, \tag{1.1}$$

$$y(0) = A, \tag{1.2}$$

$$y(1) = B. \tag{1.3}$$

This is a kind of problem known as a singular perturbation. Systematic techniques exist for constructing asymptotic approximations of the solution for small ε. As is so often the case in applied mathematics, the basic ideas suggesting the technique and its implementation arise from detailed study of various model and actual problems with a view to precisely what geometrical, analytical, and physical properties are important in the various parts of the solution. For example, what terms in a differential equation play the dominant roles in the various parts of the domain of interest? In the present problem it is clearly the case that for special forms of the coefficients $a(x)$ and $b(x)$ the problem can be solved exactly in terms of known elementary and special functions. If several such special solutions are found and sketched for small ε, the underlying reasons for our techniques and classifications with regard to properties of $a(x)$ and $b(x)$ should become clear. In fact, for $a(x)$ and $b(x)$ constant we find that the typical situation for small ε is that which is pictured in Figure 1.1.

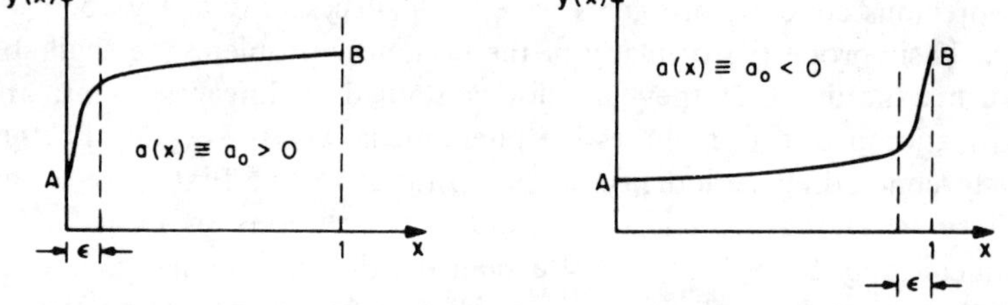

FIGURE 1.1

Since $0 < \varepsilon \ll 1$, we might suspect initially that the term $\varepsilon y''$ will be small enough to neglect in some sort of first approximation. If we do this, we see immediately that the order of our differential equation drops from second to first, and thus, in general, both boundary conditions cannot be satisfied. Hence, it must be the case that there is a part or parts of the interval on which the "curvature" y'' is so large that the product $\varepsilon y''$ cannot be neglected. We see from Figure 1.1 that this is indeed the case in a small interval (called the boundary layer) near $x = 0$ if $a(x) \equiv$ constant $a_0 > 0$ and near $x = 1$ if $a(x) \equiv$ constant $a_0 < 0$. We shall now implement our observations by developing the singular perturbation method in this case.

Assume that

$$a(x) > 0 \quad \text{for } 0 \leqslant x \leqslant 1. \tag{1.4}$$

(The case where $a(x) < 0$ on $0 \leqslant x \leqslant 1$ is handled the same way except that instead of a boundary layer near $x = 0$, there will be a boundary layer near $x = 1$.) We shall consider the case where $a(x)$ vanishes somewhere in $0 \leqslant x \leqslant 1$ later. In the outer region (i.e., away from boundary layers) we neglect $\varepsilon y''$ in the first approximation. Then,

$$a(x)y' + b(x)y = 0, \tag{1.5}$$

$$y(1) = B, \tag{1.6}$$

the solution of which is

$$y(x) = B \exp\left(\int_x^1 \frac{b(t)}{a(t)}\, dt\right). \tag{1.7}$$

Clearly, in general, this will not satisfy the condition that $y(0) = A$.

To assess the situation in the boundary layer near $x = 0$ we shall magnify this small interval by means of the change of variable

$$\tilde{x} = x/\phi(\varepsilon), \tag{1.8}$$

where $\phi(\varepsilon)$ is as yet unspecified except that we require $\phi(\varepsilon) \to 0$ as $\varepsilon \to 0$. Let $y(x) = y(\phi(\varepsilon)\tilde{x}) \equiv w(\tilde{x})$. Furthermore, near $x = 0$, we have by means of a Taylor's series that

$$a(x) = a(\phi(\varepsilon)\tilde{x}) = a_0 + a_1\tilde{x} + a_2\tilde{x}^2 + \dots,$$

$$b(x) = b(\phi(\varepsilon)\tilde{x}) = b_0 + b_1\tilde{x} + b_2\tilde{x}^2 + \dots.$$

Then, equation (1.1) becomes

$$\frac{\varepsilon}{\phi^2(\varepsilon)} w'' + [a_0 + a_1\tilde{x} + \dots]\frac{1}{\phi(\varepsilon)} w' + [b_0 + b_1\tilde{x} + \dots]w = 0. \tag{1.9}$$

We shall assume that in the stretched variable $\tilde{x}$ significant changes in w occur in distances of order unity and that the relative importance of the various terms in (1.9) is measured explicitly by the size of their coefficients. We wish to retain the dominant terms in (1.9) in such a way that we account for the effect of the second derivative in the boundary layer. If $\phi(\varepsilon) > \varepsilon$, then the dominant term is $a_0 w'$, which is not in keeping with our assumption that the w'' must come into play in a major way in the layer. If $\phi(\varepsilon) < \varepsilon$, then the dominant term is simply the second derivative, but then this couples no information from the other terms which govern in the outer region. (We shall see below when we discuss "matching" that this is not a possibility on other grounds.) If $\phi(\varepsilon) = \varepsilon$, then both w'' and a coupling to other terms are achieved, and we obtain the dominant terms in (1.9) as

$$w'' + a_0 w' = 0. \tag{1.10}$$

The solution of (1.10) satisfying $w(0) = A$ is

$$w(\tilde{x}) = A + c(e^{-a_0\tilde{x}} - 1), \tag{1.11}$$

where c is a constant of integration.

We now assume that our outer expansion (1.7) and inner expansion (1.11) are mutually valid in some common interval, and we evaluate the constant c by requiring that both formulas represent the same function as $\varepsilon \to 0$ in this common region. To accomplish this we shall write both the outer and inner expansions in terms of a common "intermediate" variable x_η which represents a point in the interval where both are valid descriptions of the solution. Thus, we let $x_\eta = x/\eta(\varepsilon)$ where $\eta(\varepsilon)$ is some function of ε such that $\eta(\varepsilon) \to 0$ as $\varepsilon \to 0$ and such that $\varepsilon \ll \eta(\varepsilon) < 1$. Then,

$$x = \eta(\varepsilon)x_\eta \to 0 \quad \text{as } \varepsilon \to 0 \text{ for fixed } x_\eta,$$

$$\tilde{x} = x/\varepsilon = (\eta(\varepsilon)/\varepsilon)x_\eta \to \infty \quad \text{as } \varepsilon \to 0 \text{ for fixed } x_\eta.$$

Hence, the effect of fixing our position at $x = x_\eta$ and letting $\varepsilon \to 0$ is equivalent to letting $x \to 0$ and $\tilde{x} \to \infty$. Alternatively, fixing our position at $x = x_\eta$ and letting $\varepsilon \to 0$ is equivalent to requiring that the outer expansion as $x \to 0$ equal the inner expansion as $\tilde{x} \to \infty$. In other words, this inner expansion as we proceed out of the boundary layer must equal the outer expansion as we proceed into the boundary layer. Now,

$$y \to B \exp\left(\int_0^1 \frac{b(t)}{a(t)}\, dt\right) \quad \text{as } x \to 0,$$

and $w \to A - c$ as $\tilde{x} \to \infty$. Thus, in order to match we must require that

$$c = A - B \exp\left(\int_0^1 \frac{b(t)}{a(t)}\, dt\right).$$

(Note that if $a_0 < 0$, then the boundary layer solution (1.11) would grow exponentially as $\tilde{x} \to \infty$, and matching would be impossible. In the case that $a_0 < 0$, however, as our sketches of Figure 1.1 show, the boundary layer is at $x = 1$. We then magnify near $x = 1$ by setting $\tilde{x} = (1 - x)/\varepsilon$, and we find exponential decay in the boundary layer solution. Matching then proceeds just as we have done above.)

In summary, then, we have that to order ε,

$$y \sim B \exp\left(\int_x^1 \frac{b(t)}{a(t)}\, dt\right) \quad \text{as } \varepsilon \to 0, \tag{1.12}$$

outside the boundary layer, and in the boundary layer

$$y \sim B \exp\left(\int_0^1 \frac{b(t)}{a(t)}\, dt\right) + \left[A - B \exp\left(\int_0^1 \frac{b(t)}{a(t)}\, dt\right)\right] \exp\left(-\frac{a_0 x}{\varepsilon}\right)$$

$$\text{as } \varepsilon \to 0. \tag{1.13}$$

The inner and outer expansions have some terms in common, namely those terms which are matched. If the two expansions are added and their common part subtracted, then we obtain an expansion uniformly valid to $O(\varepsilon)$ on the entire interval $0 \leqslant x \leqslant 1$. This is

$$y(x) \sim B \exp\left(\int_x^1 \frac{b(t)}{a(t)}\, dt\right)$$
$$+ \left[A - B \exp\left(\int_0^1 \frac{b(t)}{a(t)}\, dt\right)\right] \exp\left(-\frac{a_0 x}{\varepsilon}\right). \tag{1.14}$$

A more detailed exposition of this method together with its logical extension for generating higher order correction terms (i.e., asymptotic series to n terms) can be found in references [**1**]–[**3**].

If we consider some problems with exact solutions and then find asymptotic expansions of those solutions for small ε, we find that those expansions agree with the ones we obtain with our procedure in every case. This certainly gives us confidence in our method, but it is no rigorous proof that the expansions we generate for all our problems are indeed asymptotic expansions of the solution. Throughout this paper we will continually raise many fundamental questions of a mathematical nature. These will be pointed out, but for the most part we shall convince ourselves of the validity of our results by other means. However, there are many outstanding questions which can be resolved only by a rigorous study of the problem, and in these cases an adoption of the results of the formalism can sometimes lead to error. We shall now illustrate one such situation with the following example due to G. F. Carrier [**4**]:

$$\varepsilon u'' + 2(1 - x^2)u + u^2 = 1, \qquad -1 < x < 1, 0 < \varepsilon \ll 1, \tag{1.15}$$
$$u(-1) = 1, \tag{1.16}$$
$$u(1) = 0. \tag{1.17}$$

Our goal is to obtain some feeling for the nature of the solution or solutions for small $\varepsilon > 0$.

Here again we believe that there are large parts of the interval $-1 \leqslant x \leqslant 1$ on which the term $\varepsilon u''$ is negligible in the first approximation. In these regions the solution (the outer solution or outer expansion) is obtained from $2(1 - x^2)u + u^2 = 1$. Hence, there are two candidates for the outer solution, namely,

$$u_1(x) = (x^2 - 1) + \sqrt{(1 - x^2)^2 + 1}\,, \tag{1.18}$$
$$u_2(x) = (x^2 - 1) - \sqrt{(1 - x^2)^2 + 1}\,. \tag{1.19}$$

These are illustrated in Figure 1.2. Since neither $u_1(x)$ nor $u_2(x)$ satisfy the boundary conditions, we expect boundary layers at both

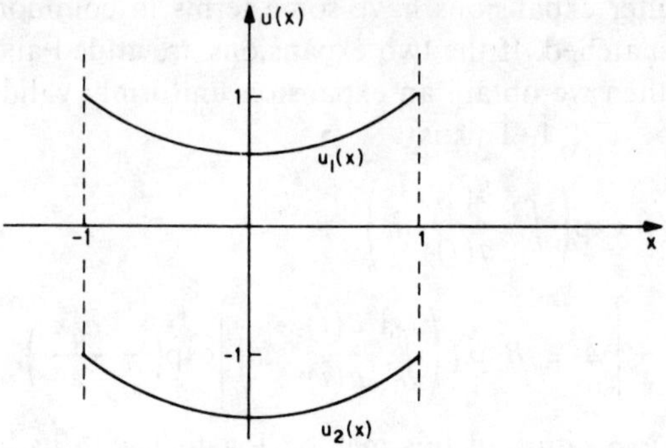

FIGURE 1.2

endpoints. Thus, we anticipate small regions (boundary layers) near the endpoints at $x = \pm 1$ where the curvature of the solution is so great that the solution goes from zero at the endpoint to match with the outer solution. We shall now proceed to show that, in fact, we can construct such boundary layer solutions (the inner solutions or inner expansions) to match with $u_2(x)$ but that such inner solutions *cannot* be constructed to match with $u_1(x)$.

To assess the behavior near $x = 1$ we shall magnify this region by letting $\tilde{x} = (x - 1)/\phi(\varepsilon)$, where $\phi(\varepsilon)$ is as yet unspecified except that we require that $\phi(\varepsilon) \to 0$ as $\varepsilon \to 0$. Let $u(x) = u(1 + \phi(\varepsilon)\tilde{x}) \equiv w(\tilde{x})$. Then, the equation (1.15) becomes

$$\varepsilon w''/\phi^2(\varepsilon) + 2\left(-2\phi(\varepsilon)\tilde{x} - \phi^2(\varepsilon)\tilde{x}^2\right)w + w^2 = 1. \tag{1.20}$$

It is clear that in order to account for the effect of the curvature u'' and at the same time couple this to the outer solution we must choose $\phi(\varepsilon) = \varepsilon^{1/2}$. With this choice the dominant terms in (1.20) are given by

$$w'' + w^2 = 1. \tag{1.21}$$

Similarly, if we magnify the region near $x = -1$ by $\tilde{x} = (x + 1)/\psi(\varepsilon)$, we find that $\psi(\varepsilon) = \varepsilon^{1/2}$ and the boundary layer equation is again given by (1.21).

Thus, the boundary layers are of width $O(\varepsilon^{1/2})$, and in the boundary layers the inner solutions will satisfy $w'' + w^2 = 1$. We must now study this boundary layer equation to see if it possesses solutions which satisfy the boundary conditions at $\tilde{x} = 0$ and which match with u_1 and/or u_2 as $\tilde{x} \to \pm\infty$. More precisely, if we are to construct inner solutions matching both $u_1(x)$ and $u_2(x)$, then we require:

left-hand boundary layer to match $u_1 \Rightarrow$ we need a solution such that $w(0) = 0$ and $\lim_{\tilde{x}\to\infty}[w(\tilde{x})] = 1$ since $u_1(-1) = 1$;

right-hand boundary layer to match $u_1 \Rightarrow$ we need a solution such that $w(0) = 0$ and $\lim_{\tilde{x}\to-\infty}[w(\tilde{x})] = 1$ since $u_1(1) = 1$;

left-hand boundary layer to match $u_2 \Rightarrow$ we need a solution such that $w(0) = 0$ and $\lim_{\tilde{x}\to\infty}[w(\tilde{x})] = -1$ since $u_2(-1) = -1$;

right-hand boundary layer to match $u_2 \Rightarrow$ we need a solution such that $w(0) = 0$ and $\lim_{\tilde{x}\to-\infty}[w(\tilde{x})] = -1$ since $u_2(1) = -1$.

To see if such solutions exist we shall study the (w, w') plane for $w'' + w^2 = 1$. Let $v = w'$. Then, (1.21) is equivalent to

$$v' = 1 - w^2, \qquad w' = v,$$

or

$$dv/dw = (1 - w^2)/v. \tag{1.22}$$

Equation (1.22) has singular points at $(w, v) = (\pm 1, 0)$. Upon linearizing about each of the singular points, we find that the point $(-1, 0)$ is a saddle point and the point $(1, 0)$ is a center and that the phase plane trajectories look as illustrated in Figure 1.3. An infinite number of trajectories pass through lines $w = \pm 1$. For matching,

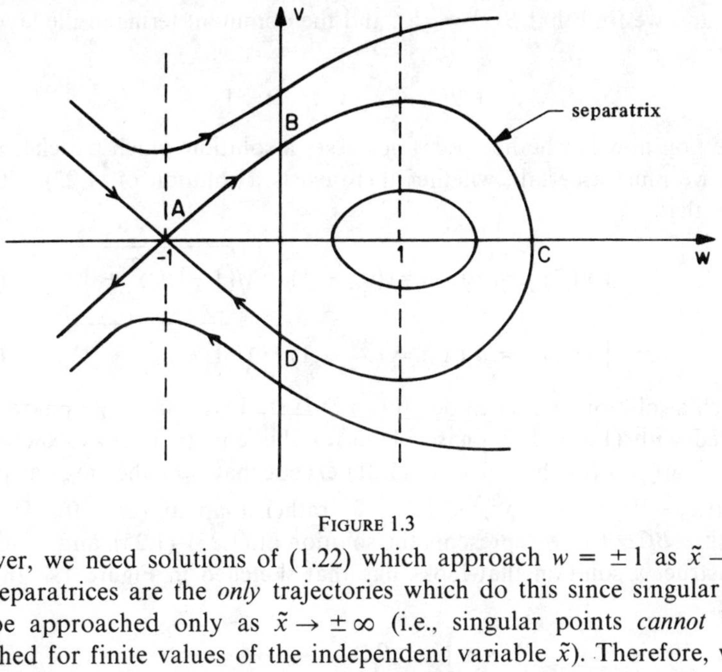

FIGURE 1.3

however, we need solutions of (1.22) which approach $w = \pm 1$ as $\tilde{x} \to \pm\infty$. The separatrices are the *only* trajectories which do this since singular points can be approached only as $\tilde{x} \to \pm\infty$ (i.e., singular points *cannot* be approached for finite values of the independent variable $\tilde{x}$). Therefore, we can immediately conclude that there are not solutions w which approach $+1$ as $\tilde{x} \to \pm\infty$. Thus, we *cannot* construct inner solutions to match $u_1(x)$, and we suspect that no such solution exists.

We can, however, construct boundary layer solutions matching $u_2(x)$. We now do this. The part AB of the separatrix satisfies $w = 0$ and $\lim_{\tilde{x}\to-\infty}[w(\tilde{x})] = -1$. We can make $w = 0$ for $\tilde{x} = 0$ since if $w(x)$ is a solution of (1.21) then so also is $w(\tilde{x} + k)$ for any constant k. Thus, we can construct a right-hand inner solution matching u_2. Similarly, the part DA of the separatrix provides us with a left-hand boundary layer solution. Hence, we expect that there does exist a solution of (1.15)–(1.17) which looks like $u_2(x)$ on $-1 + \varepsilon^{1/2} < x$

$< 1 - \varepsilon^{1/2}$ and that in boundary layers of thickness $O(\varepsilon^{1/2})$ near both endpoints the solution quickly grows from $u_2(x)$ to zero at $x = \pm 1$.

We have obtained a good idea of the nature of our solution without actually solving the inner problem. Although it is not necessary to do so, we can solve the inner problem analytically in this problem. The solution of $w'' + w^2 = 1$ satisfying $w(0) = 0$ and $w \to -1$ as $x \to -\infty$ is

$$w(x) = -1 + 3\,\mathrm{sech}^2\left(x/\sqrt{2} + \mathrm{arctanh}\, 2/3\right),$$

with a similar formula for the inner solution near $x = -1$.

In this same problem let us now raise the question of the possibility of the existence of internal layers, that is, a narrow region not adjacent to the boundary in which the solution undergoes rapid changes. Suppose we have such an internal layer at some point $x = x_0$. Proceeding just as above, we magnify this layer by setting $\tilde{x} = (x - x_0)/\beta(\varepsilon)$ and $u(x) = u(x_0 + \beta(\varepsilon)\tilde{x}) \equiv y(\tilde{x})$, and we find that $\beta(\varepsilon) = \varepsilon^{1/2}$ and the dominant terms in the layer are given by

$$y'' + 2(1 - x_0^2)y + y^2 = 1. \tag{1.23}$$

The question now is whether (1.23) possesses a solution which matches $u_2(x)$. That is, we must ascertain whether there exists a solution of (1.23) with the property that

$$\lim_{\tilde{x}\to\infty}\left[y(\tilde{x})\right] = u_2(x_0) = (x_0 - 1) - \sqrt{\left(1 - x_0^2\right)^2 + 1}\,, \tag{1.24}$$

$$\lim_{\tilde{x}\to-\infty}\left[y(\tilde{x})\right] = u_2(x_0) = (x_0 - 1) - \sqrt{\left(1 - x_0^2\right)^2 + 1}\,. \tag{1.25}$$

That such a solution exists can be proved by a study of the (y, y') phase plane associated with (1.23). The phase plane for this equation looks exactly like that of Figure 1.3 for the equation (1.21) except that now the singular points are at $((x_0 - 1) \pm ((1 - x_0^2)^2 + 1)^{1/2}, 0)$ rather than at $(\pm 1, 0)$. The full separatrix $ABCDA$ now represents the solution of (1.23)–(1.25), and hence, we can construct a solution that looks like that sketched in Figure 1.4. In fact, what our

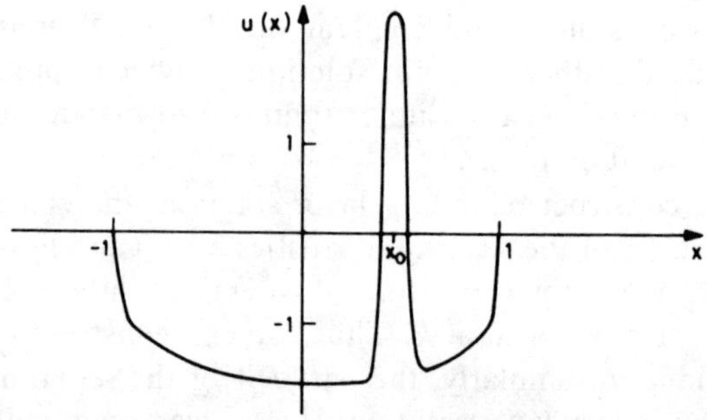

FIGURE 1.4

formalism has shown is that we can construct internal boundary layers at *any* point x_0 in the interval $(-1, 1)$, and so we have constructed solutions with any number of internal layers at any arbitrary points in $(-1, 1)$. This particular problem can be analyzed rigorously by phase plane techniques, and indeed, it is found that many solutions do exist but not as many as we have constructed with arbitrary placement of interior layers. We must conclude that our formalism leads to spurious solutions and that in this case we need either a rigorous mathematical study or some additional information on which to judge our constructions. An elegant and very nice beginning on the mathematical theory of these types of singular perturbation problems for both ordinary and partial differential equations has recently been given by P. C. Fife [**5**]–[**7**].

APPENDIX. To gain some facility with the technique for identifying and constructing boundary layer type expansions, the reader should work out some problems. Some particularly good problems (some of which are quite difficult) which illustrate points not covered in this paper are now simply listed. They can all be found worked out in the text of J. D. Cole [**1**].

1. $\varepsilon y'' + \sqrt{x}\, y' - y = 0$ subject to $y(0) = 0, y(1) = e^2$. The boundary layer is of width $O(\varepsilon^{2/3})$, and several terms in the boundary layer solution are needed to match one term in the outer solution.

2. $u'' + u'/x + uu' = 0$ subject to $u(\varepsilon) = 0$, $u(\infty) = 1$. There is a boundary layer of width $O(\varepsilon)$ near $x = \varepsilon$. Although all our examples had the small parameter multiplying the highest derivative, this is not necessary for a singular perturbation problem as shown by this example. Also, this example requires terms like $\varepsilon^i (\ln \varepsilon)^j$ in the asymptotic sequence.

3. $\varepsilon y'' + yy' - y = 0$ subject to $y(0) = A$, $y(0) = B$.Depending on the values of A and B there are many different kinds of solutions. Among the more interesting are the ones with internal boundary layers of the standard shock type which occur in fluid and gas dynamics.

REFERENCES

1. J. D. Cole, *Perturbation methods in applied mathematics*, Blaisdell, Waltham, Mass., 1968. MR **39** #7841.

2. A. H. Nayfeh, *Perturbation methods*, Wiley, New York, 1973.

3. R. E. O'Malley, *Introduction to singular perturbations*, Academic Press, New York, 1974.

4. G. F. Carrier and C. E. Pearson, *Ordinary differential equations*, Blaisdell, Waltham, Mass., 1968.

5. P. C. Fife, *Transition layers in singular perturbation problems*, J. Differential Equations **15** (1974), 77–105. MR **48** #9002.

6. P. C. Fife and W. M. Greenlee, *Interior transition layers for elliptic boundary value problems with a small parameter*, Uspehi Mat. Nauk (1974).

7. P. C. Fife, *Two point boundary value problems admitting interior transition layers* (to appear).

II. **Some special problems.** We shall introduce several current problems in singular perturbation theory by means of two special examples. Certain problems arising in the study of heat and mass transfer in porous catalysts [**1**]–[**4**] generate both initial and boundary value problems involving equations of the form

$$\varepsilon u'' - f(u, u')u' + g(u) = 0 \qquad (0 < \varepsilon \ll 1). \tag{2.1}$$

Peculiar physical effects involving local hot spots and local regions of instability (in a sense to be explained below) are observed experimentally. We believe that we have identified one possible mechanism for such occurrences, and this is reflected in a special mathematical situation first observed by N. Levinson [**5**]. Perhaps, the best way to introduce both the physics and the mathematics is to present Levinson's example; we now do this.

Consider the initial value problem

$$\varepsilon^2 u'' - 4u'/\left(3 + u'^4\right) + u = 0 \qquad (0 < \varepsilon \ll 1), \tag{2.2}$$

$$u(0) = 8/19, \tag{2.3}$$

$$u'(0) = 2. \tag{2.4}$$

Here again our goal is to obtain some feeling for the nature of the solution for $x > 0$. If we assume that in some outer region (i.e., region where $\varepsilon u''$ is negligible) the solution is given asymptotically by $u(x) \sim u_0(x) + \dots$, then clearly $u_0(x)$ satisfies the reduced problem (i.e., problem with $\varepsilon = 0$)

$$4u_0'/\left(3 + u_0'^4\right) = u_0, \tag{2.5}$$

$$u_0(0) = 8/19. \tag{2.6}$$

Note that the initial value $u_0'(0) = 2$ is compatible with (2.6) and the equation (2.5) evaluated at $x = 0$. Thus, we examine the solution of (2.5) which starts with initial values $u_0(0) = 8/19$, $u_0'(0) = 2$. Since $u_0'(0) = 2 > 0$, the solution u_0 increases from its initial value $u_0(0) = 8/19$. Now, $4a/(3 + a^4)$ is a decreasing function of a for increasing $a > 1$ so that when u_0 in (2.5) increases, u_0' must decrease. Thus, $u_0(x)$ increases with decreasing slope, and as $u_0'(x) \to 1-$, $u_0'(x) \to 1+$. Furthermore, $u_0(x)$ reaches the value of 1 for a finite value of x, which we denote by x_1, because

$$x_1 = \int_0^{x_1} dx = \int_{8/19}^{1} \frac{du_0}{u_0'} < \int_{8/19}^{1} du_0 = \frac{11}{19}.$$

Hence, $u_0(x_1) = 1$ and $u_0'(x_1) = 1$. This solution *cannot* be continued beyond $x = x_1$ because if $u_0(x)$ were continued, it would have to increase since $u_0'(x_1) = 1 > 0$. However, since $4a/(3 + a^4) \leqslant 1$ for all real a, this is impossible. Therefore, we conclude that the solution $u_0(x)$ of the reduced problem exists only in the interval $0 \leqslant x \leqslant x_1$.

Let us now study the full problem (2.2)–(2.4). Obviously, fundamental existence and uniqueness theorems guarantee that the solution $y(x)$ is continuable from any finite values of u and u'. We shall see that $u(x)$ is very close to $u_0(x)$ for $0 \leqslant x \leqslant x_1$, and beyond the point $x = x_1$ the solution $u(x)$ goes into rapid oscillation of amplitude approximately 1 and period $2\pi\varepsilon$ as illustrated in Figure 2.1. The oscillatory part of the solution on $x > x_1$

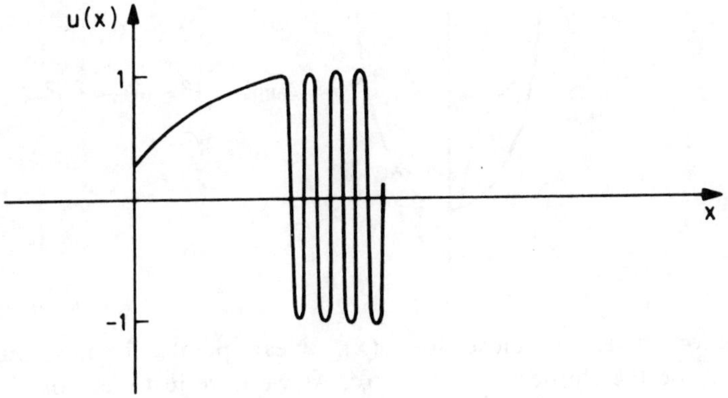

FIGURE 2.1

can be considered unstable since small changes in the initial conditions affect the phase.

To prove that $u(x)$ behaves as stated we shall study the (u, u') phase plane associated with (2.2). Let $v = u_2$. Then, (2.2) is equivalent to

$$\varepsilon^2 v' = 4v/(3 + v^4) - u, \tag{2.7}$$

$$u' = v. \tag{2.8}$$

It will also be convenient to express (2.7) and (2.8) in polar coordinate variables r and θ given by $u = r\cos\theta$, $v = \varepsilon^{-1} r \sin\theta$, so that $r^2 = u^2 + \varepsilon^2 v^2$, $\tan\theta = \varepsilon v/u$. In these variables equations (2.7), (2.8) become

$$\frac{dr^2}{dx} = \frac{8v^2}{3 + v^4} \geqslant 0, \qquad \frac{d\theta}{dx} = \frac{1}{\varepsilon}\left[\frac{4v}{3 + v^4}\,\frac{\cos\theta}{r} - 1\right]. \tag{2.9}$$

Since $|4v/(3 + v^4)| \leqslant 1$, we see that if $r > 1$, then $d\theta/dx < 0$. This fact together with the fact that $dr^2/dx \geqslant 0$ implies that for $r > 1$, a point (u, v) on the trajectory moves clockwise with r increasing. Now, the solution of our problem (2.2)–(2.4) starts at the point A of Figure 2.2 and proceeds along a trajectory from A to B very close to the curve $u = 4v/(3 + v^4)$. This is the part of

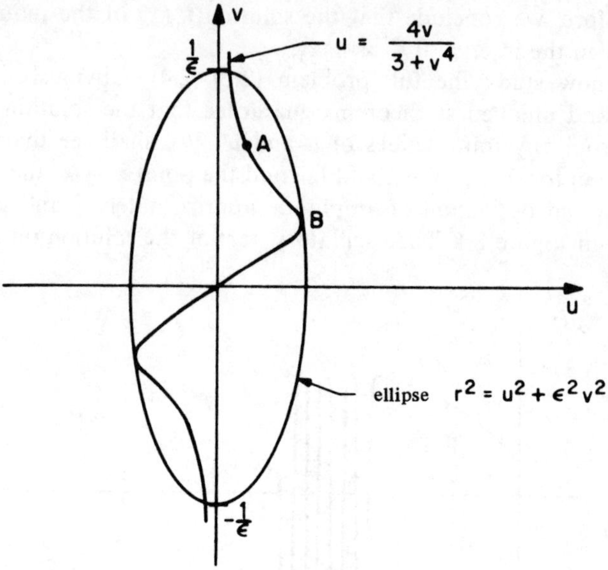

FIGURE 2.2

the solution which is close to $u_0(x)$. Near point B the trajectory passes outside the ellipse $r^2 = u^2 + \varepsilon^2 v^2$. As we have just seen for all points (u, v) outside the ellipse we have $dr^2/dx \geqslant 0$, $d\theta/dx < 0$, and thus, once the trajectory reaches such a point it then moves clockwise with r increasing. Therefore, past point B (corresponding to $x > x_1$) the solution $u(x)$ of (2.2)–(2.4) represents an increasing oscillation. To prove that it is a rapid oscillation of amplitude approximately 1 for small ε and period $2\pi\varepsilon$ requires certain precise estimates, the details of which can be found in Levinson's paper [**5**].

To the author's knowledge, Levinson's paper [**5**] is the only study made of this type of situation, namely the case where the solution of the reduced problem does not exist on the entire domain where a solution of the original problem is sought. (It should be noted that this kind of problem is basically different from those involving relaxation oscillations or interior layers, for example, where solutions of the reduced problem are discontinuous but nevertheless exist almost everywhere.) D. S. Cohen [**6**] has recently investigated a similar problem and its implication with regard to certain types of bifurcation in chemical systems.

For our second problem we consider the equations

$$u'' = v, \tag{2.10}$$

$$\varepsilon v'' + u'v' = 0 \qquad (0 < \varepsilon \ll 1), \tag{2.11}$$

subject to the conditions

$$u(0) = 0 = u(1), \tag{2.12}$$

$$0 \leqslant v(0) = v_0 < v_1 = v(1). \tag{2.13}$$

This problem was studied by F. W. Dorr and S. V. Parter [7] as a special case of more general problems which they also treated. To obtain some feeling for the nature of the solution we make the following observations: Equation (2.11) implies

$$v'(x) = c\exp(-u(x)/\varepsilon), \tag{2.14}$$

where c is a constant. Thus, $v'(x)$ is of one sign, and clearly since $0 \leqslant v_0 < v_1$, then we must have $c > 0$. Therefore, $v(x)$ is always positive. Thus, equation (2.10) implies $u''(x)$ is always positive, and this fact together with the conditions (2.12) implies that $u(x) \leqslant 0$ so that $u'(x)$ varies from negative to positive on $0 \leqslant x \leqslant 1$. Furthermore, (2.14) and (2.12) imply that $v'(0) = v'(1)$, and this together with the facts that $v' > 0$ and $v'' = -u'v'/\varepsilon$ implies that v'' varies from positive to negative on $0 \leqslant x \leqslant 1$ which means that v changes from convex to concave on $0 \leqslant x \leqslant 1$. Therefore, the solutions of (2.10), (2.11) must look as sketched in Figure 2.3. In an asymptotic expansion as $\varepsilon \to 0$ it is clear that the first term in

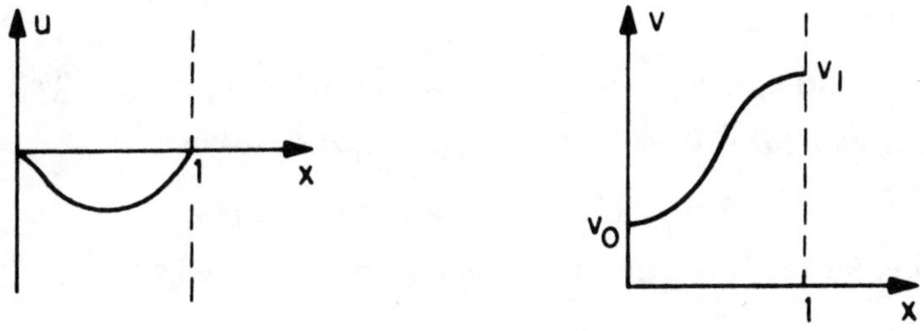

FIGURE 2.3

the outer expansion will come from $u'v' = 0$. Since we cannot have either $u \equiv 0$ or $u' \equiv 0$, then we must have $v' \equiv 0$. Hence, the first term in the outer expansion is $v \equiv$ constant. Thus, for small ε the solutions must look as sketched in Figure 2.4. To lowest order there is an interior layer in $v(x)$ at some point $x = \alpha$ in

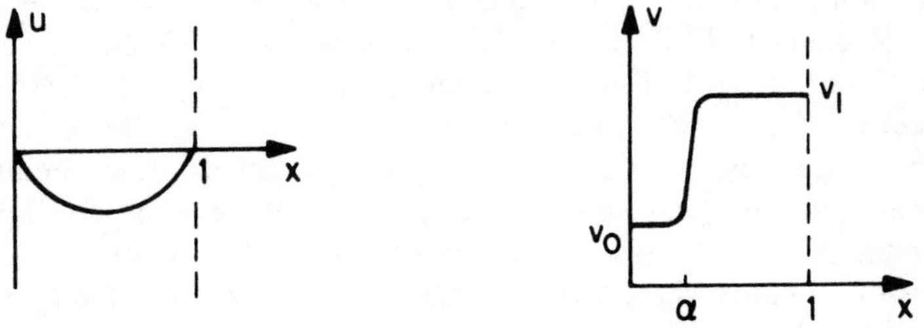

FIGURE 2.4

$0 \leqslant x \leqslant 1$, and u is smooth. We shall now find the approximations for u, v, and α.

Clearly,

$$v(x) = \begin{cases} v_0 & \text{for } 0 \leqslant x < \alpha, \\ v_1 & \text{for } \alpha < x \leqslant 1, \end{cases} \tag{2.15}$$

where α is such that $u'(\alpha) = 0$. Then, (2.10) implies that

$$u'' = v = \begin{cases} v_0 & \text{for } 0 \leqslant x < \alpha, \\ v_1 & \text{for } \alpha < x \leqslant 1, \end{cases} \tag{2.16}$$

the solution of which is

$$u(x) = \begin{cases} \frac{1}{2} v_0 x^2 + c_1 x + c_2 & \text{for } 0 \leqslant x < \alpha, \\ \frac{1}{2} v_1 x^2 + c_3 x + c_4 & \text{for } \alpha < x \leqslant 1, \end{cases} \tag{2.17}$$

where c_i, $i = 1, \ldots, 4$, are constants. Now, the conditions (2.12) immediately give us that $c_2 = 0$ and $c_4 = -\frac{1}{2} v_1 - c_3$, so that

$$u(x) = \begin{cases} \frac{1}{2} v_0^2 + c_1 x & \text{for } 0 \leqslant x < \alpha, \\ \frac{1}{2} v_1^2 + c_3 x - \frac{1}{2} v_1 - c_3 & \text{for } \alpha < x \leqslant 1. \end{cases} \tag{2.18}$$

We now impose the requirements that both u and u' be continuous at $x = \alpha$ to find that

$$c_3 = c_1 + \alpha(v_0 - v_1), \qquad c_1 = \tfrac{1}{2}\alpha^2(v_0 - v_1) + \alpha(v_1 - v_0) - \tfrac{1}{2} v_1. \tag{2.19}$$

Finally, from $u'(\alpha) = 0$ we find that $v_0 \alpha + c_1 = 0$, so that

$$\alpha = 1/(1 + \sqrt{d}\,) \quad \text{where } d_1 = v_0 / v_1. \tag{2.20}$$

Therefore, for the first terms in the outer expansion, we have

$$v(x) = \begin{cases} v_0 & \text{for } 0 \leqslant x < \alpha, \\ v_1 & \text{for } \alpha < x \leqslant 1, \end{cases}$$

$$u(x) = \begin{cases} \frac{1}{2} v_0 x^2 + c_1 x & \text{for } 0 \leqslant x \leqslant \alpha, \\ \frac{1}{2} v_1 x^2 + c_3 x - \frac{1}{2} v_1 - c_3 & \text{for } \alpha \leqslant x \leqslant 1, \end{cases}$$

where c_3, c_1, and α are given by (2.19) and (2.20). Note that $u(x)$ and $u'(x)$ are continuous on $0 \leqslant x \leqslant 1$. By appropriate scaling we can match a boundary layer solution to $v(x)$ in an interval of order $O(\varepsilon^{1/2})$ centered at $x = \alpha$.

The problem (1.1)–(1.3) of §I exhibits the same type of interior layer when the coefficient $a(x)$ vanishes at some point in $0 \leqslant x \leqslant 1$. (Here the coefficient u' of v' in (2.11) vanishes at the point $x = \alpha$ inside $0 \leqslant x \leqslant 1$.) With linear equations such problems are usually called turning point problems because in linear wave propagation problems these points correspond to the turning points of rays. Similar phenomena also occur in nonlinear problems of all kinds where they are called simply interior or internal layers. In particular, when derivatives appear nonlinearly in the equations the formal methods and the mathematical theory are far from complete. Some problems of this nature are discussed by D. S. Cohen **[8]**, and an excellent discussion of the current status of linear turning point problems is given by B. J. Matkowsky **[9]**.

References

1. V. W. Weekman, Jr. and R. L. Gorring, *Influence of volume change on gas-phase reactions in porous catalysts*, J. Catal. **4** (1965), 260–270.

2. V. W. Weekman, Jr., *Combined effect of volume change and internal heat and mass transfer on gas-phase reactions in porous catalysts*, J. Catal. **5** (1966), 44–54.

3. V. Hlavacek and M. Marek, *Effect of heat and mass transfer inside catalyst particles on the heterogeneous catalytic reaction*, Chem. Reaction Eng., (1971) Proc. Fourth European Symposium, Brussels, 1968.

4. V. Hlavacek and M. Kubicek, *Transient heat and mass transfer in a porous catalyst*. III, J. Catal. **22** (1971), 364–370.

5. N. Levinson, *An ordinary differential equation with an interval of stability, a separation point, and an interval of instability*, J. Math. Physics **28** (1949), 215–222. MR **11**, 722.

6. D. S. Cohen, *Localized instability in chemically reacting systems*, Rocky Mountain Math. J. (to appear).

7. F. W. Dorr and S. V. Parter, *Singular perturbations of nonlinear boundary value problems with turning points*, J. Math. Anal. Appl. **29** (1970), 273–293. MR **41** #7227.

8. D. S. Cohen, *Singular perturbation of nonlinear two-point boundary value problems*, J. Math. Anal. Appl. **43** (1973), 151–160.

9. B. J. Matkowsky, *On boundary layer problems exhibiting resonance*, SIAM Rev. **17** (1975), 82–100.

III. **Some problems involving partial differential equations.** Our goal will be to solve some problems and pose some more unsolved problems in chemical reactor theory involving a nonlinear partial differential equation of the form

$$u_t = \varepsilon \operatorname{div}(D \operatorname{grad} u) + \mathbf{c} \cdot \operatorname{grad} u + g(u, \lambda), \qquad 0 < \varepsilon \ll 1, \tag{3.1}$$

where D may be either constant or depend on u, $\mathbf{c}$ is a prescribed constant vector, and λ denotes a parameter. We shall complete the problem with the specification of g and the appropriate initial and boundary conditions later in this chapter. First, however, before becoming involved in the various complexities of the actual problems, let us attempt to develop our techniques and obtain some feeling for the possible structure involved by considering the simple model linear problem given by

$$\varepsilon(u_{xx} + u_{yy}) - u_x - bu_y = 0, \qquad b > 0, \quad (x, y) \text{ in } R, \tag{3.2}$$

$$u = f, \qquad (x, y) \text{ on } B. \tag{3.3}$$

Thus, for small ε we shall solve (3.2) in some region R, such as that illustrated in Figure 3.1, subject to the boundary condition that u is prescribed everywhere on the boundary B of R. Thus, f is a known function. A complete exposition of the technique we shall use can be found in the text of J. D. Cole [**1**], and we follow his presentation.

First, let us determine the properties of the reduced (or outer) equation. (We shall see that the outer expansion will describe the behavior essentially inside the region, and the inner expansion will describe the behavior near some parts of the boundary. Hence, "outer" and "inner" have no geometrical significance with respect to the exterior or interior of the region R. They are simply terms we employ to indicate the manipulations we perform.) If $u_0(x, y)$

denotes the first term in the asymptotic expansion of $u(x, y)$ as $\varepsilon \to 0$, then

$$\partial u_0/\partial x + b\,\partial u_0/\partial y = 0. \tag{3.4}$$

It should be clear (or in any event it follows immediately from the theory of first order partial differential equations) that (3.4) is equivalent to

$$dx/ds \equiv 1, \quad dy/ds = b, \quad du_0/ds = 0. \tag{3.5}$$

Hence,

$$u_0(x, y) \equiv \text{constant on lines } bx - y = \text{constant}. \tag{3.6}$$

The lines $bx - y =$ constant (the characteristics for the first order partial differential equation (3.4)) are called subcharacteristics of the equation (3.2).

The possible structure of the solution of (3.2), (3.3) should now suggest itself. To lowest order in ε, inside the region R the solution $u(x, y)$ is constant on the subcharacteristics $bx - y =$ constant. Since the boundary values are not, in general, the same at both ends of a given subcharacteristic, then we should expect a boundary layer either along MSN or along MTN (see Figure 3.1). In order to study the boundary layer and its matching to $u_0(x, y)$ it is convenient to introduce the orthogonal coordinates (ξ, η) as follows: Let

$$\xi = bx - y, \qquad \eta = x + by. \tag{3.7}$$

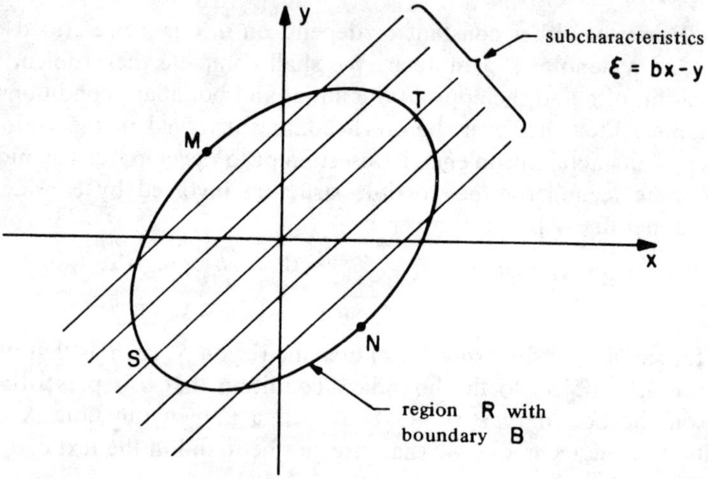

FIGURE 3.1

Then, equation (3.2) transforms to

$$\varepsilon(u_{\xi\xi} + u_{\eta\eta}) - u_\eta = 0, \qquad (\xi, \eta) \text{ in } R. \tag{3.8}$$

See Figure 3.2. The boundary condition (3.3) can now be denoted as

$$u(\xi, \eta) = \begin{cases} f_U(\xi) & \text{for } (\xi, \eta) \text{ on } MTN, \\ f_L(\xi) & \text{for } (\xi, \eta) \text{ on } MSN. \end{cases} \tag{3.9}$$

In the (ξ, η) variables the outer problem becomes $\partial u_0 / \partial \eta = 0$ implies that u_0 is some function of $\xi = bx - y$. Suppose we now try to put a boundary layer along MTN as shown in Figure 3.2. Then, the outer solution u_0 will take on the lower boundary values, and hence,

$$u_0 = f_L(\xi), \tag{3.10}$$

so that to lowest order $u \sim u_0 = f_L(\xi)$ in R minus the boundary layer along MTN.

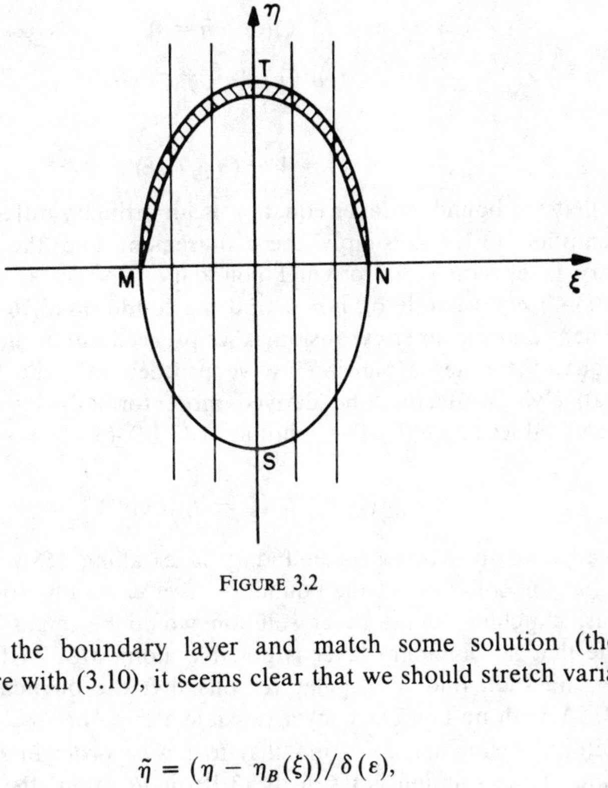

FIGURE 3.2

To magnify the boundary layer and match some solution (the inner expansion) there with (3.10), it seems clear that we should stretch variables as follows: Let

$$\tilde{\eta} = (\eta - \eta_B(\xi))/\delta(\varepsilon), \tag{3.11}$$

where $\delta(\varepsilon)$ is as yet unspecified except that we require $\delta(\varepsilon) \to 0$ as $\varepsilon \to 0$, and where $\eta = \eta_B(\xi)$ is the equation for the boundary MTN. Let $u(\xi, \eta) = u(\xi, \eta_B(\xi) + \delta(\varepsilon)\tilde{\eta}) \equiv w(\xi, \tilde{\eta})$. Then, equation (3.8) becomes

$$\varepsilon w_{\xi\xi} - \frac{2\varepsilon}{\delta(\varepsilon)} \frac{d\eta_B}{d\xi} w_{\xi\tilde{\eta}} + \frac{\varepsilon}{\delta^2(\varepsilon)} \left[\left(\frac{d\eta_B}{d\xi} \right)^2 + 1 \right] w_{\tilde{\eta}\tilde{\eta}} - \frac{1}{\delta(\varepsilon)} w_{\tilde{\eta}} = 0. \tag{3.12}$$

In keeping with the principles adopted in §I, we see that retaining the dominant terms in such a way as to anticipate matching with the outer solution requires that we take $\delta(\varepsilon) = \varepsilon$ to obtain

$$\left[(d\eta_B/d\xi)^2 + 1\right]w_{\tilde{\eta}\tilde{\eta}} - w_{\tilde{\eta}} = 0. \tag{3.13}$$

We must note that this is under the assumption that $d\eta_B/d\xi$ is finite everywhere. If $d\eta_B/d\xi$ is infinite anywhere, then the boundary lies along a subcharacteristic there, and we must modify our analysis. We shall consider this situation, but first let us proceed assuming that the boundary is nowhere subcharacteristic (i.e., that $d\eta_B/d\xi$ is nowhere infinite). Hence, our inner (or boundary layer) problem is

$$K(\xi)w_{\tilde{\eta}\tilde{\eta}} - w_{\tilde{\eta}} = 0, \tag{3.14}$$

$$w = f_U(\xi) \quad \text{on } \tilde{\eta} = 0, \tag{3.15}$$

$$w \to f_L(\xi) \quad \text{as } \tilde{\eta} \to -\infty, \tag{3.16}$$

where

$$K(\xi) = 1 + (d\eta_B/d\xi)^2. \tag{3.17}$$

Notice that the boundary layer equation is an ordinary differential equation. The condition (3.15) is simply the requirement that the solution in the boundary layer satisfy the original boundary conditions on the boundary $\eta = \eta_B(\xi)$, or equivalently on $\tilde{\eta} = 0$, and the condition (3.16) is the matching requirement that the inner expansion as we proceed out of the boundary layer must equal the outer expansion as we proceed into the boundary layer. (Alternatively, (3.16) could be derived more formally by means of intermediate variables as in §I.) The solution of (3.14)–(3.17) is

$$w = f_L(\xi) + \left[f_U(\xi) - f_L(\xi)\right]e^{\tilde{\eta}/K(\xi)}. \tag{3.18}$$

If we had tried to match in a boundary layer along MSN, we would have found that the solutions of the boundary layer equation grow exponentially, and thus, matching to the outer solution would be impossible. Hence, we conclude that no boundary layer is possible along MSN. (If we take $b < 0$, however, we shall find for similar reasons that the boundary layer is now along MSN with no boundary layer possible along MTN.)

Therefore, in summary, we find that to lowest order in ε the asymptotic expansion of the solution is given by (3.10) in R minus the boundary layer and by (3.18) in the boundary layer along MTN. (In this particular problem the boundary layer solution is also the first term of a uniformly valid expansion in the entire domain.)

We shall now study the situation when a part of the boundary lies along a subcharacteristic so that $d\eta_B/d\xi$ is infinite there. A typical situation is illustrated in Figure 3.3 where a part of the boundary lies along $\xi = \xi_s$. The boundary condition (3.3) can now be

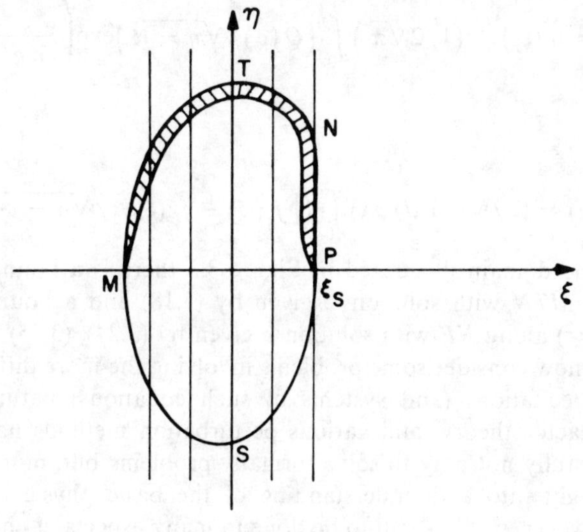

FIGURE 3.3

written as

$$u(\xi, \eta) = \begin{cases} f_U(\xi) & \text{for } (\xi, \eta) \text{ on } MTN, \\ f_L(\xi) & \text{for } (\xi, \eta) \text{ on } MSP, \\ f_s(\eta) & \text{for } (\xi, \eta) \text{ on } NP. \end{cases} \tag{3.19}$$

As we approach the subcharacteristic part of the boundary $\xi = \xi_s$, the outer solution (3.10) approaches the value $f_L(\xi_s)$, so that the boundary condition (3.19) is not satisfied. Hence, we must put a boundary layer along NP, but now we should expect that derivatives in ξ are large so that stretching should be done in the ξ-variable in such a way that we balance $\partial^2/\partial\xi^2$ and $\partial/\partial\eta$. Thus, we let

$$\tilde{\xi} = (\xi - \xi_s)/\sqrt{\varepsilon} \tag{3.20}$$

and we let $u(\xi, \eta) = u(\xi_s + \sqrt{\varepsilon}\,\tilde{\xi}, \eta) \equiv v(\tilde{\xi}, \eta)$. Then, the dominant terms in (3.2) become

$$v_{\tilde{\xi}\tilde{\xi}} = v_{\eta}. \tag{3.21}$$

Note that this is a partial differential equation. In fact it is the well-known heat equation. (It is a general "rule of thumb" that nonsubcharacteristic boundary layers generate ordinary differential equations and subcharacteristic boundary layers generate partial differential equations.) In order that $v(\tilde{\xi}, \eta)$ take on the prescribed boundary values and match to the outer solution, we require that

$$v(0, \eta) = f_s(\eta), \tag{3.22}$$

$$v(-\infty, \eta) = f_L(\xi_s). \tag{3.23}$$

The solution of (3.21) satisfying (3.22) and (3.23) is

$$v(\xi, \eta) = f_L(\xi_s) + (1/2\sqrt{\pi}) \int_0^\eta \left(Q(\alpha)/\sqrt{\eta - \alpha}\right) \exp\left(\frac{-\tilde{\xi}^2}{4(\eta - \alpha)}\right) d\alpha, \tag{3.24}$$

where

$$Q(\alpha) = (2/\sqrt{\pi})(d/d\alpha) \int_0^\alpha \left([f_s(\omega) - f_L(\xi_s)]/\sqrt{\alpha - \omega}\right) d\omega. \tag{3.25}$$

Thus, for the domain illustrated in Figure 3.3 there is a boundary of width $O(\varepsilon)$ along MTN with solution ω given by (3.18) and a boundary layer of width $O(\varepsilon^{1/2})$ along NP with solution v given by (3.24), (3.25).

We shall now consider some problems involving the more difficult equation (3.1). Such equations (and systems of such equations) naturally occur in chemical reactor theory, and various perturbation methods have been used very successfully not only to solve formally problems but, more importantly, to give insight into and understanding of the basic physical processes involved. However, much is still to be done in many aspects of chemical reactor theory. Some idea of some problems and results together with many references can be found in [**2**]–[**4**].

Now, we consider the following steady state model problem:

$$\varepsilon \Delta u + u_y - g(\lambda, u) = 0 \quad \text{in } D, \tag{3.26}$$

$$u = h \quad \text{on } \Gamma, \tag{3.27}$$

$$u_y(x, 0) + bu(x, 0) = f(x). \tag{3.28}$$

Here D is the interior of a semicircle bounded by the x-axis and the semicircle Γ in the half-plane $y > 0$, and h represents the prescribed values of u on Γ. This is a somewhat idealized and simplified problem arising from the chemical reactor theory when the geometry is not a tube. We present it because it retains all the features of the actual problems without the complicated algebraic and analytical calculations of those problems. The different boundary conditions on different parts of the boundary usually arise in the following two ways: (i) For practical reasons various parts of the boundary are insulated or not insulated for heat transfer, thus necessitating different forms of boundary conditions to account for this. (ii) Different boundary conditions account for mass inflow and mass outflow (corresponding to feeding in and draining out the various reactants) at different positions on the boundary.

We ask the following questions: For what values of the parameters λ and b do solutions exist? How many solutions are there? Are they stable? Are there solutions of the time dependent equations which are not steady? Construct asymptotic formulas for the solution for small ε. What does the response (or bifurcation) diagram look like? (The response diagram is simply the graph of some norm $\|u\|$ of the solution vs. a parameter λ or b.) Finally, the question of mathematical justification arises. Especially in situations where singular perturbation methods yield multiple solutions, some rigorous study would be desirable since, as we have seen, spurious solutions can sometimes be constructed.

Our perturbation technique indicates that there is a boundary layer along the semicircular part of the boundary and that in the rest of the domain the first term in the asymptotic expansion of a solution is given by

$$u_y - g(\lambda, u) = 0, \tag{3.29}$$

$$u_y(x, 0) + bu(x, 0) = f(x). \tag{3.30}$$

Evaluating the equation (3.29) at $y = 0$, we find that (3.29) and (3.30) together imply that

$$g(\lambda, u(x, 0)) = f(x) - bu(x, 0). \tag{3.31}$$

Clearly, the solutions $u(x, 0)$ of the algebraic equation (3.31) provide the proper initial conditions for the equation (3.29), the solution of which is the first term of the outer expansion of a solution (3.26)–(3.28). Furthermore, there are as many solutions of (3.26)–(3.28) as there are solutions $u(x, 0) \equiv \alpha_i(x)$, $i = 1, \ldots, N$, of the equation (3.31). Hence, if there are N solutions $\alpha_i(x)$, $i = 1, \ldots, N$, the first term in the outer expansion of each solution is given by

$$u_y - g(\lambda, u) = 0, \tag{3.32}$$

$$u(x, 0) = \alpha_i(x). \tag{3.33}$$

Finally, the boundary layer (or inner) expansion can be constructed just as we did for the problem (3.2), (3.3).

In the simple situation that $f(x) \equiv$ constant f_0 we can easily sketch the response (or bifurcation) diagram as follows: Since $f(x)$ is constant, the solutions $u(x, 0)$ of (3.31) will be constants. Figure 3.4a illustrates the roots of this equation for the nonlinearity $g(\lambda, u)$ which has the form sketched for λ fixed. Figure 3.4a shows how the roots change with b; clearly each intersection of the line $f_0 - bu(x, 0)$ with $g(\lambda, u(x, 0))$ is a solution $u(x, 0) = \alpha_i$, $i = 1, \ldots, N$. If we arbitrarily take our norm $\|u\|$ as $\|u\| = \max_x[u(x, 0)]$, then $\|u_i\| = \alpha_i$, and Figure 3.4b shows how the roots change with b (i.e., Figure 3.4b is the response of bifurcation diagram). Figures 3.5a and 3.5b represent the corresponding situations when we vary f_0 with b fixed.

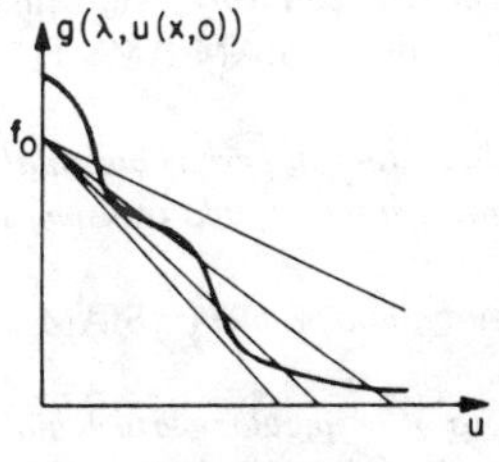

FIGURE 3.4a

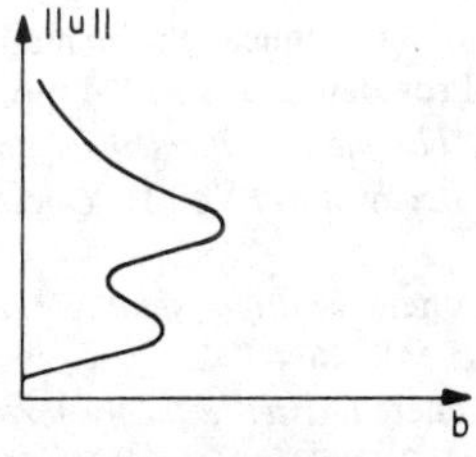

FIGURE 3.4b

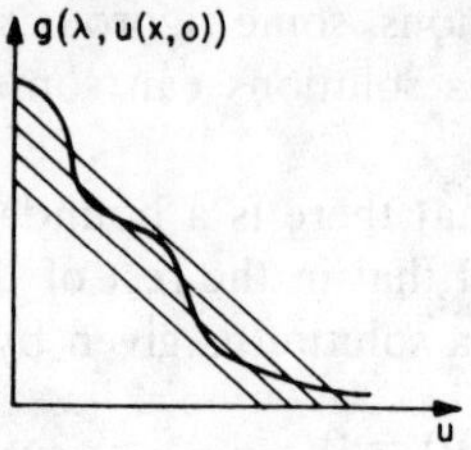

FIGURE 3.5a

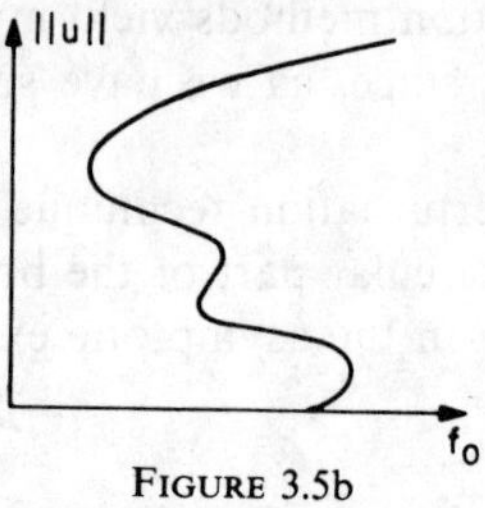

FIGURE 3.5b

For the one-dimensional version (3.26) where Δu is simply u_{yy} and (3.26) reduces to an ordinary differential equation, rigorous proofs of the existence of multiple solutions have been given by D. S. Cohen [**5**] and H. B. Keller [**6**], and stability of the solutions has been analyzed by D. S. Cohen [**2**]. Analogous results for the partial differential equations have not yet been given either for the model problems or the actual reactor problems. Especially nice would be a rigorous proof of the existence of multiple solutions. To further complicate matters in many of the reactor problems the diffusion coefficient D of (3.1) depends on temperatures and/or concentrations of the reactants. Thus, for (3.1) the coefficient D would depend on u. The operation div(D grad u) would then generate nonlinear gradients since

$$\operatorname{div}(D(u) \operatorname{grad} u) \equiv D(u)\Delta u + D'(u)(\nabla u)^2.$$

In addition, it is often the case that the physically relevant solutions for various parameter ranges are oscillatory with no stable steady state so that the full time dependent equations must be investigated. Finally, packed bed and moving bed reactor problems generate equations with all these complications in addition to further complexity due to localized instabilities and localized oscillations for reactions in the catalyst zone.

REFERENCES

1. J. D. Cole, *Perturbation methods in applied mathematics*, Blaisdell, Waltham, Mass., 1968. MR **39** #7841.

2. D. S. Cohen, *Multiple solutions of nonlinear partiál differentil equations*, Nonlinear Problems in the Physical Sciences and Biology, Lecture Notes in Math., vol. 322, Springer-Verlag, Berlin and New York, 1972, pp. 15–77.

3. N. R. Amundson, *Nonlinear problems in chemical reactor theory*, Mathematical Aspects of Chemical and Biochemical Problems in Quantum Chemistry, SIAM-AMS Proc., vol. 8, Amer. Math. Soc., Providence, R.I., 1974, pp. 59–84.

4. R. Aris, *The mathematical theory of diffusion and reaction in permeable catalysts*; Vol. I: *The theory of the steady state*; Vol. II: *Questions of uniqueness, stability, and transient behavior*, Oxford Press, 1975.

5. D. S. Cohen, *Multiple solutions of singular perturbation problems*, SIAM J. Math. Anal. **3** (1972), 72–82. MR **46** #9473.

6. H. B. Keller, *Existence theory for multiple solutions of a singular perturbation problem*, SIAM J. Math. Anal. **3** (1972), 86–92. MR **46** #9474.

IV. **Multi-scale expansion procedures**. Just as was true for singular perturbation methods, the multi-scale technique is a name for a large class of methods applicable in widely disparate situations, but there are common conceptual properties. Thus, rather than attempt any general definition (which would be of questionable value anyway) we shall present the ideas via several examples.

In order to be able to consider problems of current interest in the space allotted here we have chosen problems of considerably more complexity than is needed simply to illustrate the ideas and manipulations commonly used in the technique. A good feeling for the basic ideas and appreciation of the analytical manipulations necessary to carry out the method can be obtained from the text of J. D. Cole [**1**] or the lectures of J. Kevorkian [**2**].

Several recent problems in the theory of both chemical and nuclear reactors and the theory of chemical and biochemical reactions, at one stage or another, involve the study of the bifurcation of periodic solutions in an autonomous system of the form

$$dX/dt = F(X, \lambda), \tag{4.1}$$

where

$$x = \begin{bmatrix} x_1 \\ \cdot \\ \cdot \\ \cdot \\ x_n \end{bmatrix}, \qquad F(X) = \begin{bmatrix} f_1(x_1, \ldots, x_n) \\ \cdot \\ \cdot \\ \cdot \\ f_n(x_1, \ldots, x_n) \end{bmatrix},$$

and where $\lambda = (\lambda_1, \ldots, \lambda_m)$ denotes the various parameters occurring. In the applications it is important to obtain asymptotic formulas for the oscillations, rates of decay onto the oscillatory state, asymptotic frequency shifts, and information concerning stability. We shall discuss many of the current applications and still outstanding problems after we pose the mathematical problem.

Critical points (steady states in the physical problems mentioned) are given by the solutions of $F(X, \lambda) = 0$. Suppose $X_0(\lambda)$ is such a solution, that is, suppose that

$$F(X_0(\lambda), \lambda) = 0. \tag{4.2}$$

To determine the (linearized) stability of a steady state $X_0(\lambda)$, we linearize (4.1) about this steady state. In the standard manner we obtain a system of linear ordinary differential equations with constant coefficients, the solutions of which are linear combinations of exponentials of the form $\exp(\mu_i t)$, where the μ_i are the eigenvalues of the matrix

$$F_X(X_0(\lambda), \lambda) = \begin{bmatrix} \dfrac{\partial f_1}{\partial x_1}(x_{10}, \ldots, x_{n0}) & \cdots & \dfrac{\partial f_n}{\partial x_1}(x_{10}, \ldots, x_{n0}) \\ \vdots & \cdot & \vdots \\ \dfrac{\partial f_1}{\partial x_n}(x_{10}, \ldots, x_{n0}) & \cdots & \dfrac{\partial f_n}{\partial x_n}(x_{10}, \ldots, x_{n0}) \end{bmatrix}. \tag{4.3}$$

Thus, based on linear theory the steady state is stable if all eigenvalues have negative real parts and unstable if at least one eigenvalue has a positive real part.

Suppose that we wish to determine the stability of the steady state $X_0(\lambda)$ as a function of λ for some scalar λ, and suppose that for some range of values of λ the eigenvalues of (4.3) consist of two complex conjugate numbers $\alpha(\lambda) \pm i\beta(\lambda)$ with all the other eigenvalues being negative real numbers. (Other possibilities are discussed later.) Suppose, further, that for some value of λ, say λ_0, we have $\alpha(\lambda_0) = 0$ with $\alpha(\lambda) > 0$ for $\lambda < \lambda_0$ and $\alpha(\lambda) < 0$ for $\lambda > \lambda_0$. Hence, as a result of varying the parameter λ, the quantity $\alpha(\lambda)$ changes in sign from negative to positive (at the stability boundary $\lambda = \lambda_0$), and linearized stability theory predicts that the steady state loses its stability via an exponentially growing function of time t. This (linearized) exponentially growing function cannot represent the solution for very long because clearly the nonlinear terms must then become important. If, in fact, this exponentially growing function tends to a stable oscillatory solution, then growth on another time scale must come into play so that in some sense the perturbation from the unstable steady state should exhibit a more or less typical multi-time scale representation; That is, for the solution of (4.1) we expect a representation of the form

$$X(t) = \begin{bmatrix} x_1(t) \\ \cdot \\ \cdot \\ \cdot \\ x_n(t) \end{bmatrix} = \begin{bmatrix} A_1(\tau)P_1(t^*) \\ \cdot \\ \cdot \\ \cdot \\ A_n(\tau)P_n(t^*) \end{bmatrix},$$

where the $P_k(t^*)$ represent a periodic oscillation on a so-called "fast time" t^* and the $A_k(\tau)$ represent "slow time" modulation which perhaps approach constant values as $t \to \infty$. We shall now apply a multi-time scale (or "two-timing") perturbation technique in this way. Our method will produce a representation of the solution which easily interpretable physically and from which the stability of the solution is immediately resolved without recourse to further analysis. Change of stability and time evolution of the solution in somewhat different problems have been studied in this way by D. S. Cohen [3], B. J. Matkowsky [4], and J. P. Keener and D. S. Cohen [5]–[7].

On the stable side ($\lambda > \lambda_0$) of the stability boundary ($\lambda = \lambda_0$) all solution trajectories in phase space approach the steady state $X_0(\lambda)$. As λ is varied so that we cross the stability boundary, all solution trajectories leave the steady state and approach a limit cycle in phase space. Thus, $\lambda = \lambda_0$ is a bifurcation point in the sense that stable time periodic solutions bifurcate (or branch) from the unstable equilibrium solution. In to order exhibit this phenomenon for our equations we shall investigate the bifurcation of periodic solutions in the system

$$dX/dt = PX + \varepsilon^2 AX + g(X), \tag{4.4}$$

where

$$P = \begin{bmatrix} 0 & \beta & 0 & \cdot & \cdot & \cdot & \cdot & 0 \\ -\beta & 0 & 0 & \cdot & \cdot & \cdot & \cdot & 0 \\ 0 & 0 & -\mu_3 & 0 & \cdot & \cdot & \cdot & 0 \\ \cdot & & 0 & \cdot & & & & \cdot \\ \cdot & & & & \cdot & & & \cdot \\ \cdot & & & & & \cdot & & \cdot \\ \cdot & & & & & & \cdot & \cdot \\ 0 & \cdots & 0 & \cdot & \cdot & \cdot & \cdot & \mu_n \end{bmatrix}, \tag{4.5}$$

$$g(X) = \begin{bmatrix} g_1(x_1, \ldots, x_n) \\ \vdots \\ g_n(x_1, \ldots, x_n) \end{bmatrix}, \tag{4.6}$$

and where $0 < \varepsilon \ll 1$, β and μ_i $(i = 2, \ldots, n)$ are constants, $A = (a_{ij})$ is a constant matrix, and the nonlinear functions $g_i(X)$, $i = 1, \ldots, n$, contain no linear terms in X near $X = 0$, that is,

$$g_i(0) = g_{ix_1}(0) = \cdots = g_{ix_n}(0) = 0, \qquad i = 1, \ldots, n. \tag{4.7}$$

The transformations which reduce the system (4.1) to the form (4.4)–(4.6) will be explicitly demonstrated immediately following our study of (4.4)–(4.6). Furthermore, we shall see that to treat analytically the problem posed with the multi-scale method it is necessary first to transform (4.1) to some system of the form of (4.4).

To avoid considerable algebraic manipulations we shall carry out our procedure only for a three-dimensional version of (4.1) and (4.4). The calculations for the n-dimensional system can be done similarly, and in fact a judicious use of a computer (in a given problem) can greatly aid the manipulations as shown by Cohen and Keener [7]. Thus, we now consider (4.4) where

$$X = \begin{pmatrix} x \\ y \\ z \end{pmatrix}, \quad P = \begin{bmatrix} 0 & \beta & 0 \\ -\beta & 0 & 0 \\ 0 & 0 & -\mu \end{bmatrix}, \quad g(X) = \begin{bmatrix} g_1(x, y, z) \\ g_2(x, y, z) \\ g_3(x, y, z) \end{bmatrix}.$$

The only necessary tool we shall need in carrying out the two-timing formalism is an elementary fact which we shall state in the form of an easily referenced lemma.

LEMMA. *The general solution of*

$$du/dt + v = m \sin t + n \cos t, \qquad dv/dt - u = p \sin t + q \cos t,$$

is

$$u(t) = A \sin t + B \cos t + \tfrac{1}{2}(m - q)t \sin t + \tfrac{1}{2}(n + p)t \cos t + \tfrac{1}{2}(n - p)\sin t,$$

$$v(t) = -A \cos t + B \sin t + \tfrac{1}{2}(n + p)t \sin t - \tfrac{1}{2}(m - q)t \cos t + \tfrac{1}{2}(m + q)\sin t.$$

Thus, in order to suppress secular terms (i.e., in order to have solutions bounded for all $t \geqslant 0$) it is sufficient to require $m - q = 0$ and $n + p = 0$.

Now, we assume that

$$x = x(t^*, \varepsilon) = \varepsilon x_1(t^*, \tau) + \varepsilon^2 x_2(t^*, \tau) + \dots, \tag{4.8}$$

$$y = y(t^*, \varepsilon) = \varepsilon y_1(t^*, \tau) + \varepsilon^2 y_2(t^*, \tau) + \dots, \tag{4.9}$$

$$z = z(t^*, \varepsilon) = \varepsilon z_1(t^*, \tau) + \varepsilon^2 z_2(t^*, \tau) + \dots, \tag{4.10}$$

or equivalently,

$$X = X(t^*, \tau) = \sum_{n=1}^{\infty} \varepsilon^n X_n(t^*, \tau), \quad \text{where } X_n = \begin{bmatrix} x_n \\ y_n \\ z_n \end{bmatrix}, \tag{4.11}$$

and where the "slow time" τ and the "fast time" t^* are defined by

$$\tau = \varepsilon^2 t, \tag{4.12}$$

$$t^* = \left(1 + \varepsilon\omega_1 + \varepsilon^2\omega_2 + \dots\right)t. \tag{4.13}$$

We shall now require that the ω_i and the other unknowns which will occur shall be chosen according to the principle that we suppress secular terms in such a way that we generate a self-consistent procedure for determining bounded functions $X_i(t^*, \tau)$ with modulation only on the slow time scale. We shall now carry out this procedure.

With the definitions of equations (4.12) and (4.13), we find that

$$d/dt = \left(1 + \varepsilon\omega_1 + \varepsilon^2\omega_2 + \dots\right) \partial/\partial t^* + \varepsilon^2\, \partial/\partial\tau.$$

Thus, upon substituting equations (4.8)–(4.13) into equation (4.4) and equating coefficients of like powers of ε, we obtain

$$\partial X_1/\partial t^* = PX_1, \tag{4.14}$$

$$\partial X_2/\partial t^* = PX_2 = -\omega_1\, \partial X_1/\partial t^* + G(X_1), \tag{4.15}$$

$$\frac{\partial X_3}{\partial t^*} - PX_3 = AX_1 - \omega_2 \frac{\partial X_1}{\partial t^*} - \omega_1 \frac{\partial X_2}{\partial t^*} + F(X_1, X_2) - \frac{\partial X_1}{\partial \tau}, \tag{4.16}$$

where $G(X_1)$ and $F(X_1, X_2)$ represent higher order terms for which the precise formulas will be given later in equations (4.18) and (4.24), (4.25). The solution of equation (4.14) is

$$x_1(t^*, \tau) = R(\tau)\sin(\beta t^* + \varphi(\tau)),$$
$$y_1(t^*, \tau) = R(\tau)\cos(\beta t^* + \varphi(\tau)),$$
$$z_1(t^*, \tau) = B(\tau)e^{-\mu^*}, \tag{4.17}$$

where the unknown $R(\tau)$, $\varphi(\tau)$ and $B(\tau)$ will be determined at a later stage of the perturbation procedure. Using equation (4.17), we find that

$$\begin{aligned} G_i(X_1) &= \tfrac{1}{2}\, g_{ixx}(0)R^2 \sin^2\psi + g_{ixy}(0)R^2 \sin\psi\cos\psi \\ &\quad + \tfrac{1}{2}\, g_{iyy}(0)R^2\cos\psi + (\text{terms involving } e^{-\mu t^*} \text{ as a factor}) \\ &= \tfrac{1}{4}R^2\big[\, g_{ixx}(0) + g_{iyy}(0)\big] + \tfrac{1}{4}R^2\big[\, g_{iyy}(0) - g_{ixx}(0)\big]\cos\partial\psi \\ &\quad + \tfrac{1}{2}R^2 g_{ixy}(0)\sin 2\psi + (\text{exponentially decaying terms}), \end{aligned}$$

where

$$\psi \equiv \beta t^* + \varphi(\tau). \tag{4.19}$$

We shall be interested in finding only the leading term of equation (4.11), and thus, we do not need to retain the exponentially decaying terms in equation (4.18) since these terms can never give rise to secular terms in the calculation of $R(\tau)$ and $\varphi(\tau)$. Using our lemma to suppress secular terms in equation (4.15), we see immediately that we must require that $\omega_1 = 0$. Then, the solution of (4.15) is found to be

$$\begin{aligned} x_2(t^*, \tau) &= a\sin 2\psi + b\cos 2\psi + c + D(\tau)\sin(\beta t^* - \delta(\tau)) \\ &\quad + (\text{exponentially decaying terms}), \end{aligned} \tag{4.20}$$

$$\begin{aligned} y_2(t^*, \tau) &= d\sin 2\psi + e\cos 2\psi + f + D(\tau)\cos(\beta t^* - \delta(\tau)) \\ &\quad + (\text{exponentially decaying terms}), \end{aligned} \tag{4.21}$$

$$\begin{aligned} z_2(t^*, \tau) &= l\sin 2\psi + h\cos 2\psi + j \\ &\quad + (\text{exponentially decaying terms}), \end{aligned} \tag{4.22}$$

where

$$\begin{aligned} a &= (6\beta)^{-1}R^2\big[\, g_{1yy}(0) - g_{1xx}(0) - g_{2xy}(0)\big], \\ b &= (3\beta)^{-1}R^2\big[\, \tfrac{1}{4}\, g_{2xx}(0) - \tfrac{1}{4}\, g_{2yy}(0) - g_{1xy}(0)\big], \\ c &= (4\beta)^{-1}R^2\big[\, g_{2xx}(0) + g_{2yy}(0)\big], \\ d &= (6\beta)^{-1}R^2\big[\, g_{1xy}(0) + g_{2yy}(0) - g_{2xx}(0)\big], \\ e &= (3\beta)^{-1}R^2\big[\, \tfrac{1}{4}\, g_{1yy}(0) - \tfrac{1}{4}\, g_{1xx}(0) - g_{2xy}(0)\big], \\ f &= -(4\beta)^{-1}R^2\big[\, g_{1xx}(0) + g_{1yy}(0)\big], \\ l &= \big(2(\mu^2 + 4\beta^2)\big)^{-1}R^2\big[\, \mu g_{3xy}(0) + \beta g_{3yy}(0) - \beta g_{3xx}(0)\big], \\ h &= (\mu^2 + 4\beta^2)^{-1}R^2\big[\, \tfrac{1}{4}\,\mu g_{3yy}(0) - \tfrac{1}{4}\,\mu g_{3xx}(0) - \beta g_{3xx}(0)\big], \\ j &= (4\mu)^{-1}R^2\big[\, g_{3xx}(0) + g_{3yy}(0)\big], \end{aligned} \tag{4.23}$$

and where $D(\tau)$ and $\delta(\tau)$ must be determined at a later stage. Upon substituting (4.17)–(4.23) into equation (4.16), we find, after a considerable amount of algebraic and trigonometric manipulation, that the first two equations of (4.16) become

$$\begin{aligned}\frac{\partial x_3}{\partial t^*} - \beta y_3 = {} & R^2 \sin\psi - R' \sin\psi - \varphi' R \cos\psi - \omega_2 \beta R \cos\psi \\ & + R^3(L_1 + P_1)\sin\psi + R^3(M_1 + Q_1)\cos\psi \\ & + (\text{terms in } e^{-\mu t^*} \text{ and higher harmonics}),\end{aligned} \tag{4.24}$$

$$\begin{aligned}\frac{\partial y_3}{\partial t^*} + \beta x_3 = {} & R^2 \cos\psi + \varphi' R \sin\psi - R' \cos\psi + \omega_2 \beta R \sin\psi \\ & + R^3(L_2 + P_2)\sin\psi + R^3(M_2 + Q_2)\cos\psi \\ & + (\text{terms in } e^{-\mu t^*} \text{ and higher harmonics}),\end{aligned} \tag{4.25}$$

where, for $i = 1, 2$,

$$\begin{aligned}L_i = {} & \frac{1}{6\beta} g_{ixx}(0)\left[\frac{5}{4} g_{2xx}(0) + \frac{7}{4} g_{2yy}(0) + g_{1xy}(0)\right] \\ & + \frac{1}{3\beta} g_{ixy}(0)\left[-\frac{7}{8} g_{1xx}(0) - \frac{5}{8} g_{1yy}(0) + \frac{1}{4} g_{2xy}(0)\right] \\ & + \frac{1}{12\beta} g_{iyy}(0)\left[g_{1xy}(0) + g_{2yy}(0) - g_{2xx}(0)\right] \\ & + \frac{1}{4\mu} g_{ixz}(0)\left[g_{3xx}(0) + g_{3yy}(0)\right] \\ & + \frac{1}{\mu^2 + 4\beta^2} g_{ixz}(0)\left[\frac{1}{2}\beta g_{3xy}(0) + \frac{1}{8}\mu g_{3xx}(0) - \frac{1}{8}\mu g_{3yy}\right] \\ & + \frac{1}{4(\mu^2 + 4\beta^2)} g_{iyz}(0)\left[\beta g_{3yy}(0) - \beta g_{3xx}(0) + \mu g_{3xy}(0)\right],\end{aligned}$$

$$\begin{aligned}M_i = {} & \frac{1}{12\beta} g_{ixx}(0)\left[g_{1yy}(0) - g_{1xx}(0) - g_{2xy}(0)\right] \\ & + \frac{1}{3\beta} g_{ixy}(0)\left[-\frac{1}{4} g_{1xy}(0) + \frac{7}{8} g_{2yy}(0) + \frac{5}{8} g_{2xx}(0)\right] \\ & + \frac{1}{6\beta} g_{iyy}(0)\left[-\frac{5}{4} g_{1yy}(0) - \frac{7}{4} g_{1xx}(0) - g_{2xy}(0)\right] \\ & + \frac{1}{4\mu} g_{iyz}(0)\left[g_{3xx}(0) + g_{3yy}(0)\right] \\ & + \frac{1}{4(\mu^2 + 4\beta^2)} g_{ixz}(0)\left[\mu g_{3xy}(0) + \beta g_{3yy}(0) - \beta g_{3xx}(0)\right] \\ & + \frac{1}{\mu^2 + 4\beta^2} g_{iyz}(0)\left[\frac{1}{8}\mu g_{3yy}(0) - \frac{1}{8}\mu g_{3xx}(0) - \frac{1}{2}\beta g_{3xy}(0)\right],\end{aligned}$$

$$P_i = \frac{1}{8} g_{ixxx}(0) + \frac{1}{8} g_{ixyy}(0),$$

$$Q_i = \frac{1}{8} g_{ixx}(0) + \frac{1}{8} g_{iyyy}(0).$$

Now, the application of our lemma to the system (4.24), (4.25) implies that to suppress secular terms we must require that

$$2dR/d\tau = (a_{11} + a_{22})R + (L_1 + M_2 + P_1 + Q_2)R^3, \tag{4.26}$$

$$2Rd\varphi/d\tau = (a_{12} - a_{21})R - 2\omega_2 \beta R + (M_1 - L_2 + Q_1 - P_2)R^3. \tag{4.27}$$

The periodic nature of the solutions $x_1(t^*, \tau)$ and $y(t^*, \tau)$ can now be established from a consideration of equation (4.26) alone. The solution of (4.26) is easily found to be given by

$$R^2(\tau) = \frac{\alpha}{\gamma} \frac{1}{1 + ke^{-\alpha\tau}}, \tag{4.28}$$

where k is a constant and where

$$\alpha = a_{11} + a_{22}, \qquad \gamma = -L_1 - M_2 - P_1 - Q_2. \tag{4.29}$$

Thus, when $\alpha > 0$ (and $\gamma > 0$), equations (4.17) and (4.28) show that the solution $X_1(t^*, \tau)$ approaches a limit cycle in the (x_1, y_1)-plane of amplitude

$$\frac{\alpha^{1/2}}{\gamma} \varepsilon = \left(\frac{a_{11} + a_{22}}{-L_1 - M_2 - P_1 - Q_2} \right)^{1/2} \varepsilon. \tag{4.30}$$

On the other hand, if $\alpha < 0$ (and $\gamma > 0$), the oscillatory solution decays to zero as $\tau \to \infty$. (The solution with $\gamma < 0$ will not be of interest to us.)

Note that the stability of the oscillatory solution follows immediately by writing equation (4.26) in the form

$$dR^2/d\tau = \gamma R^2(\alpha/\gamma - R^2), \tag{4.31}$$

from which we can immediately conclude that for $R^2(\tau)$ less (greater) than its steady state value α/γ we have $dR^2/d\tau$ greater (less) than zero, implying motion towards the steady state (i.e., stability).

Since $\omega_1 = 0$, we see from (4.13) that the first nontrivial correction to the asymptotic frequency shift is given by the term $\varepsilon^2\omega_2$. The quantity ω_2 can be found from (4.27) as follows: As $\tau \to \infty$, we expect that $\varphi(\tau)$ approaches a constant value which implies that $d\varphi/d\tau \to 0$. Thus, as $\tau \to \infty$, equation (4.27) implies

$$(a_{12} - a_{21})R(\infty) - 2\omega_2 \beta R(\infty) + (M_1 - L_2 + Q_1 - P_2)R^3(\infty) = 0. \tag{4.32}$$

Also, (4.28) yields

$$R^2(\infty) = \alpha/\gamma. \tag{4.33}$$

Hence, from (4.32) and (4.33) we obtain

$$\omega_2 = \frac{1}{2\beta}\left[a_{12} - a_{21} + (a_{11} + a_{22})\frac{L_2 + P_2 - M_1 - Q_1}{L_1 + P_1 + M_2 + Q_2}\right].$$

We shall now employ our results to study oscillations and their stability in a nuclear reactor model with the reactivity a function of two effective temperatures. We consider the following model of Vreeke and Sandquist [**8**]:

$$dN/dt = \rho N, \tag{4.34}$$

$$dT_1/dt = \omega_1(N - T_1), \tag{4.35}$$

$$dT_2/dt = \omega_2(N - T_2), \tag{4.36}$$

where $\omega_1, \omega_2 > 0$, and

$$\rho = -\alpha_1(T_1 - 1) - \alpha_2(T_2 - 1). \tag{4.37}$$

Here, in normalized variables, N is the neutron density, T_1 is the temperature of the fuel, T_2 is the temperature associated with the moderator and coolant, ρ is the reactivity, ω_i $(i = 1, 2)$ are heat transfer coefficients, and α_i $(i = 1, 2)$ are effective neutron life-time parameters.

Equations (4.34)–(4.36) have two equilibrium states (i.e., critical points), namely the steady-state shut-down point $(N, T_1, T_2) = (0, 0, 0)$ and the steady-state operating equilibrium point $(N, T_1, T_2) = (1, 1, 1)$. Vreeke and Sandquist [**8**] have studied the linearized stability of both points, and they present numerical calculations (via projections on various planes) of oscillatory solutions near the operating equilibrium point. In order to obtain a deeper qualitative feeling for the process by which these oscillations appear in higher order systems we shall now apply our two-timing method. For algebraic simplicity we shift this operating point $(N, T_1, T_2) = (1, 1, 1)$ to the origin with the transformation

$$n = N - 1, \qquad \theta_1 = T_1 - 1, \qquad \theta_2 = T_2 - 1. \tag{4.38}$$

Then, (4.34)–(4.36) become

$$dn/dt = (-\alpha_1\theta_1 - \alpha_2\theta_2)(1 + n), \tag{4.39}$$

$$d\theta_1/dt = \omega_1 n - \omega_1\theta_1, \tag{4.40}$$

$$d\theta_2/dt = \omega_2 n - \omega_2\theta_2. \tag{4.41}$$

The characteristic equation of the linearized system in the neighborhood of the steady-state operating point $(n, \theta_1, \theta_2) = (0, 0, 0)$ is

$$\lambda^3 + (\omega_1 + \omega_2)\lambda^2 + (\omega_1\omega_2 + \alpha_1\omega_1 - \alpha_2\omega_2)\lambda + \omega_1\omega_2(\alpha_1 + \alpha_2) = 0. \tag{4.42}$$

It is easy to show by a number of standard techniques that the linearized stability boundary is given by

$$\omega_1^2\alpha_1 + \omega_2^2\alpha_2 = -\omega_1\omega_2(\omega_1 + \omega_2). \tag{4.43}$$

For fixed ω_1, ω_2 this represents a straight line in the (α_1, α_2)-plane. Thus, as we cross this line due to small perturbations in α_1 and/or α_2 the steady-state equilibrium point loses its stability point on the stability boundary given by

$\omega_1 = 1, \omega_2 = 2, \alpha_1 = 10, \alpha_2 = -4$. Fix ω_1 and ω_2, and let $\alpha_1 = 10 - 26\nu^2$ and $\alpha_2 = -4 - 26\delta^2$. Then, (4.39)–(4.41) can be written

$$\frac{d}{dt}\begin{bmatrix} n \\ \theta_1 \\ \theta_2 \end{bmatrix} = \begin{bmatrix} 0 & -10+26\nu^2 & 4+26\delta^2 \\ 1 & -1 & 0 \\ 2 & 0 & -2 \end{bmatrix}\begin{bmatrix} n \\ \theta_1 \\ \theta_2 \end{bmatrix} + \begin{bmatrix} (-10+26\nu^2)n\theta_1 + (4+26\delta^2)n\theta_2 \\ 0 \\ 0 \end{bmatrix}. \tag{4.44}$$

Now, make the transformation

$$\begin{bmatrix} n \\ \theta_1 \\ \theta_2 \end{bmatrix} = \begin{bmatrix} 2 & 4 & 2 \\ 2 & 0 & -1 \\ 3 & 1 & -4 \end{bmatrix}\begin{pmatrix} x \\ y \\ z \end{pmatrix} \equiv Q\begin{pmatrix} x \\ y \\ z \end{pmatrix},$$

so that

$$\begin{pmatrix} x \\ y \\ z \end{pmatrix} = \frac{1}{26}\begin{bmatrix} 1 & 18 & -4 \\ 5 & -14 & 6 \\ 2 & 10 & -8 \end{bmatrix}\begin{bmatrix} n \\ \theta_1 \\ \theta_2 \end{bmatrix} \equiv Q^{-1}\begin{bmatrix} n \\ \theta_1 \\ \theta_2 \end{bmatrix}.$$

Under this transformation, the equations (4.44) become

$$\frac{d}{dt}\begin{pmatrix} x \\ y \\ z \end{pmatrix} = \begin{bmatrix} 2\nu^2 + 3\delta^2 & 2+\delta^2 & -\nu^2 - 4\delta^2 \\ -2 + 10\nu^2 + 15\delta^2 & 5\delta^2 & -5\nu^2 - 20\delta^2 \\ 4\nu^2 + 6\delta^2 & 2\delta^2 & -3 - 2\nu^2 - 8\delta^2 \end{bmatrix}\begin{pmatrix} x \\ y \\ z \end{pmatrix} + \frac{1}{26}\begin{bmatrix} f \\ 5f \\ 2f \end{bmatrix},$$

where

$$\begin{aligned} f = (-10 + 26\mu^2)&(4x^2 + 8xy + 2xz - 4yz - 2z^2) \\ &+ (4 + 26\delta^2)(6x^2 + 14xy - 2xz + 4y^2 - 14yz - 8z^2). \end{aligned}$$

Two different situations have now been reduced to the form of system (4.4). First, we let $\omega_1 = 1$, $\omega_2 = 2$, $\alpha_1 = 10$, and cross the stability boundary by varying α_2 according to $\alpha_2 = -4 - 26\delta^2$. In this case we see by comparing system (4.45) with system (4.4) that $\varepsilon^2 = \delta^2$, $\beta = 2$, $\mu = 3$, $g_1 = f/26$, $g_2 = 5f/26$, and $g_3 = f/13$. In the second case we let $\omega_1 = 1$, $\omega_2 = 2$, $\alpha_2 = -4$, and cross the stability boundary by varying α_1 according to $\alpha_1 = 10 - 26\nu^2$. In this case $\varepsilon^2 = \nu^2$. For the system (4.45) it is a simple calculation to show that in both cases $L_1 = -352/507$, $M_2 = 16064/2535$, $P_1 = Q_2 = 0$, so that $\gamma = M_2 - Q_2 - P_1 - L_1 = 16064/2535 + 352/507 > 0$. Hence, we conclude that when we cross the stability boundary, the steady-state operating point $(n, \theta_1, \theta_2) = (0, 0, 0)$ loses its stability, and stable oscillatory solutions (limit cycles in phase space) appear. Note, in particular, that stability of the oscillation implies that all (sufficiently small) initial conditions lead to solutions which approach this limit cycle, and thus, any perturbation of the reactor from its steady-state operating equilibrium point which crosses the stability boundary forces the reactor into the stable oscillatory state.

REFERENCES

1. J. D. Cole, *Perturbation methods in applied mathematics*, Blaisdell, Waltham, Mass., 1968. MR **39** #7841.

2. J. Kevorkian, *The two variable expansion procedure for the approximate solution of certain nonlinear differential equations*, Lectures in Appl. Math., vol. 7, part III, Amer. Math. Soc., Providence, R. I., 1966, 206–275. MR **34** #5295.

3. D. S. Cohen, *Multiple solutions and periodic oscillations in nonlinear diffusion processes*, SIAM J. Appl. Math. **25** (1973), 640–654.

4. B. J. Matkowsky, *A simple nonlinear dynamic stability problem*, Bull. Amer. Math. Soc. **76** (1970), 620–625. MR **41** #2194.

5. D. S. Cohen and J. P. Keener, *Oscillatory processes in the theory of particulate formation in supersaturated chemical solutions*, SIAM J. Appl. Math. **28** (1975), 307–318.

6. J. P. Keener and D. S. Cohen, *Nonlinear oscillations in a reactor with two temperature coefficients*, Nuclear Sci. and Eng. **56** (1975), 354–359.

7. D. S. Cohen and J. P. Keener, *Multiplicity and stability of oscillatory states in a continuous stirred tank reactor with exothermic consecutive reactions $A \to B \to C$*, Chem. Eng. Sci. **31** (1976), 115–122.

8. S. A. Vreeke and G. M. Sandquist, *Phase space analysis of reactor kinetics*, Nuclear Sci. Eng. **42** (1970), 295–305.

V. **Some current problems**. This section will be devoted to a discussion of some current problems in bifurcation and perturbation theory. We shall discuss specific problems from various fields such as the theory of chemical and biochemical reactors, chemical and nuclear reactors, combustion, diffusion through membranes and porous media, Joule heating, and soil mechanics. The rest of this section consists of the details of one of the problems to be considered and forms the basis of a research paper [**1**] to be published by D. S. Cohen.

1. *Introduction*. We shall study bifurcation and stability for nonlinear ordinary differential systems of arbitrary dimension when an equilibrium solution loses its stability by virtue of *two* pairs $(\alpha(\lambda) \pm i\beta(\lambda), \gamma(\lambda) \pm i\delta(\lambda))$ of complex conjugate eigenvalues of the linearized system *simultaneously* crossing the imaginary axis. Such a situation is not all uncommon, and we shall give applications where this situation is indeed the usual phenomenon encountered. In these situations considerably different and more diverse behavior can occur than in the simpler bifurcation at a simple complex eigenvalue (Hopf bifurcation). As we shall see, the complexity of bifurcating solutions will result principally from the simple fact that superpositions such as $\sin \beta t + \sin \delta t$ are not periodic if β and δ are incommensurate. The extension of our theory to account for arbitrary numbers of pairs of complex conjugate eigenvalues will be clear.

We consider the system

$$dY/dt = F(Y, \lambda), \tag{5.1}$$

where

$$Y = \begin{bmatrix} y_1 \\ \cdot \\ \cdot \\ \cdot \\ y_n \end{bmatrix}, \qquad F(Y) = \begin{bmatrix} f_1(y_1, \ldots, y_n) \\ \vdots \\ f_n(y_1, \ldots, y_n) \end{bmatrix}, \tag{5.2}$$

and where λ is a parameter. Equilibrium solutions (or steady states or critical points) are given by the solutions of $F(Y, \lambda) = 0$. Suppose that $Y_0(\lambda)$ is such a solution, that is, suppose that

$$F(Y_0(\lambda), \lambda) = 0. \tag{5.3}$$

To determine the (linearized) stability of the steady state $Y_0(\lambda)$, we linearize (5.1) around this steady state. In the standard manner we obtain a system of linear ordinary differential equations with constant coefficients, the solutions of which are linear combinations of exponentials of the form $\exp(\mu_i t)$, where the μ_i are the eigenvalues of the matrix

$$F_Y(Y_0(\lambda), \lambda) = (\partial f_i / \partial y_j(Y_0(\lambda))). \tag{5.4}$$

Hence, based on linear theory the steady state $Y_0(\lambda)$ is stable (or attracting) if all eigenvalues have negative real parts and unstable if at least one eigenvalue has a positive real part.

Suppose now that (5.4) possesses two simple complex conjugate pairs of eigenvalues given by

$$\mu_{1,2} = \alpha(\lambda) \pm i\beta(\lambda), \qquad \mu_{3,4} = \gamma(\lambda) \pm i\delta(\lambda), \tag{5.5}$$

and all other eigenvalues $\mu_i(\lambda)$, $i \geqslant 5$, are distinct negative real numbers for all λ. Suppose further that for som $\lambda = \lambda_0$ we have

$$\begin{aligned} &\alpha(\lambda_0) = 0, \qquad \alpha'(\lambda_0) \neq 0, \\ &\gamma(\lambda_0) = 0, \qquad \gamma'(\lambda_0) \neq 0, \\ &\alpha(\lambda)\gamma(\lambda) \geqslant 0 \quad \text{for all } \lambda. \end{aligned} \tag{5.6}$$

Thus, the real parts of $\mu_{1,2}$ and $\mu_{3,4}$ are both of the same sign and simultaneously change sign. As λ varies in such a way that $\alpha(\lambda)$ and $\gamma(\lambda)$ change from negative to positive, the equilibrium solution $Y_0(\lambda)$ loses its stability, and new (possibly quite complicated) time dependent solutions bifurcate from the branch of steady states at $\lambda = \lambda_0$. We shall study these bifurcating solutions in the following subsections.

Our main result, a representation for the form of the bifurcating solutions, is stated and derived in part 2. Two quantities (which we call amplitude modulations) appear in our representation, and we give the general equations (called the modulation equations) which determine these quantities. In specific problems these amplitude modulations give quite detailed information about the bifurcating solutions and their stability. Thus, in part 3 we study the general modulation equations. Finally, in part 4 we give some specific applications.

There has been a considerable amount of recent interest and research in bifurcation theory in many different contexts. In particular, a considerable literature exists, and many of the papers treat bifurcation at a single simple complex conjugate pair. An excellent reference, together with many references, is the mongraph of D. H. Sattinger [**2**]. More recent results, concerned with bifurcation at a simple eigenvalue, have been given by M. G. Crandall and P. H. Rabinowitz [**3**] and D. D. Joseph and D. A. Nield [**4**], and results and further references dealing with bifurcation from a multiple *real* eigenvalue can be found in the work of H. B. Keller and W. F. Langford [**5**]. However, our present results are believed to be the first governing the situation of bifurcation from multiple complex eigenvalues. Our methods were suggested by the chemical and nuclear reactor problems recently treated by D. S. Cohen and J. P. Keener [**6**]–[**8**].

2. *The bifurcation theory*. We shall now derive representations for the bifurcating solutions. Specifically, we derive the following:

MAIN RESULT. *Let* $\beta_0 = \beta(\lambda_0)$, $\delta_0 = \delta(\lambda_0)$, *let* $\hat{\mu} = \min_{5 \leqslant i \leqslant n}[|\mu_i|]$, *and let*

$$\varepsilon^2 = \begin{cases} \lambda - \lambda_0 & \text{if } \alpha'(\lambda_0) > 0, \\ \lambda_0 - \lambda & \text{if } \alpha'(\lambda_0) < 0. \end{cases} \tag{5.7}$$

Then, if $|\beta_0 - 2\delta_0|$, $|2\beta_0 - \delta_0|$, *and* $|\beta_0 - \delta_0|$ *are all* $O(1)$ *in* ε, *we have*

$$\begin{aligned} Y(t, \lambda) = Y_0(\lambda_0) &+ \varepsilon\left[C_1 R(\tau)\sin\Omega + C_2 R(\tau)\cos\Omega + O(e^{-\hat{\mu}t})\right] \\ &+ \varepsilon\left[D_1 S(\tau)\sin N + D_1 S(\tau)\cos N + O(e^{-\hat{\mu}t})\right] + O(\varepsilon^2), \end{aligned} \tag{5.8}$$

where C_i, D_i, $i = 1, 2$, *are constant vectors, where*

$$\Omega = \left(1 + \omega_2\varepsilon^2\right)\beta_0 t + \phi(\tau), \tag{5.9}$$

$$N = \left(1 + \nu_2\varepsilon^2\right)\delta_0 t + \psi(\tau), \tag{5.10}$$

$$\tau = \varepsilon^2 t, \tag{5.11}$$

and where $R(\tau)$, $S(\tau)$, $\phi(\tau)$, *and* $\psi(\tau)$ *are scalar functions determined by differential equations of the form*

$$\begin{aligned} R' &= k_{11}R + k_{12}S + k_{13}R^3 + k_{14}R^2S + k_{15}RS^2 + k_{16}S^3, \\ R\phi' + \beta_0 w_2 R &= k_{21}R + k_{22}S + k_{23}R^3 + k_{24}R^2S + k_{25}RS^2 + k_{26}S^3, \\ S' &= k_{31}R + k_{32}S + k_{33}R^3 + k_{34}R^2S + k_{35}RS^2 + k_{36}S^3, \\ S\psi' + S_0\nu_2 S &= k_{41}R + k_{42}S + k_{43}R^3 + k_{44}R^3S + k_{44}RS^2 + k_{46}S^3, \end{aligned} \tag{5.12}$$

or equivalently,

$$\begin{bmatrix} R' \\ R\phi' + \beta w_2 R \\ S' \\ S\psi' + \delta v_2 S \end{bmatrix} = K \begin{bmatrix} R \\ S \\ R^3 \\ R^2 S \\ RS^2 \\ S^3 \end{bmatrix}, \tag{5.13}$$

where $K = (k_{ij})$ is a constant 4×6 matrix. The representation (5.8) is valid at least for times of order $O(1/\varepsilon^2)$.

REMARKS. (1) Note that in sharp constrast with the case of (Hopf) bifurcation at a simple complex eigenvalue, we do not generally obtain periodic bifurcating solutions in the present case because the superposition of harmonics in Ω and N is not periodic if β_0 and δ_0 are incommensurate.

(2) The conditions that $|\beta_0 - 2\delta_0|$, $|2\beta_0 - \delta_0|$, and $|\beta_0 - \delta_0|$ all be of order $O(1)$ in ε is essential; otherwise, the representation (5.8) is not valid. These conditions are a manifestation of the so-called problem of small divisors. In the present context the small divisor problem occurs as a purely mathematical defect in most analytical treatments and is neither inherent in the physics described by the equations nor a property of the mathematical solutions. A good historical survey (starting with the attempts of Poincaré and Whittaker) of the small divisor problem together with a formal method for systematically overcoming the difficulties has been given by H. A. Kabakow [**9**]. We shall use Kabakow's method to eliminate all possible small divisors occurring in all terms of order $O(\varepsilon^2)$, thus insuring that the representation (5.8) is uniformly valid on intervals $0 \leqslant t \leqslant T(\varepsilon)$ where $T(\varepsilon) = O(1/\varepsilon^2)$. Kabakow's method can, in fact, be used to remove the imposed conditions that $|\beta_0 - \delta_0|$ be $O(1)$. Then, a slightly different form for (5.8) would occur. (See the remarks after equation (2.28).)

(3) Stability of the bifurcating solutions is determined by the asymptotic behavior of $R(\tau)$ and $S(\tau)$ as $\tau \to \infty$. If these approach constant values or oscillate, then the bifurcating solutions are stable.

(4) The first and third equations of (5.12) decouple from the rest, and thus, the amplitude modulations $R(\tau)$ and $S(\tau)$, and their implications for stability, can be studied by phase-plane techniques for example. We shall consider these modulation equations in part 3.

(5) Clearly, to lowest order $O(\varepsilon)$ the frequency shifts ω_2 and ν_2 in (5.9), (5.10) can be neglected. We have included them because later in part 3 we shall show how to determine them quite easily in certain cases and thereby obtain the first nontrivial correction to the frequency.

(6) It will be quite clear from our derivation of the representation (5.8) how to write the appropriate representation for bifurcating solutions in the general case when $Y_0(\lambda)$ loses its stability due to j pairs of complex eigenvalues simultaneously crossing the imaginary axis.

We shall now derive the results which we have just stated. In order to apply our methods it is first necessary to transform the system (5.1) to a certain canonical form. We now do this. Let $Z(t, \lambda) = Y(t, \lambda) - Y_0(\lambda)$, and expand the right-hand side of (5.1) in a Taylor's series around the equilibrium solution $Y_0(\lambda)$ to write (5.1) as

$$dZ/dt = F_Y(Y_0(\lambda), \lambda)Z + H(Z), \tag{5.14}$$

where $H(Z)$ denotes the remainder after the linear terms and hence contains no linear terms near $Y_0(\lambda)$. Now, let $Z = TX$ where the (constant) matrix T is yet to be specified. Then, (5.14) becomes

$$dX/dt = T^{-1}F_Y(Y_0(\lambda), \lambda)TX + T^{-1}H(TX). \tag{5.15}$$

At $\lambda = \lambda_0$ we can choose T such that

$$T^{-1}F_Y(Y_0(\lambda_0), \lambda_0)T = \begin{bmatrix} 0 & \beta_0 & 0 & 0 & 0 & \cdot & \cdot & \cdot & 0 \\ -\beta_0 & 0 & 0 & 0 & \cdot & & & & \cdot \\ 0 & 0 & 0 & \delta_0 & \cdot & & & & \cdot \\ 0 & 0 & -\delta_0 & 0 & & & & & \cdot \\ 0 & \cdot & \cdot & \cdot & \mu_5(\lambda_0) & & & & \\ & & & & & \cdot & & & \\ & & & & & & \cdot & & \\ & & & & & & & \cdot & \\ 0 & \cdot & \cdot & \cdot & & & & & \mu_n(\lambda_0) \end{bmatrix} \equiv P. \tag{5.16}$$

The parameter ε defined in (5.7) insures that we are slightly into the region where the steady state $Y_0(\lambda)$ is unstable. Therefore, since $F_Y(Y_0(\lambda), \lambda) = F_Y(Y_0(\lambda_0), \lambda_0) + O(|\lambda - \lambda_0|)$, we can choose T such that

$$T^{-1}F_Y(Y_0(\lambda), \lambda)T = P + \varepsilon^2 A \tag{5.17}$$

where $A = (a_{ij})$. The matrix A can be singular or even identically zero. (For specific applied problems the matrix T has been explicitly given by D. S. Cohen and J. P. Keener [**6**]–[**8**] in cases where there is a transition to instability due to a single complex conjugate pair crossing the imaginary axis, and we exhibit a specific T in the applications given in §4.) Therefore, we choose T such that (5.17) is satisfied where P is given in (5.16). Then, the system (5.14) becomes

$$dX/dt = PX + \varepsilon^2 AX + G(X), \tag{5.18}$$

where $G(X) = T^{-1}H(X)$ and where

$$G(X) = \begin{bmatrix} g_1(X) \\ \cdot \\ \cdot \\ g_n(X) \end{bmatrix}, \qquad g_i(0) = \frac{\partial g_i}{\partial x_j}(0) = 0, \qquad i, j = 1, \ldots, n. \tag{5.19}$$

We shall now investigate the bifurcation from $X \equiv 0$ of time-dependent solutions in the system (5.18), (5.19). Since $Y(t, \lambda) = Y_0(\lambda) + TX$, the elements y_i of Y are just linear combinations of the elements x_i of X.

As we cross the stability boundary $\lambda = \lambda_0$ from the stable to the unstable region, linearized stability theory predicts that the equilibrium solution $X \equiv 0$ loses its stability via an exponentially growing function of time t. This (linearized) exponentially growing function cannot represent the solution for very long because clearly the nonlinear terms must then become important. If, in fact, this exponentially growing function tends to a stable time-dependent solution, then growth on another time scale must come into play so that in some sense the perturbation from the unstable steady state should exhibit a more or less typical multi-scale representation; that is, to lowest order we expect a representation of the form

$$X(t) = \begin{bmatrix} A_1(\tau)P_1(t_1^*) \\ \vdots \\ A_n(\tau)P_n(t_n^*) \end{bmatrix}$$

where the $P_k(t_k^*)$ represent periodic oscillation on a so-called "fast time" t_k^* and the $A_k(\tau)$ represent "slow time" modulation. More specifically, the form of the matrix P strongly suggests that *to lowest order in* ε the components $x_1(t)$ and $x_2(t)$ of $X(t)$ are oscillatory with fast time $t_\beta^* = \beta t$ (or a small perturbation of βt), $x_3(t)$ and $x_4(t)$ are oscillatory with fast time $t_\delta^* = \delta t$ (or a small perturbation of δt), and all other $x_i(t)$, $i = 5, \ldots, n$, decay exponentially. We shall now implement this hermetic reasoning with a formal perturbation technique.

(*Note*. It should now be clear that the transformation of the original problem to the form (5.18) is crucial. Suppose that $x_{1,2}(t)$ and $x_{3,4}(t)$ are indeed periodic with angular frequencies β and δ respectively and that all other $x_i(t)$, $i = 5, \ldots, n$, decay exponentially. Then, the $y_i(t)$, which are linear combinations of the $x_i(t)$, are generally not periodic functions because, for example, $\cos \beta t + \cos \delta t$ is not periodic if β and δ are incommensurate.)

The only necessary tool we shall need in carrying out the n-dimensional multi-timing formalism is an elementary fact for a specific two-dimensional system which we shall state in the form of an easily referenced lemma.

LEMMA. *The general solution of*

$$du/dt + v = m \sin t + n \cos t, \qquad dv/dt - u = p \sin t + q \cos t,$$

is

$$\begin{aligned} u(t) &= A \sin t + B \cos t + \tfrac{1}{2}(m - q)t \sin t \\ &\quad + \tfrac{1}{2}(n + p)t \cos t + \tfrac{1}{2}(n - p)\sin t, \\ v(t) &= -A \cos t + B \sin t + \tfrac{1}{2}(n + p)t \sin t \\ &\quad - \tfrac{1}{2}(m - q)t \cos t + \tfrac{1}{2}(m + q)\sin t. \end{aligned}$$

Thus, in order to suppress secular terms (i.e., in order to have solutions bounded for all $t \geqslant 0$) it is sufficient to require $m - q = 0$ and $n + p = 0$.

In order to completely determine the first term in an asymptotic expansion of $X(t)$ for small ε, we shall see that we must actually carry out the formalism to three terms. To find an asymptotic expansion valid to lowest order it is sufficient now to assume that

$$X(t) = \sum_{n=1}^{3} \varepsilon^n X_n = \sum_{n=1}^{3} \varepsilon^n \begin{bmatrix} x_{1n}(t_\beta^*, \tau) \\ x_{2n}(t_\beta^*, \tau) \\ x_{3n}(t_\delta^*, \tau) \\ x_{4n}(t_\delta^*, \tau) \\ x_{5n}(t_n^*, \tau) \\ \vdots \\ x_{nn}(t_n^*, \tau) \end{bmatrix} \tag{5.20}$$

where

$$\tau = \varepsilon^2 t, \quad t_\beta^* = (1 + \omega_2 \varepsilon^2)\beta t,$$
$$t_\delta^* = (1 + \nu_2 \varepsilon^2)\delta t, \quad t_k^* = \mu_k t, \quad k = 5, \ldots, n. \tag{5.21}$$

We wish to reiterate the fact that the assumed form (5.20), (5.21) is sufficient to find the lowest order terms x_{1i} $(i = 1, \ldots, n)$. To determine higher order terms much more general multi-time dependence must be assumed to overcome small divisor problems, etc. (See Kabakow [**9**].)

With the definitions (5.21), we find that $dx_{ik}/dt = (\partial x_{ik}/\partial t_i^*)(dt_i^*/dt) + \varepsilon^2 \partial x_{ik}/\partial \tau$. Thus, upon substituting (5.20), (5.21) into (5.18) and equating coefficients of like powers of ε, we obtain from the coefficients of the first power of ε that

$$\partial X_1/\partial t^* = P X_1 \tag{5.22}$$

where

$$\partial X_1/\partial t^* = \begin{bmatrix} \partial x_{11}/\partial t_\beta^* \\ \vdots \\ \partial x_{n1}/\partial t_n^* \end{bmatrix}. \tag{5.23}$$

The solution of (5.22) is

$$x_{11} = R(\tau)\sin(t_\beta^* + \phi(\tau)), \qquad x_{21} = R(\tau)\cos(t_\beta + \phi(\tau)), \tag{5.24}$$

$$x_{31} = S(\tau)\sin(t_\delta + \psi(\tau)), \qquad x_{41} = S(\tau)\sin(t_\delta^* + \psi(\tau)), \tag{5.25}$$

$$x_{k1} = C_k(\tau)e^{t_k^*}, \qquad k = 5, \ldots, n, \tag{5.26}$$

where the unknown functions $R(\tau)$, $S(\tau)$, $\phi(\tau)$, $\psi(\tau)$ and $C_k(\tau)$ are to be determined at a later stage of the perturbation procedure.

Since $t_k^* = \mu_k t$ with $\mu_k < 0$, all the x_{k1}, $k = 5, \ldots, n$, decay exponentially. Thus, shortly after bifurcation the dominant terms are x_{11}, x_{21}, x_{31}, and x_{41}. Since all components of $Y(t, \lambda)$ to lowest order are linear combinations of the x_{i1}, $i = 1, \ldots, n$, it suffices now to consider only the steps in the perturbation procedure which determine $R(\tau)$, $S(\tau)$, $\phi(\tau)$, and $\psi(\tau)$. We shall now see that due to the specific form of our transformed system (5.18), the equations for these quantities decouple from the rest of the system.

It will be convenient to carry out the derivation in two stages. First, we shall proceed for the special situation where $G(X)$ of (5.18), (5.19) is given by

$$g_i(X) = \sum_{k,l=1}^{n} b_{kl}^{(i)} x_k x_l \equiv X^T B_i X, \qquad i = 1, \ldots, n. \tag{5.27}$$

It will then be an easy matter to return to our derivation for the general $G(X)$ at the second stage because we shall be able to use certain identities and tricks arising from the case where $G(X)$ is the quadratic form (5.27). A special example, with all the necessary algebraic and trigonometric manipulations worked out, is given in the Appendix so that the reader can follow the ensuing calculations in this concrete case if he wishes.

Upon substituting (5.20), (5.21) into (5.18) with $G(X)$ given by (5.27) and equating coefficients of like powers of ε, we find that the coefficients of ε yield (5.22), the solution of which is given by (5.24)–(5.26). Upon equating the coefficients of ε^2 and ε^3 and retaining only those terms which are needed to determine $R(\tau)$, $S(\tau)$, $\phi(\tau)$, and $\psi(\tau)$, we obtain

$$\partial \hat{X}_2/\partial t^* - \hat{P}\hat{X}_2 = V(X), \tag{5.28}$$

$$\frac{\partial \hat{X}_3}{\partial t^*} - \hat{P}\hat{X}_3 = -M\frac{\partial \hat{X}_2}{\partial t^*} - \frac{\partial \hat{X}_1}{\partial \tau} + \hat{A}\hat{X}_1 + W(X) \tag{5.29}$$

where

$$M = \begin{bmatrix} \omega_2 & 0 & 0 & 0 \\ 0 & \omega_2 & 0 & 0 \\ 0 & 0 & \nu_2 & 0 \\ 0 & 0 & 0 & \nu_2 \end{bmatrix},$$

$$V(X) = \begin{bmatrix} v_1(X) \\ \vdots \\ v_4(X) \end{bmatrix}, \qquad W(X) = \begin{bmatrix} w_1(X) \\ \vdots \\ w_4(X) \end{bmatrix}, \tag{5.30}$$

$$v_i(X) = \sum_{k,l=1}^{n} b_{kl}^{(i)} x_{k1} x_{l1} \equiv X_1^T B_i X, \qquad i = 1, \ldots, 4, \tag{5.31}$$

$$w_i(X) = \sum_{k,l=1}^{n} b_{kl}^{(i)}(x_{k2}x_{l1} + x_{k1}x_{l2})$$
$$\equiv X_2^T B_i X_1 + X_1^T B_i X_2, \quad i = 1, \ldots, 4, \tag{5.32}$$

and where the superscript cap ($\hat{\ }$) is used to denote the truncation of an n-vector or an $n \times n$ matrix to a 4-vector or a 4×4 matrix respectively. For example,

$$\frac{\partial \hat{X}_2}{\partial t^*} = \begin{bmatrix} \partial x_{12}/\partial t_\beta^* \\ \cdot \\ \cdot \\ \cdot \\ \partial x_{42}/\partial t^*\delta \end{bmatrix}, \quad \hat{P} = \begin{bmatrix} 0 & \beta_0 & 0 & 0 \\ -\beta_0 & 0 & 0 & 0 \\ 0 & 0 & 0 & \delta_0 \\ 0 & 0 & -\delta_0 & 0 \end{bmatrix}. \tag{5.33}$$

We now solve system (5.28). From (5.24)–(5.26) and (5.31), we see that the right-hand side of (5.28) contains decreasing exponentials (i.e., those with $e^{t_k^*}$, $k = 5, \ldots, n$, as factors) plus purely trigonometric terms of the forms

$$R^2 \sin^2 \Omega, \quad R^2 \cos^2 \Omega, \quad RS \sin \Omega \sin N, \quad RS \sin \Omega \cos N,$$
$$S^2 \sin^2 N, \quad S^2 \cos^2 N, \quad RS \cos \Omega \sin N, \quad RS \cos \Omega \cos N,$$

where Ω and N are defined in (5.9) and (5.10). (See equations (A-10)–(A-13) of the Appendix.) Upon using common trigonometrical identities such as $\cos^2 A = \frac{1}{2} + \frac{1}{2} \cos 2A$ and $\sin A \cos B = \frac{1}{2}\sin(A + B) + \frac{1}{2}\sin(A - B)$, we find that the purely trigonometrical terms on the right-hand side of (5.28) are all of the form

$$R^2 \genfrac{}{}{0pt}{}{\sin}{\cos} 2\Omega, \quad S^2 \genfrac{}{}{0pt}{}{\sin}{\cos} 2\Omega, \quad RS \genfrac{}{}{0pt}{}{\sin}{\cos} (\Omega - N), \quad RS \genfrac{}{}{0pt}{}{\sin}{\cos} (\Omega + N). \tag{5.34}$$

(See equations (A-10)–(A-13) of the Appendix.) Hence, if $\beta_0 \neq 2\delta_0$, $2\beta_0 \neq \delta_0$, there are no terms which correspond to forcing at a resonance, and thus, the solution of (5.28) contains constants (R^2 and S^2), terms of the form (5.34), and decaying exponentials. (This corresponds to equations (A-14)–(A-17) of the Appendix.) It is at this stage that we require that $|\beta_0 - 2\delta_0|$ and $|2\beta_0 - \delta_0|$ be of order $O(1)$ in ε. To see this, suppose that they were $O(\varepsilon)$, for example. Then, the $O(\varepsilon^2)$ terms in our perturbation expansion (5.20) would really be of order $O(\varepsilon)$, and our expansion would not be uniformly valid. (In our special example, this is illustrated with equations (A-14)–(A-17) of the Appendix.) Kabakow's method [9] allows us to investigate the situation even when $|\beta_0 - 2\delta_0|$ and $|2\beta_0 - \delta_0|$ are $O(\varepsilon)$, but we shall not do so here.

Finally, we examine the system (5.29). From (5.32) we see that resonance producing terms (secular terms) arise only from products of the terms (5.24), (5.25) and the terms of the form (5.34) (equations (A-18) and (A-19) of the Appendix). Thus, for example, the first two equations of (5.29) are of the form

$$\frac{\partial x_{13}}{\partial t_\beta^*} - x_{23} = \frac{1}{\beta}(\sin \Omega)\Big\{- dR/d\tau + (\ )R + (\ )S$$

$$+ (\ )R^3 + (\ )R^2S + (\ )RS^2 + (\ )S^3$$

$$+ \frac{1}{\beta}(\cos \Omega) - \beta w_2 R - R d\phi/d\tau + (\ )R + (\ )S$$

$$+ (\ )R^3 + (\ )R^2S + (\ )RS^2 + (\ )S^3$$

$$+ \text{(other nonresonance producing terms)}, \tag{5.35}$$

$$\frac{\partial x_{23}}{\partial t_\beta^*} + x_{13} = \frac{1}{\beta}(\sin \Omega)\, \beta w_2 R + R d\phi/d\tau + (\ )R + (\ )S$$

$$+ (\ )R^3 + (\ )R^2S + (\ )RS^2 + (\ )S^3$$

$$+ \frac{1}{\beta}(\cos \Omega) - dR/d\tau + (\ )R + (\ )S$$

$$+ (\ )R^3 + (\ )R^2S + (\ )RS^2 + (\ )S^3$$

$$+ \text{(other nonresonance producing terms)}. \tag{5.36}$$

Now, the application of our lemma to the system (5.35), (5.36) implies that to suppress secular terms we must require that

$$R' = (\ )R + (\ )S + (\ )R^3 + (\ )R^2S + (\ )RS^2 + (\ )S^3,$$

$$R\phi' + \beta w_2 R = (\ )R + (\ )S + (\ )R^3 + (\ )R^2S + (\ )RS^2 + (\ )S^3.$$

Similar considerations clearly apply to the third and fourth equations of (5.29), and we then obtain the equations for S and ψ. Therefore, R, S, ϕ, and ψ are determined by equations of the form (5.12), and the amplitude modulation equations (i.e., those for R and S) decouple from the others. Thus, our derivation is complete. (Note that the vectors C_1, C_2, D_1, D_2 in (5.8) are determined by the matrix transformation T, and initial conditions are used to determine the constants of integration arising from solving the modulation equations (5.12).)

It should be clear that the generalization from the quadratic form to the general case proceeds in the same way because the Taylor's series expansion of $G(X)$ is a power series first containing a quadratic form whose coefficients are second derivatives of the $g_i(X)$.

3. *The modulation equations and stability considerations*. First, we shall consider the amplitude modulation equations, namely, the first and third equations of (5.12) for the quantities $R(\tau)$ and $S(\tau)$. The twelve constants k_{1i}, k_{3i} $(i = 1, \ldots, 6)$ can take on all real values depending in any specific case on the nonlinearity $F(Y, \lambda)$ in (5.1) and thus, any general treatment of these equations to account for all possibilities would be prohibitively lengthy. However, since phase-plane considerations are clearly possible, we can make some general statements. For example, it is clear what implications the

appearance of the more usual structures such as spiral points, saddle points, limit cycles, etc. have with regard to analytical form, growth or decay rates, and stability for $R(\tau)$ and $S(\tau)$ and thus for $Y(t, \lambda)$ as given by (5.8). We shall now illustrate some of these situations for two special examples which occur in the theory of chemical reactions.

Suppose that

$$k_{1i}, k_{3i} > 0 \quad \text{for } i = 1, 2,$$
$$k_{1i}, k_{3i} < 0 \quad \text{for } i = 3, \ldots, 6, \tag{5.37}$$

and suppose that the only critical point is the point $(R, S) = (0, 0)$. Multiply the first equation of (5.12) by R and the third equation of (5.12) by S and add the equations to obtain

$$\begin{aligned} \tfrac{1}{2}(d/d\tau)(R^2 + S^2) = {} & k_{11}R^2 + (k_{12} + k_{31})RS + k_{32}S^2 + k_{13}R^4 \\ & + (k_{14} + k_{33})R^3S + (k_{15} + k_{34})R^2S^2 \\ & + (k_{16} + k_{35})RS^3 + k_{36}S^4. \end{aligned} \tag{5.38}$$

Now, on some sufficiently small circle surrounding the origin the quadratic terms are the dominant terms on the right-hand side of (5.38), and on some sufficiently large circle (completely containing the small one) the quartic terms dominate. Hence, (5.37) implies that $d/d\tau\,(R^2 + S^2)$ is positive on the small circle and negative on the large circle. Since there is no singular point in the annular region between the circles, the fundamental Poincaré-Bendixson theory implies that there exists a stable limit cycle contained inside the annulus. Therefore, according to (5.8) the bifurcating solutions consist of the superposition of two components oscillating on fast time scales $\beta_0 t$ and $\delta_0 t$ and modulated by amplitudes periodically changing on the slow time scale $\varepsilon^2 t$.

A totally different situation is encountered with the following example: Suppose that the nonlinearity $F(Y, \lambda)$ is such that the amplitude modulation equations possess a saddle point at the origin and two stable spiral points, one in the first quadrant and the other, symmetric with respect to the origin, in the third quadrant. (Note that the result of the transformation $R \to -R$, $S \to -S$ leaves the equations invariant, and thus, the trajectories in the phase plane are symmetric with respect to the origin.) Therefore, depending on the initial conditions, $R(\tau)$ and $S(\tau)$ approach one or the other of the spiral points, and the bifurcating solution approaches a time dependence consisting of the superposition of sines and cosines of Ω and N. If the spiral points occur on an axis, then one component asymptotically vanishes, and the bifurcating solution eventually becomes periodic at the surviving frequency.

In the case that $R(\tau)$ and $S(\tau)$ approach finite limits as $\tau \to \infty$ we can find the $O(\varepsilon^2)$ frequency shifts ω_2 and ν_2 directly from the equations (5.12) as follows: We expect that $\phi(\tau)$ and $\psi(\tau)$ approach finite limits as $\tau \to \infty$. Then, $\phi'(\tau), \psi'(\tau) \to 0$ as $\tau \to \infty$. Therefore, the second and third equations of (5.12) become algebraic equations for ω_2 and ν_2 in terms of the limiting values $R(\infty)$ and $S(\infty)$.

4. *Applications.* The author was led to developing the methods for the results obtained by considering various problems and observations occurring in chemical and biochemical reactions. In particular, the appearance of complicated and peculiar time-dependent states as the result of varying the concentration of one reactant **[10]**–**[11]** cannot be accounted for by bifurcation at a single simple complex eigenvalue. The chemical kinetics is given by equations of the form (5.1) where the nonlinearity in most cases is a quadratic form in the unknowns. Thus, the vector Y of concentrations is governed by (5.1) where the right-hand side contains linear and quadratic terms in the concentrations. This corresponds to accounting for all reactions on the basis of single and bimolecular reactions. (Trimolecular and higher type reactions are thought to be extremely rare or nonexistent. In any event, our theory could handle them because we have developed it for a general nonlinearity.) A typical situation is worked out in the Appendix.

In the chemical kinetics the bifurcation parameter λ can depend on many other parameters and groupings of parameters, and it can easily happen that complex eigenvalues *simultaneously* cross the imaginary axis. However, even in the classical mechanics of particle motion these occurrences are not only common but in some situations the usual case, and perhaps this would be a good context in which to illustrate this. We shall do so with the following simple example: Consider two coupled harmonic oscillators nonlinearly coupled and with a slight amount of damping. Hence, consider

$$\ddot{y} + \omega^2 y - \varepsilon\alpha\dot{y} + f(y, z) = 0, \tag{5.39}$$

$$\ddot{z} + \gamma^2 z - \varepsilon\beta\dot{z} + g(y, z) = 0. \tag{5.40}$$

Let $u = y$ and $v = z$, and then (5.39), (5.40) can be written as

$$dU/dt = BU + \varepsilon DU + F(U), \tag{5.41}$$

where

$$U = \begin{bmatrix} u \\ y \\ v \\ z \end{bmatrix}, \quad B = \begin{bmatrix} 0 & -\omega^2 & 0 & 0 \\ 1 & 0 & 0 & 0 \\ 0 & 0 & 0 & -\gamma^2 \\ 0 & 0 & 1 & 0 \end{bmatrix},$$

$$D = \begin{bmatrix} \alpha & 0 & 0 & 0 \\ 0 & 0 & 0 & 0 \\ 0 & 0 & \beta & 0 \\ 0 & 0 & 0 & 0 \end{bmatrix}, \quad F(U) = \begin{bmatrix} -f(y, z) \\ 0 \\ -g(y, z) \\ 0 \end{bmatrix}.$$

Let $U = TX$ with T constant. Then, (5.41) becomes

$$dX/dt = T^{-1}BTX + \varepsilon T^{-1}DTX + T^{-1}F(TX). \tag{5.42}$$

Now, let

$$T = \begin{bmatrix} 1 & \omega & 0 & 0 \\ 1 & -1/\omega & 0 & 0 \\ 0 & 0 & 1 & \gamma \\ 0 & 0 & 1 & -1/\gamma \end{bmatrix}.$$

Then, simple straightforward calculations yield

$$T^{-1}BT = \begin{bmatrix} 0 & \omega & 0 & 0 \\ -\omega & 0 & 0 & 0 \\ 0 & 0 & 0 & \gamma \\ 0 & 0 & -\gamma & 0 \end{bmatrix}, \tag{5.43}$$

$$T^{-1}DT = \frac{\omega\gamma}{(1+\omega^2)(1+\gamma^2)} \cdot \begin{bmatrix} \frac{\alpha}{\omega}\left(\gamma + \frac{1}{\gamma}\right) & \alpha\left(\gamma + \frac{1}{\gamma}\right) & 0 & 0 \\ \alpha\left(\gamma + \frac{1}{\gamma}\right) & \alpha\omega\left(\gamma + \frac{1}{\gamma}\right) & 0 & 0 \\ 0 & 0 & \frac{\beta}{\gamma}\left(\omega + \frac{1}{\omega}\right) & \beta\left(\omega + \frac{1}{\omega}\right) \\ 0 & 0 & \beta\left(\omega + \frac{1}{\omega}\right) & \beta\gamma\left(\omega + \frac{1}{\omega}\right) \end{bmatrix}. \tag{5.44}$$

Therefore, from (5.42)–(5.44) we see that (5.41) can be written as

$$dX/dt = PX + \varepsilon AX + G(X),$$

where $P = T^{-1}BT$, $A = T^{-1}DT$, and $G(X) = T^{-1}G(TX)$. This is clearly of the form (5.18) which was the canonical starting form for all our analyses. As both damping coefficients α and β change from negative to positive, the complex eigenvalues of the system simultaneously cross the imaginary axis.

APPENDIX. We consider the problem

$$dX/dt = PX + \varepsilon^2 AX + G(X), \tag{A-1}$$

where

$$X = \begin{bmatrix} x_1 \\ \cdot \\ \cdot \\ \cdot \\ x_4 \end{bmatrix}, \quad P = \begin{bmatrix} 0 & \beta & 0 & 0 \\ -\beta & 0 & 0 & 0 \\ 0 & 0 & 0 & \delta \\ 0 & 0 & -\delta & 0 \end{bmatrix}, \quad A = \begin{bmatrix} a_{11} & 0 & 0 & 0 \\ a_{12} & 0 & 0 & 0 \\ 0 & 0 & a_{33} & 0 \\ 0 & 0 & a_{43} & 0 \end{bmatrix}, \tag{A-2}$$

$$G(X) = \begin{bmatrix} b_{11}x_1^2 + b_{13}x_1x_3 \\ b_{21}x_1x_2 + b_{24}x_2x_4 \\ b_{31}x_1x_3 \\ b_{41}x_1x_4 + b_{43}x_3x_4 \end{bmatrix}. \tag{A-3}$$

Upon substituting (5.20), (5.21) into (A-1) and equating coefficients of like powers of ε, we obtain

$$\partial X_1/\partial t^* = PX, \tag{A-4}$$

$$\frac{\partial X_2}{\partial t^*} - PX_2 = \begin{bmatrix} b_{11}x_{11}^2 + b_{13}x_{11}x_{31} \\ b_{21}x_{11}x_{21} + b_{24}x_{21}x_{41} \\ b_{31}x_{11}x_{31} \\ b_{41}x_{11}x_{41} + b_{43}x_{31}x_{41} \end{bmatrix}, \tag{A-5}$$

$$\frac{\partial X_3}{\partial t^*} - PX_3 = -M\frac{\partial X_2}{\partial t^*} - \frac{\partial X_1}{\partial \tau} + AX$$

$$+ \begin{bmatrix} b_{11}x_{11}x_{12} + b_{13}(x_{11}x_{32} + x_{12}x_{31}) \\ b_{21}(x_{11}x_{22} + x_{12}x_{21}) + b_{24}(x_{21}x_{42} + x_{22}x_{41}) \\ b_{31}(x_{11}x_{32} + x_{12}x_{31}) \\ b_{41}(x_{11}x_{42} + x_{12}x_{41}) + b_{43}(x_{31}x_{42} + x_{32}x_{41}) \end{bmatrix}, \tag{A-6}$$

where M is given by (5.30). The solution of (A-4) is

$$x_{11} = R(\tau)\sin\Omega, \qquad x_{21} = R(\tau)\cos\Omega, \tag{A-7}$$

$$x_{31} = S(\tau)\sin N, \qquad x_{41} = S(\tau)\cos N, \tag{A-8}$$

where

$$\Omega = (1 + \omega_2\varepsilon^2)\beta t + \phi(\tau), \qquad N = (1 + \nu_2\varepsilon^2)\phi t + \psi(\tau). \tag{A-9}$$

The system (A-5) then becomes

$$\frac{\partial x_{12}}{\partial t_\beta^*} - x_{22} = \frac{1}{\beta}b_{11}R^2\sin 2\Omega + \frac{1}{\beta}b_{13}RS\sin\Omega\sin N$$

$$= \frac{1}{2\beta}b_{11}R^2(1 - \cos 2\Omega)$$

$$+ \frac{1}{2\beta}b_{13}RS[\cos(\Omega - N) - \cos(\Omega + N)], \tag{A-10}$$

$$\frac{\partial x_{22}}{\partial t_\beta^*} + x_{12} = \frac{1}{\beta}b_{21}R^2\sin\Omega\cos\Omega + \frac{1}{\beta}b_{24}RS\cos\Omega\cos N$$

$$= \frac{1}{2\beta}b_{21}R^2\sin^2\Omega + \frac{1}{2\beta}b_{24}RS[\cos(\Omega + N) + \cos(\Omega - N)], \tag{A-11}$$

$$\frac{\partial x_{32}}{\partial t_\delta^*} - x_{42} = \frac{1}{\delta}b_{31}RS\sin\Omega\sin N$$

$$= \frac{1}{2\delta}RS[\cos(\Omega - N) - \cos(\Omega + N)], \tag{A-12}$$

$$\frac{\partial x_{42}}{\partial t_\delta^*} + x_{32} = \frac{1}{\delta} b_{41} RS \sin \Omega \sin N + \frac{1}{\delta} b_{43} S^2 \sin N \cos N$$

$$= \frac{1}{2\delta} b_{41} RS [\sin(\Omega + N) + \sin(\Omega - N)] + \frac{1}{2\delta} b_{43} S^2 \sin 2N. \quad \text{(A-13)}$$

We now require that $\beta \neq 2\delta$ and $2\beta \neq \delta$ so that the terms $\cos(\Omega - N)$ in (A-10)–(A-12) and $\sin(\Omega - N)$ in (A-13) do not contribute secular terms. Then, the solution of (A-10)–(A-13) is given by

$$x_{12} = \frac{-1}{5\beta}(b_{11} + \tfrac{1}{2} b_{21}) R^2 \sin 2\Omega + \frac{\beta b_{24} RS}{2[\beta^2 - (\beta - \delta)^2]} \cos(\Omega - N)$$
$$+ \frac{\beta b_{24} RS}{2[\beta^2 - (\beta + \delta)^2]} \cos(\Omega + N) - \frac{(\beta - \delta) b_{13} RS}{2[(\beta - \delta)^2 - \beta^2]} \sin(\Omega - N)$$
$$- \frac{(\beta + \delta) b_{13} RS}{2[(\beta + \delta)^2 - \beta^2]} \sin(\Omega + N), \quad \text{(A-14)}$$

$$x_{22} = \frac{-1}{2\beta} b_{11} R^2 - \frac{1}{5\beta}(\tfrac{1}{2} b_{11} - b_{21}) R^2 \cos 2\Omega$$
$$+ \frac{\beta b_{13} RS}{2[(\beta - \delta)^2 - \beta^2]} \cos(\Omega - N)$$
$$- \frac{\beta b_{13} RS}{2[(\beta + \delta)^2 - \beta^2]} \cos(\Omega + N) - \frac{(\beta - \delta) b_{24} RS}{2[\beta^2 - (\beta - \delta)^2]} \sin(\Omega - N)$$
$$- \frac{(\beta + \delta) b_{24} RS}{2[\beta^2 - (\beta + \delta)^2]} \sin(\Omega + N), \quad \text{(A-15)}$$

$$x_{42} = \frac{-1}{3\delta} b_{43} S^2 \sin 2N + \frac{[(\beta - \delta) b_{31} - \delta b_{41}]}{2[(\beta - \delta)^2 - \delta^2]} RS \sin(\Omega - N)$$
$$- \frac{[(\beta + \delta) b_{31} + \delta b_{41}]}{2[(\beta + \delta)^2 - \delta^2]} RS \sin(\Omega + N), \quad \text{(A-16)}$$

$$x_{42} = \frac{-2}{3\delta} b_{43} S^2 \cos 2N + \frac{[\delta b_{31} - (\beta - \delta) b_{41}]}{2[(\beta - \delta)^2 - \delta^2]} RS \cos(\Omega - N)$$
$$- \frac{[\delta b_{31} - (\beta + \delta) b_{41}]}{2[(\beta + \delta)^2 - \delta^2]} RS \cos(\Omega + N). \quad \text{(A-17)}$$

Now, upon substituting (A-7)–(A-9) and (A-14)–(A-17) into (A-6) and using the standard trigonometrical addition formulas, we find after a considerable amount of algebraic and trigonometric manipulation that if $\beta \neq \delta$ the first two equations of system (A-6) become

$$\frac{\partial x_{13}}{\partial t_\beta^*} - x_{23} = \frac{1}{\beta}(\sin\Omega)\left[-R' + a_{11}R + \frac{\beta^2 b_{13}b_{24}}{\delta(2\beta-\delta)(2\beta+\delta)}RS^2\right]$$

$$+ \frac{1}{\beta}(\cos\Omega)\left[-\beta w_2 R - R\phi' + \frac{1}{5}b_{11}\left(-b_{11}+\frac{1}{2}b_{21}\right)R^3\right.$$

$$\left.-\frac{b_{13}^2(2\beta^2-\delta^2)}{2\delta(2\beta-\delta)(2\beta+\delta)}RS^2\right]$$

$$+ \text{(other nonresonance producing terms)}, \tag{A-18}$$

$$\frac{\partial x_{23}}{\partial t_\beta^*} + x_{13} = \frac{1}{\beta}(\sin\Omega)\,\beta w_2 R + R\phi' + a_{11}R - \frac{1}{2\beta}b_{11}b_{21}R^3$$

$$+ \frac{1}{10\beta}b_{21}\left(\frac{3}{2}b_{11} - \frac{1}{2}b_{21}\right)R^3$$

$$-\frac{b_{24}(2\beta^2-\delta^2)}{2\delta(2\beta-\delta)(2\beta+\delta)}RS^2$$

$$+\frac{1}{\beta}(\sin\Omega)\left[-R' - \frac{b_{24}b_{11}}{2\beta}R^2S - \frac{\beta^2 b_{13}b_{24}}{2\delta(2\beta-\delta)(2\beta+\delta)}RS^2\right]$$

$$+ \text{(other nonresonance producing terms)}. \tag{A-19}$$

Now, the application of our lemma to the system (A-18), (A-19) implies that to suppress secular terms we must require that

$$R' = k_{11}R + k_{14}R^2S + k_{15}RS^2, \tag{A-20}$$

$$R\phi' + \beta w_2 R = k_{21}R + k_{23}R^3 + k_{25}RS^2, \tag{A-21}$$

where

$$k_{11} = k_{21} = \frac{a_{11}}{2}, \quad k_{14} = \frac{-b_{24}b_{11}}{4\beta}, \quad k_{15} = \frac{\beta^2 b_{13}b_{24}}{4(2\beta-\delta)(2\beta+\delta)},$$

$$k_{23} = \frac{b_{21}}{10\beta} = \left(\frac{3}{2}b_{11} - \frac{1}{2}b_{21}\right) - \frac{b_{11}}{5}\left(-b_{11}+\frac{1}{2}b_{21}\right),$$

$$k_{25} = \frac{(b_{13}^2 - b_{24}^2)(2\beta^2-\delta^2)}{2\delta(2\beta-\delta)(2\beta+\delta)}. \tag{A-22}$$

Equations (A-20), (A-21) are of the form given in (2.6), and quite clearly by similar manipulations on the second two equations of system (A-6) we obtain the equations for S and ψ which are also of the form given in (5.12).

References

1. D. S. Cohen, *Bifurcation from multiple complex eigenvalues*, J. Math. Anal. Appl. (to appear).

2. D. H. Sattinger, *Topics in stability and bifurcation theory*, Lecture Notes in Math., vol. 309, Springer-Verlag, New York and Berlin, 1973.

3. M. G. Crandall and P. H. Rabinowitz, *Bifurcation, perturbation of simple eigenvalues, and linearized stability*, Arch. Rational Mech. Anal. **52** (1973), 161–180. MR **49** #5962.

4. D. D. Joseph and D. A. Nield, *Stability of bifurcating time-periodic and steady solutions of arbitrary amplitude* (to appear).

5. H. B. Keller and W. F. Langford, *Iterations, perturbations and multiplicities for nonlinear bifurcation problems*, Arch. Rational Mech. Anal. **48** (1972), 83–108. MR **49** #1252.

6. D. S. Cohen and J. P. Keener, *Oscillatory processes in the theory of particulate formation in super-saturated chemical solutions*, SIAM J. Appl. Math. **28** (1975), 307–318.

7. J. P. Keener and D. S. Cohen, *Nonlinear oscillations in a reactor with two temperature coefficients*, Nuclear Sci. and Eng. **56** (1975), 354–359.

8. D. S. Cohen and J. P. Keener, *Multiplicity and stability of oscillatory states in a continuous stirred tank reactor with exothermic consecutive reactions $A \to B \to C$*, Chem. Eng. Sci. **31** (1976), 115–122.

9. H. A. Kabakow, *A perturbation procedure for nonlinear oscillations* (*The dynamics of two oscillators with weak nonlinear coupling*), Ph. D. Thesis, California Institute of Technology, 1968.

10. K. Pye and B. Chance, *Sustained sinusoidal oscillations of reduced pyridic nucleotide in a cell-free extract of saccharomyces carlsbergensis*, Proc. Nat. Acad. Sci. U.S.A. **55** (1966), 888–894.

11. B. Hess and A. Borteux, *Oscillatory phenomena in biochemistry*, Annual Review of Biochemistry **40** (1971), 237–258.

Department of Applied Mathematics, California Institute of Technology, Pasadena, California 91125

Lectures in Applied Mathematics
Volume 16, 1977

Introduction to the Asymptotic Analysis of Stochastic Equations

George C. Papanicolaou[1]

1. Introduction. The purpose of these notes is to bring to the attention of people interested in stochastic problems some tools for their analysis. It is easy to pose stochastic problems in applied science and engineering and it is also clear that they are more realistic than their deterministic counterparts. It is not easy, however, to analyze these problems. We shall attempt to show here that a class of interesting stochastic problems admit satisfactory solutions by using perturbation theory, in fact, perturbation theory in much the same format as for deterministic problems, multiple scales [**1**], the Fredholm alternative [**2**], etc.

We shall also attempt to show what it is about the stochastic problems under consideration that permits this analysis. Loosely speaking, it is the difference in the time scales of the observables as compared to the time scales of the noise fluctuations. Basic references for the material of §3, our main subject, are [**3**], [**4**], [**5**]. References to works on specific topics, mathematical or physical, are given in this section. The material of §2 is a general review on stochastic processes. The class of Markov processes plays the role of "canonical problems" for the problems we seek to analyze, i.e., the principal approximation will be Markovian and frequently a diffusion process. The assumption is, of course, that these processes are relatively easy to understand and their analysis reduces to reasonably standard mathematical problems, such as solving (deterministic) differential equations.

§3 deals with the asymptotics for ordinary differential equations with random coefficients following the basic work of Stratonovich [**3**]. Every result stated in this section has some mathematical theory behind it; there are few problems that are not understood here. Naturally, implementing the suggested methods to specific problems is quite challenging and more effort is needed in this direction.

Asymptotics for stochastic partial differential equations are, naturally, less well developed. A good deal of work has been done in connection with waves in random media and the emergence of transport theory as a first approximation [**7**], [**8**]. The analysis of the stochastic parts of the problem is formally similar to the one of §3 but there are many other issues that must be considered. A mathematical theory is lacking at present.

The formalism of §3 seems to apply also to a variety of problems in modern physics; see for example [**4**] and other work by M. Lax, [**5**], [**6**].

AMS (*MOS*) *subject classifications* (1970). Primary 60–02, 60H10.

[1]Partially supported by an Alfred P. Sloan Foundation Fellowship and by the Air Force Office of Scientific Research under Grant AFOSR-71-2013.

REFERENCES

1. D. S. Cohen, *Lecture notes in perturbation theory*, Lectures in Appl. Math., vol. 16, Amer. Math. Soc., Providence, R.I., 1976, pp.

2. J. B. Keller, *Perturbation theory*, Lecture Notes, Dept. of Mathematics, Michigan State University, East Lansing, Michigan, 1968.

3. R. S. Stratonivič, *Topics in the theory of random noise*. Vols. 1, 2, "Sovetskoe Radio", Moscow, 1961; English transl., Gordon and Breach, New York, 1963. MR **28** #1660.

4. M. Lax, *Classical noise*. IV. *Langevin methods*, Rev. Modern Phys. **38** (1966), 561–566.

5. F. Haake, *Statistical treatment of open systems by generalized master equations*, Springer-Tracts in Modern Physics, no. 66, Springer-Verlag, Berlin and New York, 1973.

6. H. Haken, *Cooperative phenomena in systems far from thermal equilibrium and in nonlinear systems*, Rev. Modern Phys. **47** (1975), 67–121.

7. A. S. Gurvič and V. I. Tatarskii, *Coherence and intensity fluctuations of light in the turbulent atmosphere*, Radio Science **10** (1975), 3–14.

8. Yu.N. Barabanenkov, A. G. Vinogradov, Yu.A. Kravtsov and V. I. Tatarskiĭ, *Application of theory of multiple scattering to the radiation transfer equation for a statistically inhomogeneous medium*, Radiofisica **15** (1972), 1852–1860.

2. Review of stochastic processes, diffusions and stochastic calculus. References to the subsections that follow are [**1**]–[**6**] at the end of this section.

(i) *Stochastic processes, conditional expectations, Markov property*. A real-valued stochastic process is, at first, a consistent family of probability distribution functions, that is, for any $n \geqslant 1$ and $0 \leqslant t_0 < t_1 < \cdots < t_n < \infty$, functions

$$F_{t_1, t_2, \ldots, t_n}(x_1, x_2, \ldots, x_n)$$

with $x_i \in R$, $i = 1, 2, \ldots, n$, are given which represent the probability that the process at t_1 is less than x_1, at t_2 less than $x_2, \ldots,$ at t_n less than x_n. With this information at hand a probability space $(\Omega, \mathfrak{F}, P)$ is constructed where the set of elementary events Ω is the collection of real-valued functions on $[0, \infty)$, the events $\mathfrak{F}$ are the σ-algebra generated by cylinder sets and P is the probability measure on $\mathfrak{F}$ induced by the given distribution functions. Cylinder sets are sets of functions defined by $\{x(\cdot) \in \Omega;\ x(t_1) \in A_1, x(t_2) \in A_2, \ldots, x(t_n) \in A_n\}$ where $A_1, A_2, \ldots, A_n$ are Borel subsets of R and $0 \leqslant t_1 < t_2 < \cdots < t_n < \infty$.

The σ-algebra generated by the cylinder sets with the restriction that $0 \leqslant t_0 < t_1 < \cdots < t_n \leqslant t$ is denoted by $\mathfrak{F}_t$. We say that $\mathfrak{F}_t$ is the collection of events that occurs up to time t. Thus, along with $(\Omega, \mathfrak{F}, P)$ we have also an increasing family $\mathfrak{F}_t$, $t \geqslant 0$, of σ-algebras and the process, denoted by $\{x(t), t \geqslant 0\}$, is said to be adapted to $\mathfrak{F}_t$, by construction in this case.

Processes that are vector valued or take values in more general spaces can be considered without additional complication.

Under mild regularity conditions on the two-dimensional distributions $F_{t_1, t_2}(x_1, x_2)$, probabilities of events that depend on uncountably many t points can be computed using only a denumerable (dense) set of t points. The process is said to be separable in this case. If $\omega \in \Omega$ let $x(t, \omega)$ denote the value of ω (a function on $[0, \infty)$) at time t. Again, under mild restrictions on $F_{t_1, t_2}(x_1, x_2)$ (stochastic continuity) it follows that $x(t, \omega)$ can be considered a random variable on $[0, \infty) \times \Omega$. In particular, if integrals exist, we may for P-almost all ω integrate $x(t, \omega)$ as a function of t. [**3**, Chapter III] gives a good account of these facts.

If Y is a random variable on $(\Omega, \mathfrak{F}, P)$ taking real values, its expectation is denoted by

$$E\{Y\} = \int_\Omega Y(\omega)\, P(d\omega).$$

Since Ω is the space of functions on $[0, \infty)$, Y is a functional. We wish to define the conditional expectation of Y given $\mathfrak{F}_t$, i.e., given information about the process $\{x(t), t \geqslant 0\}$ up to time t. This conditional expectation is denoted by $E\{Y|\mathfrak{F}_t\}$ and it is a random variable that depends on the past up to time t, i.e., it is an $\mathfrak{F}_t$ measurable random variable on $(\Omega, \mathfrak{F}, P)$. Its defining perperty is that if B is any event in $\mathfrak{F}_t$ then

$$\int_B E\{Y|\mathfrak{F}_t\}\, P(d\omega) = \int_B Y\, P(d\omega)$$

or

$$E\{\chi(B) E\{Y|\mathfrak{F}_t\}\} = E\{\chi(B) Y\}$$

where $\chi(B)$ is the characteristic function of the set B. The random variable $E\{Y|\mathfrak{F}_t\}$ is easily shown to exist but it is defined only for almost all $\omega \in \Omega$. Two basic properties of conditional expectations are these. If Z is an $\mathfrak{F}_t$ measurable random variable then $E\{ZY|\mathfrak{F}_t\} = ZE\{Y|\mathfrak{F}_t\}$ and if $s < t$ (so that $\mathfrak{F}_s \subset \mathfrak{F}_t$), $E\{Y|\mathfrak{F}_s\} = E\{E\{Y|\mathfrak{F}_t\}|\mathfrak{F}_s\}$. The latter is called the iterated expectation formula. A succinct presentation of the above facts is in [**3**, Chapter I].

It is clear that a general stochastic process is a complicated object. The measure P requires for its construction an enormous amount of information contained in the finite-dimensional distributions. To obtain useful information, special classes of processes are introduced such as Gaussian processes, stationary processes, martingales, Markov processes, etc. Their study is the main objective of the theory of stochastic processes.

Gaussian processes are characterized by their mean function $E\{X(t)\} = m(t)$ and their covariance

$$\rho(t, s) = E\{(X(t) - m(t))(X(s) - m(s))\},$$

since if $0 \leqslant t_1 < t_2 < \cdots < t_n < \infty$, the joint characteristic function of $X(t_1), X(t_2), \ldots, X(t_n)$ is given by

$$\exp\left[-\frac{1}{2} \sum_{p,q=1}^{n} \rho(t_p, t_q) z_p z_q + i \sum_{p=1}^{n} m(t_p) z_p\right],$$

where $(z_1, z_2, \ldots, z_n) \in R^n$. Stationary processes are characterized by the fact that the joint distribution of $X(t_1 + h), X(t_2 + h), \ldots, X(t_n + h)$ is independent of h, i.e., the statistics are translation invariant. Martingales (see (vii) ahead) are defined by $\sup_{t \geqslant 0} E\{|X(t)|\} < \infty$, and $E\{X(t + s)|\mathfrak{F}_t\}$

$= X(s)$ for almost all realizations. The significance of such processes will become apparent later. Finally, Markov processes, which are of principal interest to us, are characterized by the property that if A is any event in $\mathfrak{F}$ but not in $\mathfrak{F}_t$ then

$$P\{A|\mathfrak{F}_t\} = E\{\chi(A)|\mathfrak{F}_t\}$$

is a function of $X(t)$ only. In words, the conditional probability of an event in the future given the past up to the present t depends only on the present.

(ii) *Brownian motion*. This process $\{w(t), t \geqslant 0\}$ is defined as follows.

(a) $w(0) = 0$.

(b) For any $0 \leqslant s_1 < t_1 \leqslant s_2 < t_2 \leqslant \cdots < s_n \leqslant t_n < \infty$, $w(t_1) - w(s_1)$, $w(t_2) - w(s_2), \ldots, w(t_n) - w(s_n)$ are independent random variables.

(c) $w(t) - w(s)$ has Gaussian distribution with mean zero and variance $t - s$.

From the properties of the Gaussian distribution we have

$$E\{(w(t) - w(s))^{2n}\} = \int (y - x)^{2n} \frac{\exp[-(y-x)^2/2(t-s)]}{(2\pi(t-s))^{1/2}} \tag{2.1}$$
$$= c_n(t-s)^n$$

where c_n is a constant. Now a sufficient condition that a process have with probability one continuous trajectories is [3, p. 186] that there exist $\alpha, \beta > 0$ and $C < \infty$ such that

$$E\{|X(t) - X(s)|^\alpha\} \leqslant C|t-s|^{1+\beta},$$

with $t, s \in [0, \infty)$. It is clear therefore that Brownian motion has continuous trajectories with probability one. We can now take as basic probability space $(\Omega, \mathfrak{F}, P)$ the set Ω of bounded continuous functions on $[0, \infty)$ since P is concentrated on this set. The σ-algebra of cylinder sets is the same as the σ-algebra of Borel sets of Ω viewed as a metric space with $\sup_t|x(t) - y(t)|$ as metric.

It turns out that Brownian motion is nowhere differentiable with probability one [**5**]. This is a phenomenon inherited from the independent increments property; it is the root of the difficulties associated with the stochastic calculus. One can see formally from (2.1) that derivatives will not exist since

$$E\left|\frac{w(t) - w(s)}{t-s}\right| = \frac{\tilde{c}_1}{|t-s|^{1/2}}$$

where $\tilde{c}_1$ is a constant.

Since $w(t)$ is Gaussian, for any $f(x)$ bounded and continuous $u(t, x) = E\{f(w(t) + x)\}$ exists and

$$u(t, x) = \int_{-\infty}^{\infty} f(z + x) \frac{\exp[-z^2/2t]}{(2\pi t)^{1/2}}\, dz$$

so that

$$\frac{\partial u}{\partial t} = \frac{1}{2}\frac{\partial^2 u}{\partial x^2}, \qquad t > 0, u(0, x) = f(x). \tag{2.2}$$

Note that for $t > 0$, $u(t, x)$ is C^∞ in x.

Brownian motion is a process of independent increments and hence Markovian. Let $P(t, x, y)$ denote the transition probability density that $w(s + t)$ will be at y given that $w(s)$ equals x. Clearly

$$P(t, x, y) = \frac{\exp\left[-(x - y)^2/2t\right]}{(2\pi t)^{1/2}} \tag{2.3}$$

and also, $u(t, x) = \int_{-\infty}^{\infty} P(t, x, y) f(y)\, dy$. Equation (2.2) is called the backward equation associated with Brownian motion. $P(t, x, y)$ satisfies (2.2) as a function of both x and y so the forward and backward equations are the same for Brownian motion.

An important characterization of Brownian motion is the following **[2]**, **[4]**. Let Ω be the space of continuous trajectories on $[0, \infty)$ and let $\mathcal{F}_t$ be the σ-algebra generated by the trajectories up to time t. Suppose $\{w(t), t \geqslant 0\}$ is a process with the following properties.

(a) $w(\cdot) \in \Omega$, i.e., $w(\cdot)$ is continuous with probability one.

(b) $E\{w(t) - w(s) | \mathcal{F}_s\} = 0$, $t \geqslant s$.

(c) $E\{(w(t) - w(s))^2 | \mathcal{F}_s\} = t - s$, $t \geqslant s$. Then $\{w(t), t \geqslant 0\}$ is the Brownian motion process.

The point of this characterization is that the Gaussian distribution for the increments is deduced from (a), (b) and (c) above.

(iii) *Orstein-Uhlenbeck process.* Brownian motion plays a basic role in diffusion theory since a large class of diffusions can be obtained as functionals of it (see (vi) below). On the other hand as a physical process meant to describe a particle undergoing many collisions, it has many drawbacks. One is that it is not differentiable and the other is that its transition density (2.3) does not stabilize for t large to a limit that can be conrected with phenomenological thermodynamics.

To overcome these difficulties we suppose [7] that the velocity $\{v(t), t \geqslant 0\}$ of the particle satisfies Langevin's equation

$$dv(t) + \gamma v(t)dt = \alpha dw(t), \quad v(0) = v_0, \gamma > 0, \tag{2.4}$$

where $\{w(t), t \geqslant 0\}$ is Brownian motion. The stochastic differential equation (2.4) is easily solved, even though the right-hand side has no formal meaning, and

$$v(t) = e^{-\gamma t} v_0 + \alpha \int_0^t e^{-\gamma(t-s)}\, dw(s). \tag{2.5}$$

In (2.5) the integral is defined by integration by parts and this then is taken as the definition for $v(t)$ with (2.4) being merely symbolic. The constant γ is called the friction constant and α the noise intensity level; (2.4) has the form $F = ma$ with αdw the external (symbolic) force and $-\gamma v$ the frictional force.

From (2.5) we deduce that $v(t)$ is Gaussian and Markov and that its transition probability density is given by

$$P(t, v_0, v) = \frac{\exp\left[-\dfrac{\gamma(v - v_0 e^{-\gamma t})^2}{\alpha^2(1 - e^{-2\gamma t})}\right]}{\left(\pi(1 - e^{-2\gamma t})\alpha^2/\gamma\right)^{1/2}} \tag{2.6}$$

To see how (2.6) arises it is enough to compute the mean and variance of $v(t)$ from (2.5):

$$E\{v(t)\} = e^{-\gamma t} v_0,$$

$$E\left\{\left(v(t) - e^{-\gamma t} v_0\right)^2\right\} = \alpha^2 \int_0^t \int_0^t e^{-\gamma(t-s)} e^{-\gamma(t-\sigma)} E\{dw(s)dw(\sigma)\}.$$

Now $E\{w(t)w(s)\} = \min(t, s)$; hence, formally, $E\{dw(t)dw(s)\} = \delta(t - s)$ $dtds$ so that

$$E\left\{\left(v(t) - e^{-\gamma t} v_0\right)^2\right\} = \alpha^2 \int_0^t e^{-2\gamma(t-s)}\, ds = \frac{\alpha^2}{2\gamma}(1 - e^{-2\gamma t}).$$

Note that for $t \ll 1$

$$P(t, v_0, v) \sim \frac{\exp\left[-(v - v_0)^2/2\alpha^2 t\right]}{(2\pi\alpha^2 t)^{1/2}},$$

and for $t \gg 1$

$$P(t, v_0, v) \sim (\gamma/\pi\alpha^2)^{1/2} \exp\left[-(\gamma/\alpha^2)v^2\right]. \tag{2.7}$$

The friction constant γ can be related to the fluid density through which the particle is moving, its radius, assuming it spherical, etc. Thus γ is determinable from first principles. On the other hand α comes into the problem in a somewhat artificial way. To determine it we assume that for $t \gg 1$, i.e., at thermodynamic equilibrium, the velocity of the particle has the Maxwell-Gibbs distribution $ce^{-v^2/kT}$ where k is Boltzmann's constant and T is the absolute temperature. Matching (2.7) with this distribution gives

$$\alpha = \sqrt{kT\gamma} \tag{2.8}$$

and so the noise level is determined in terms of measurable quantities. Relation (2.8) is an instance of a fluctuation-dissipation theorem (α measures the fluctuation, γ the dissipation) [**8**].

From (2.6) we deduce by direct computation that

$$u(t, v_0) = \int_{-\infty}^{\infty} P(t, v_0, v) f(v)\, dv$$

satisfies

$$\frac{\partial u}{\partial t} = \frac{\alpha^2}{2} \frac{\partial^2 u}{\partial v_0^2} - \gamma v_0 \frac{\partial u}{\partial v_0}, \qquad u(0, v_0) = f(v_0)$$

which is the backward equation. P as a function of v satisfies also the forward equation

$$\frac{\partial P}{\partial t} = \frac{\alpha^2}{2}\frac{\partial^2 P}{\partial v^2} + \gamma \frac{\partial}{\partial v}(vP), \qquad P(0, v_0, v) = \delta(v_0 - v).$$

Multi-dimensional versions of the above can be obtained easily. The class of Gaussian-Markov processes, although very special, has many applications primarily because it is perhaps the only class of processes for which the transition density is explicitly known and is relatively simple to work with.

An important problem arising from (2.4) is to find a mechanical model for this equation so that in some relevant limit the "white noise" driving force on the right-hand side is deduced, rather than postulated. This problem is solved in [**9**].

(iv) *Transition functions, backward equation, diffusion.* Markov processes are in most cases completely characterized by their transition function

$$P(t, x, A) = P\{X(t+s) \in A \mid X(s) = x\}.$$

Here A is a subset of E, the state space of $\{X(t), t \geqslant 0\}$ which may be the real line, a discrete set, etc. We assume that $P(t, x, A)$ is a probability measure for each x, $0 \leqslant P(t, x, A) \leqslant 1$, $P(t, x, E) = 1$, $t \geqslant 1$, and that $P(t, x, A_x) \to 1$ as $t \to 0$ where A_x is any neighborhood of x and the limit is in a suitably uniform way in x. We also assume the Chapman-Kolmogorov relation,

$$P(t+s, x, A) = \int_E P(t, x, dy) P(s, y, A). \tag{2.9}$$

Let $\mu_0(A)$ be the initial measure for X, i.e., $\mu_0(A) = P\{X(0) \in A\}$. Then,

$$P\{X(t) \in A\} \equiv \mu_t(A) = \int_E \mu_0(dx) P(t, x, A), \qquad t > 0,$$

defines a transformation $S_t\mu_0 = \mu_t$ on probability measures on E such that, from (2.9), $S_{t+s} = S_t S_s$, i.e., S_t is a semigroup. Similarly, for $f(x)$ a bounded measurable function on E

$$T_t f(x) \equiv E\{f(X(t)) \mid X(0) = x\} = \int_E P(t, x, dy) f(y), \qquad t > 0,$$

defines a transformation that takes bounded measurable functions into bounded measurable functions and, from (2.9) again, $T_{t+s} = T_t T_s$, i.e., we have a semigroup.

It is convenient to study the process via the semigroup T_t, $t \geqslant 0$, both because a certain amount of abstraction helps clarify the issues and because it provides well-organized means for discussing the problems; see [**1**, Volume II].

Assume now that $E = R^n$ and that if $C_0(R^n)$ denotes the space of bounded continuous functions on R^n vanishing at ∞ then,

$$T_t f(x) = \int_{R^n} P(t, x, dy) f(y), \qquad t > 0, \tag{2.10}$$

is also an element of $C_0(R^n)$ and that

$$\lim_{t\downarrow 0} \|T_t f - f\| = \lim_{t\downarrow 0} \sup_{x\in R^n} |T_t f(x) - f(x)| = 0. \tag{2.11}$$

With these assumptions we have enough restrictions on $P(t, x, A)$ so that it defines a strongly continuous contraction semigroup on the Banach space $C_0(R^n)$. The contractiveness is immediate.

$$\|T_t f\| = \sup_x \left| \int P(t, x, dy) f(y) \right| \leqslant \sup_x |f(x)| = \|f\|.$$

The point of semigroup theory is that, under favorable circumstances, the semigroup T_t, hence the transition function $P(t, x, A)$ and hence the process, can be reconstructed from its infinitesimal generator A defined by

$$\lim_{h\downarrow 0} \frac{1}{h} (T_h f - f) \equiv Af$$

for all f for which the (norm) limit exists and which form the domain of A, $D(A)$. From the point of view of physical applications, one usually has A available, as a differential or integral operator suggested from first principles, and one wants to solve for the transition function, i.e., solve the backward equation which is formally

$$(d/dt)(T_t f) = A(T_t f), \qquad T_0 f = f,$$

or more carefully, for $f \in D(A)$,

$$T_t f - f - \int_0^t T_s A f \, ds = 0. \tag{2.12}$$

Since A is usually given, the question is: does it determine T_t or not? This may mean that giving a differential operator is not enough if boundary conditions are not given as well. A useful result answering this question (see [**1**, Volume II]) is the Hille-Yoshida theorem:

In order that A be the infinitesimal generator of a strongly continuous contraction semigroup on $C_0(R^n)$ it is necessary and sufficient that

(a) $D(A)$ be dense in $C_0(R^n)$;

(b) for $\lambda > 0$, $(\lambda - A)$ maps $D(A)$ onto $C_0(R^n)$;

(c) for every $f \in D(A)$ and $\lambda > 0$, $\|(\lambda - A)f\| \geqslant \lambda\|f\|$. The necessity of these conditions is easy. If for $\lambda > 0$ we let

$$R_\lambda f = \int_0^\infty e^{-\lambda t} T_t f \, dt,$$

then $\|\lambda R_\lambda f\| \leqslant \|f\|$ which is the same as (c); $R_\lambda f$ is in $D(A)$, $\lambda > 0$ and $AR_\lambda f = -f + \lambda R_\lambda f$ so (b) holds; and $\lim_{\lambda\uparrow\infty} \lambda R_\lambda f = f$ so (a) holds also. The sufficiency requires more work [**1**, Volume II]. In specific problems condition (b) is usually the most difficult to verify.

Diffusion processes are characterized by the fact that for any $\varepsilon > 0$

$$\frac{1}{t} \int_{|x-y|>\varepsilon} P(t, x, dy) \to 0 \quad \text{as } t\downarrow 0. \tag{2.13}$$

This condition alone [3, Chapter III] implies that $\{X(t), t \geqslant 0\}$ is continuous with probability one. For $\varepsilon > 0$ define $a(x)$ and $b(x)$ by (we let $E = R$ for simplicity)

$$\lim_{t\downarrow 0} \frac{1}{t} \int_{|x-y|<\varepsilon} (x-y)P(t, x, dy) = b(x), \tag{2.14}$$

$$\lim_{t\downarrow 0} \frac{1}{t} \int_{|x-y|<\varepsilon} (x-y)^2 P(t, x, dy) = a(x). \tag{2.15}$$

Then we have that $C^2(R)$, functions with two continuous derivatives, is in $D(A)$ and on $C^2(R)$

$$Af(x) = \tfrac{1}{2}a(x)\partial^2 f(x)/\partial x^2 + b(x)\partial f(x)/\partial x, \tag{2.16}$$

so that the backward equation for $u(t, x) = \int P(t, x, dy) f(y)$ is

$$\begin{aligned}\frac{\partial u(t,x)}{\partial t} &= \frac{1}{2}a(x)\frac{\partial^2 u(t,x)}{\partial x^2} + b(x)\frac{\partial u(t,x)}{\partial x},\\ t &> 0,\ u(0,x) = f(x).\end{aligned} \tag{2.17}$$

To verify these facts we note that

$$\begin{aligned} Af &= \lim_{h\downarrow 0} \frac{1}{h}(T_h f - f) = \lim_{h\downarrow 0} \frac{1}{h} \int P(h, x, dy)(f(y) - f(x)) \\ &= \lim_{h\downarrow 0} \Bigg[\frac{1}{h} \int_{|x-y|>\varepsilon} P(h, x, dy)(f(y) - f(x)) \\ &\quad + \frac{1}{h} \int_{|x-y|<\varepsilon} P(h, x, dy) \Big[f'(x)(x-y) \\ &\qquad + \frac{1}{2} f''(x)(x-y)^2 + \rho(\varepsilon) \Big] dy \Bigg], \end{aligned}$$

where $\rho(\varepsilon)$ goes to zero with ε. The hypotheses (2.13)–(2.15) now imply (2.16).

Let us note again that because no special assumptions have been made about $a(x)$ and $b(x)$, there is no a priori reason that (2.17) be a well-posed problem. So, for example, the question of existence and uniqueness of the diffusion process given its infinitesimal mean $b(x)$ and its infinitesimal variance $a(x)$ is a difficult problem with many ramifications.

(v) *Trotter-Kato theorem.* This is a general semigroup approximation theorem that gives conditions for a sequence T_t^n of semigroups to converge as $n \to \infty$ to another semigroup T_t. It is useful in proving, for example, diffusion approximations to random walks, convergence of finite difference approximations to the solution of differential equations, etc. For the kind of singular perturbations we have in mind it becomes useful after a preliminary adjustment.

Let T_t^n, $n = 1, 2, \ldots,$ be strongly continuous contraction semigroups on a Banach space, say B. Let R_λ^n, $\lambda > 0$, be the corresponding resolvents

$$R_\lambda^n = \int_0^\infty e^{-\lambda t} T_t^n \, dt.$$

Let T_t be another continuous semigroup and R_λ its resolvent. If

$$\sup_{0 \leq t \leq \tau} \|T_t^n f - T_t f\| \to 0, \qquad \tau < \infty, f \in B, \tag{2.18}$$

as $n \to \infty$ then clearly $\|R_\lambda^n f - R_\lambda f\| \to 0$, $\lambda > 0$, as $n \uparrow \infty$. The converse of this (simplified somewhat) is one aspect of the Trotter-Kato theorem: If $\|R_\lambda^n f - R_\lambda f\| \to 0$, $\lambda > 0$, $f \in B$, $n \uparrow \infty$, then the semigroups converge.

Naturally, showing that $\|R_\lambda^n f - R_\lambda f\| \to 0$, $n \uparrow \infty$, is not easy. Suppose that A_n and A are the infinitesimal generators of T_t^n and T_t respectively. Suppose there exists a set $D \subset B$ such that D is dense in B,

$$A_n f \to Af, \qquad f \in D, \tag{2.19}$$

and $(\lambda - A)D$, $\lambda > 0$, is dense in B. Then the resolvents converge also. To see this we note that, for $f \in D$,

$$T_t^n f - f - \int_0^t T_s^n A f \, ds = \int_0^t T_s^n (A_n - A) f \, ds,$$

and hence

$$\sup_{0 \leq t \leq \tau} \left\| T_t^n f - f - \int_0^t T_s^n A f \, ds \right\| \to 0.$$

From this it follows that, for $f \in D$,

$$\|f - R_\lambda^n (\lambda - A) f\| \to 0, \qquad \lambda > 0. \tag{2.20}$$

Let $g \in (\lambda - A)D$, $\lambda > 0$, and put $f = R_\lambda g$. Then (2.20) implies that $\|R_\lambda g - R_\lambda^n g\| \to 0$, $\lambda > 0$, for $g \in (\lambda - A)D$. But since R_λ, R_λ^n are bounded and $(\lambda - A)D$ is dense the result follows.

Condition (2.19) is impossible to satisfy in many problems of interest. This is because nonuniformities in the problem preclude the simple generator convergence (2.19). Kurtz **[10]** observed that it is enough that, exactly as above, for $f \in D$ there exists a sequence $f_n \in D$ such that

$$A_n f_n \to Af, \qquad f_n \to f, \tag{2.21}$$

the extra freedom allowing for the removal of nonuniformities. To see how (2.21) implies (2.18) we note that

$$T_t^n f - f - \int_0^t T_s^n A f \, ds = T_t^n (f - f_n) - (f - f_n) + \int_0^t T_s^n (A^n f_n - Af) \, ds,$$

so

$$\sup_{0 \leq t \leq \tau} \left\| T_t^n f - f - \int_0^t T_s^n A f \, ds \right\| \to 0, \qquad f \in D, \tau < \infty.$$

But this again implies (2.20) and hence, as above, the desired result (2.18).

The extension (2.21) of (2.19) may appear somewhat superficial at first but it is extremely useful as we shall see in §3 below.

(vi) *Itô's stochastic differential approach to diffusions.* Diffusion processes look locally like Brownian motion. This is so because in a small time interval Δt we have, from the above considerations, that $x(t+\Delta t)-x(t)$ is approximately Gaussian with mean $b(x(t))\Delta t$ and variance $a(x(t))\Delta t$. If $w(t)$ is the Brownian motion process we may summarize this by writing

$$x(t+\Delta t) - x(t) \sim b(x(t))\Delta t + \sigma(x(t))(w(t+\Delta t) - w(t)) \qquad (2.22)$$

where $\sigma(x) = (a(x))^{1/2}$. In the limit $\Delta t \downarrow 0$ we may write formally

$$dx(t) = b(x(t))dt + \sigma(x(t))dw(t), \qquad x(0) = x. \qquad (2.23)$$

So far this is merely a restatement of the fact that $A = \frac{1}{2}a(x)\partial^2/\partial x^2 + b(x)\partial/\partial x$ is the infinitesimal generator of $\{x(t), t \geqslant 0\}$. Can one make literal sense of this equation? If this is successful then we have a nice way to define diffusions pathwise as (nonlinear) functionals of Brownian motion via (2.23). Clearly the major difficulty is that $w(t)$ is not differentiable and this is an essential aspect of its behavior as well as the pathology that emerges in studying (2.23).

We remark that the issue of making sense of (2.23) should not be confused with what "physical" stochastic differential equations should look like. In physical problems the random coefficients are always correlated processes and hence the resulting solutions are never diffusion processes. That they may be approximated well by diffusion processes calls for the analysis of some relevant asymptotic limits. It is not a matter subject to definitions.

To continue, we rewrite (2.23) as an integral equation

$$x(t) = x + \int_0^t b(x(s))\, ds + \int_0^t \sigma(x(s))\, dw(s). \qquad (2.23')$$

We see that the first step in the analysis is the definition of the second integral on the right side of (2.23′), called an Itô stochastic integral. Integration by parts will not work since the integrand is also random. Furthermore, its definition must somehow include the intuitive idea expressed by (2.22), i.e., the Brownian increment must be in the future ($\Delta t > 0$).

Let Ω be the space of continuous trajectories on $[0, T]$, $T < \infty$, and let $\mathcal{F}_t$, $0 \leqslant t \leqslant T$, be the σ-algebra generated by the cylinder sets up to time t. A real random variable on Ω which is $\mathcal{F}_t$-measurable, denoted by $G(t) = G(t, \omega)$, $\omega \in \Omega$, is called a nonanticipating functional. Let $\mathcal{F}$ be the σ-algebra generated by cylinder sets and let P be the measure on $\mathcal{F}$ of the Brownian motion. We wish to define

$$\int_0^t G(s)\, dw(s), \qquad 0 \leqslant t \leqslant T. \qquad (2.24)$$

To begin with we define (2.24) for nonanticipating step functionals

$$\int_0^T G(s)\, dw(s) = \sum_{k=1}^{n} G(s_{k-1})[w(s_k) - w(s_{k-1})] \qquad (2.25)$$

where $s_0, s_1, \ldots, s_n$ is a partition of $[0, T]$ corresponding to the nonanticipating step functional G. The point of the definition (2.25) is that the Brownian increments are in the future, which is what we wanted in (2.22). A different definition leads to different results contrary to what is the case with ordinary integrals.

We now state the following properties of Itô's integral, assuming that $\int_0^T E\{G^2\}\,ds < \infty$.

$$\text{(a)}\ E\{\textstyle\int_0^t G\,dw\} = 0. \qquad \text{(b)}\ E\{\{\textstyle\int_0^t G\,dw\}^2\} = \textstyle\int_0^t E\{G^2(s)\}\,ds. \tag{2.26}$$

(c) $\int_0^t G\,dw$ is a.s. continuous as a function of t and it is a martingale (see (vii) below).

These properties are preserved under the limiting process necessary to define (2.24) for general nonanticipating functionals subject to $\int_0^T E\{G^2\}\,ds < \infty$; see [**4**], [**5**].

With the definition and properties of stochastic integrals under control we return to (2.23). Assuming that $b(x)$ and $\sigma(x)$ are Lipschitz continuous, uniformly in x, then, an iteration argument resembling the usual one from ODE (but more complicated) leads to the existence and uniqueness of the diffusion process $x(t)$ as a nonanticipating functional of Brownian motion.

What use can be made of this pathwise definition of diffusions? Most of what can be done depends on Itô's lemma, which is as follows.

Suppose $f(x, t)$ has the bounded continuous derivatives. Then we have the identity

$$\begin{aligned} f(x(t), t) &= f(x(s), s) + \int_s^t \left(A + \frac{\partial}{\partial\gamma}\right) f(x(\gamma), \gamma)\,d\gamma \\ &\quad + \int_s^t \sigma(x(\gamma))\,\frac{\partial f(x(\gamma), \gamma)}{\partial x}\,dw(\gamma), \qquad 0 \leqslant s \leqslant t < \infty, \end{aligned} \tag{2.27}$$

where

$$A = \tfrac{1}{2}a(x)\partial^2/\partial x^2 + b(x)\partial/\partial x, \qquad a(x) = \sigma^2(x).$$

To see what this formula says we recall that if $x(t)$ satisfies the deterministic ODE

$$dx(t)/dt = F(x(t), t), \qquad x(0) = x,$$

then

$$f(x(t), t) = f(x(s), s) + \int_s^t \left(F(x(\gamma), \gamma)\frac{\partial}{\partial x} + \frac{\partial}{\partial\gamma}\right) f(x(\gamma), \gamma)\,d\gamma$$

is an obvious identity. So Itô's formula is a generalization of this fact for the stochastic equation (2.23).

A simple application of Itô's formula is this. Suppose $u(t, x)$ is a solution of

$$\partial u(t, x)/\partial t = Au(t, x), \qquad t > 0,\ u(0, x) = f(x). \tag{2.28}$$

Apply Itô's formula to $u(t - s, x(s))$, with t a parameter. This leads to

$$f(x(t)) = u(t, x) + \int_0^t \left(A + \frac{\partial}{\partial\gamma}\right) u(t - \gamma, x(\gamma))\, d\gamma$$

$$+ \int_0^t \sigma(x(\gamma)) \frac{\partial u(t - \gamma, x(\gamma))}{\partial x}\, dw(\gamma).$$

Assuming that $\sigma \partial u/\partial x$ is bounded then $u(t, x) = E\{f(x(t))\}$, since by (2.26)(a) the last term in Itô's formula drops out.

We have deduced by pathwise means that

$$u(t, x) = E\{f(x(t))\} = \int P(t, x, dy) f(y) \tag{2.29}$$

satisfies the backward equation (2.28) provided the latter has smooth solutions. Actually, one can deduce under appropriate conditions and using stochastic calculus that $u(t, x)$ defined by (2.29) is smooth; therefore it must satisfy (2.28) [**4**], [**5**].

More interesting applications of Itô's formula are given in (viii).

(vii) *Martingales, stopping times, optional stopping theorem.* Let $(\Omega, \mathfrak{F}, P)$ be a probability space and let $\mathfrak{F}_t$, $t \geqslant 0$, be an increasing family of σ-algebras contained in $\mathfrak{F}$. A process $\{X(t), t \geqslant 0\}$ is a martingale if $\sup_{t\geqslant 0} E\{|X(t)|\} < \infty$ and $E\{X(t)|\mathfrak{F}_s\} = X(s)$ a.s., for all $t \geqslant s$. Note that Brownian motion is a martingale since $E\{w(t) - w(s)|\mathfrak{F}_s\} = 0$. Note also that stochastic integrals with, say, $|G| \leqslant C$ are martingales

$$E\left\{\int_0^t G(\sigma)\, dw(\sigma)|\mathfrak{F}_s\right\} = E\left\{\int_s^t G(\sigma)\, dw(\sigma)|\mathfrak{F}_s\right\} + E\left\{\int_0^s G(\sigma)\, dw(\sigma)|\mathfrak{F}_s\right\}$$

$$= \int_0^s G(\sigma)\, dw(\sigma).$$

It is clear therefore that Itô's formula generates martingales in profusion.

In (ii) we saw that Brownian motion can be characterized by continuity of paths and

$$E\{w(t) - w(s)|\mathfrak{F}_s\} = 0, \qquad E\{(w(t) - w(s))^2|\mathfrak{F}_s\} = t - s,$$

which says that $w(t)$ is a continuous square-integrable martingale with increasing process equal to t.

General diffusion processes can also be characterized in this martingale way, as is Brownian motion [**11**]. A number of interesting applications of this idea can be given in the area of asymptotics or limit theorems.

Aside from this connection with diffusions, we need for the next section the following two facts about martingales [**1**], [**2**], [**5**]:

(a) the optional stopping theorem,

(b) Kolmogorov's inequality.

Kolmogorov's inequality says that, for $\varepsilon > 0$,

$$P\left\{\sup_{0 \leqslant t \leqslant T} X(t) > \varepsilon\right\} \leqslant \frac{E\{X(T)^+\}}{\varepsilon}$$

where $X^+(t) = \max(0, X(t))$. Without the sup inside $P\{\ \}$ this is just Tchebyshev's inequality. The importance of having the sup inside is clear from the example of the next section.

A random variable t^* is called a stopping time if for all $t \geqslant 0$ the event $\{t^* > t\}$ belongs to $\mathfrak{F}_t$, i.e., to decide if $t^* > t$ or not, it is enough to look at information available up to time t. We denote by $\mathfrak{F}_{t^*}$ the σ-algebra of events generated by events of the form $A \cap \{t^* > t\}$ with $A \in \mathfrak{F}$ and $t \geqslant 0$. The optional stopping theorem is now as follows.

Let $s^* \leqslant t^*$ be two finite stopping times. Then

$$E\{X(t^*)|\mathfrak{F}_{s^*}\} = X(s^*),$$

i.e., the definition of the martingale is still valid if t and s are finite stopping times.

The optional stopping theorem is most useful in carrying over to stochastic problems the method of characteristics, as we do in the next section.

General Markov processes, not necessarily diffusions, can be treated, under favorable circumstances, in much the same way as diffusions; but, because the backward equation will be an integrodifferential equation now, the results are not as detailed as for diffusions. The martingale characterization of these processes is again helpful, and Itô's formula holds again in a suitably generalized form.

(viii) *Applications of Itô's formula.* As mentioned above we shall give two applications. The first one concerns Lyapounov's criterion of stability. We recall the deterministic problem.

Let $x(t)$ be the solution of

$$dx(t)/dt = F(x(t)), \qquad x(0) = x,$$

where $F(x)$ is a vector-valued function of x with $F(0) = 0$, i.e., 0 is an equilibrium point for the system. Suppose that there is a Lyapounov function $V(x)$, i.e., a positive definite function $V(x) \geqslant 0$, $V(x) = 0 \Rightarrow x = 0$, which is smooth, and

$$F(x) \cdot \partial V(x)/\partial x \leqslant 0, \quad x \text{ near zero}.$$

Then

$$V(x(t)) = V(x) + \int_0^t F(x(s)) \cdot \frac{\partial V(x(s))}{\partial x}\, ds \leqslant V(x),$$

and hence $|x(t)|$ will remain near zero for all $t \geqslant 0$ if the initial state x is sufficiently close to zero.

For stochastic (Itô) equations the problem is **[4]** to study the behavior of the diffusion $x(t)$ when

$$dx(t) = b(x(t))dt + \sigma(x(t))dw(t), \qquad x(0) = x,$$

and $b(0) = 0$, $\sigma(0) = 0$, so that $x = 0$ is again an equilibrium point (or a trap). Let A denote again the infinitesimal generator of $\{X(t), t \geqslant 0\}$. Suppose there exists a positive definite function $V(x)$, smooth near zero, such that $\mathcal{A}V(x) \leqslant 0$. Then for any $\varepsilon_1, \varepsilon_2 > 0$ we can find a $\delta > 0$ such that for $|x| < \delta$,

$$P\left\{\sup_{t \geqslant 0} |X(t)| < x + \varepsilon_1\right\} \geqslant 1 - \varepsilon_2, \tag{2.30}$$

i.e., the identically zero solution is stable in the above sense.

From Itô's formula we have

$$V(x(t)) = V(x) + \int_0^t AV(x(s))\,ds + \int_0^t \sigma(x(s)) \frac{\partial V(x(s))}{\partial x}\,dw(s),$$

and we asume here that $\sigma \partial V/\partial x$ is bounded. By hypothesis

$$0 \leqslant V(x(t)) \leqslant V(x) + \int_0^t \sigma(x(s)) \frac{\partial V(x(s))}{\partial x}\,dw(s) \equiv V(x) + f(t).$$

Now $V(x) + f(t)$ is a nonnegative martingale, and hence by Kolmogorov's inequality we have

$$P\left\{\sup_{t \geqslant 0} (V(x) + f(t)) > V(x) + \varepsilon\right\} \leqslant \frac{V(x)}{\varepsilon + V(x)} \leqslant \frac{V(x)}{\varepsilon}.$$

Thus,

$$P\left\{\sup_{t \geqslant 0} V(x(t)) \leqslant V(x) + \varepsilon\right\} \geqslant 1 - \frac{V(x)}{\varepsilon}.$$

The positive definiteness of $V(x)$ implies now the assertion (2.30).

The second application is the extension of the method of characteristics. We recall the deterministic situation.

Consider the first order equation

$$\frac{\partial u(t, x)}{\partial t} = b(x) \frac{\partial u(t, x)}{\partial x} + c(x)u(t, x) + d(x), \qquad t > 0,\ u(0, x) = f(x). \tag{2.31}$$

Let $x(t)$ be the solution of

$$dx(t)/dt = b(x(t)), \qquad x(0) = x. \tag{2.32}$$

Then we can express the solution of (2.31) as

$$u(t, x) = \exp\left\{\int_0^t c(x(s))\,ds\right\} f(x(t)) + \int_0^t \exp\left\{\int_0^s c(x(\sigma))\,d\sigma\right\} d(x(s))\,ds.$$

This is simply the method of characteristics. If (2.31) is to be solved in a region $D \subset R^n$, say, with boundary condition $u(t, x) = g(x)$ for all $x \in \partial D$ that can be reached by solutions of (3.32) with $x(0) \in D$, then we have

$$\begin{aligned} u(t, x) = {} & \chi(t < t^*)\exp\left\{\int_0^t c(x(s))\,ds\right\} f(x(t)) \\ & + \chi(t \geqslant t^*)\exp\left\{\int_0^{t^*} c(x(s))\,ds\right\} g(x(t^*)) \\ & + \int_0^{t \wedge t^*} \exp\left\{\int_0^s c(x(\sigma))\,d\sigma\right\} d(x(s))\,ds, \qquad t \wedge t^* = \min(t, t^*). \end{aligned}$$

Here t^* is the first time the solutions of (2.32) reach ∂D starting from $x \in D$, and $\chi(t < t^*) = 0$ if $t \geqslant t^*$ and one otherwise. Similarly for $\chi(t \geqslant t^*)$.

The generalization of the above to the problem

$$\frac{\partial u}{\partial t} = \frac{1}{2} a(x) \frac{\partial^2 u}{\partial x^2} + b(x) \frac{\partial u}{\partial x} + c(x)u + d(x), \qquad t > 0,$$
$$u(0, x) = f(x), x \in D, \qquad u(t, x) = g(x), x \in \partial D, \tag{2.33}$$

is immediate. Let $\{x(t),\ t \geqslant 0\}$ be the diffusion process with stochastic differential equation

$$dx(t) = b(x(t))dt + \sigma(x(t))dw(t), \qquad x(0) = x,$$

and let t^* = first time $x(t)$ hits ∂D starting from the interior, which is a stopping time. In the vector case we assume that $a(x) = \sigma(x) \cdot (\sigma(x))^T$ where $\sigma(x)$ is any Lipschitz continuous matrix. From Itô's formula and the optional stopping theorem with $t \wedge t^*$ as the upper stopping time and zero as the lower one we readily obtain the representation

$$\begin{aligned} u(t, x) = E &\left\{ \chi(t < t^*) \exp\left\{ \int_0^t c(x(s))\, ds \right\} f(x(t)) \right\} \\ + E &\left\{ \chi(t \geqslant t^*) \exp\left\{ \int_0^{t^*} c(x(s))\, ds \right\} g(x(t^*)) \right\} \\ + E &\left\{ \int_0^{t \wedge t^*} \exp\left\{ \int_0^s c(x(s))\, ds \right\} d(x(s))\, ds \right\}. \end{aligned} \tag{2.34}$$

Representations such as (2.34) are useful in many ways, and sometimes they yield very simply results that are otherwise difficult to obtain, for example, in singular perturbation results when $a(x)$ in (2.33) is replaced by $\varepsilon a(x)$ (small diffusion) and the limit $\varepsilon \downarrow 0$ is sought.

A very interesting analysis, using the Cameron-Martin formula [**4**], [**5**] not mentioned here, concerns equations of the form (2.33), with $a(x)$ replaced by $\varepsilon a(x)$ when the orbits of the deterministic problem (2.32) have a stable singular point in D and never exit from D, is given in [**12**].

REFERENCES

1. W. Feller, *An introduction to probability theory*. Vols. I, II, Wiley, New York, 1966. MR **35** #1048.

2. J. L. Doob, *Stochastic processes*, Wiley, New York; Chapman & Hall, London, 1953. MR **15**, 445.

3. I. I. Gihman and A. V. Skorohod, *The theory of stochastic processes*. I, "Nauka", Moscow, 1971; English transl., Die Grundlehren der math. Wissenschaften, Band 210, Springer-Verlag, New York and Berlin, 1974.

4. ———, *Stochastic differential equations*, "Naukova Dumka", Kiev, 1968; English transl., Ergebnisse der Mathematik und ihrer Grenzgebiete, Band 72, Springer-Verlag, New York and Berlin, 1972. MR **41** #7777.

5. H. P. McKean, Jr., *Stochastic integrals*, Probability and Math. Statist., no. 5, Academic Press, New York and London, 1969. MR **40** #947.

6. S. R. S. Varadhan, *Stochastic processes*, Courant Institute of Mathematical Sciences, New York University, New York, 1968. MR **41** #4657.

7. G. E. Uhlenbeck and L. S. Ornstein, *On the theory of Brownian motion*, Phys. Rev. **36** (1930), 823–841; Reprinted in *Selected papers on noise and stochastic processes*, N. Wax, Editor, Dover, New York, 1953.

8. R. Kubo, *Statistical mechanics*, North-Holland, Amsterdam; Interscience, New York, 1965. MR **31** #3139.

9. G. W. Ford, M. Kac and P. Mazur, *Statistical mechanics of assemblies of coupled oscillators*, J. Mathematical Phys. **6** (1965), 504–515. MR **31** #4512.

10. T. G. Kurtz, *Extensions of Trotter's operator semigroup approximation theorems*, J. Functional Analysis **3** (1969), 354–375. MR **39** #3351.

11. D. Stroock and S. R. S. Varadhan, *Diffusion processes with continuous coefficients*. I, II, Comm. Pure Appl. Math. **22** (1969), 345–400, 479–530. MR **40** #6641; #8130.

12. A. D. Ventsel' and M. I. Freidlin, *On small perturbations of dynamical systems*, Uspehi Mat. Nauk **25** (1970), no. 1 (151), 3–55 = Russian Math. Surveys **25** (1970), no. 1, 1–56. MR **42** #2123.

3. Asymptotics for initial value problems.

(i) *Formulation of the problem*. Physical problems generate stochastic processes in many ways, one of which is as solutions of differential equations with random coefficients. The prototype problem is of the form

$$dx(t)/dt = F(x(t), y(t), t) \qquad x(0) = x, \tag{3.1}$$

where $x(t)$ is an R^n-valued function, $y(t)$ is an R^m-valued given stochastic process, and F is a sufficiently regular vector-valued function. Equation (3.1) generates the process $x(t)$ from the process $y(t)$. A basic question is: given the statistical behavior of $y(t)$, the coefficients, can we deduce in some reasonably effective way the statistical behavior of the solution $x(t)$ of (3.1)?

To motivate the kind of answer we have in mind consider how equations (3.1) arise frequently in physical problems. In the differential equation (3.1) the coefficients $y(t)$ represent external influences of a complicated nature. One can always enlarge the state space to include the dynamics of $y(t)$, but this may complicate the problem too much. So one assumes that the coefficients are random and looks for results concerning (3.1) that depend as little as possible on detailed description of $y(t)$, i.e., they are model-independent. To make this statement precise we must differentiate between the characteristic time scales of evolution of the system $x(t)$ and the external noise effects $y(t)$: the latter change much faster than the former. It is in the asymptotic limit where this behavior is fully developed that interesting results about $x(t)$ emerge. We shall therefore concentrate attention on this problem and we shall use perturbation theory to analyze it.

The quantitative differentiation of time scales is effected by the introduction of a small parameter ε. Let us assume that $y(t)$ is an R^m-valued stationary stochastic process where the correlation time, typically defined by

$$\sum_{i=1}^{m} \int_0^\infty E\{y_i(0)y_i(t)\}\,dt$$

with $E\{y(t)\} = 0$, is a finite number. For the process $y^\varepsilon(t) \equiv y(t/\varepsilon^2)$ (we use ε^2

to avoid square roots later) the correlation time is of order ε^2. We seek now to analyze the process $x^\varepsilon(t)$ when ε is small where

$$dx^\varepsilon(t)/dt = F^\varepsilon(x^\varepsilon(t), y^\varepsilon(t), t), \qquad x^\varepsilon(0) = x. \tag{3.2}$$

Here $F^\varepsilon(x, y, t)$ is a vector-valued function on $R^n \times R^m \times [0, \infty)$ depending on ε. It will turn out later that F^ε can have the form

$$F = \frac{1}{\varepsilon} F_0 + F_1 + \varepsilon F_2 + \ldots$$

where, usually, $E\{F_0(x, y(t), t)\} = 0$. In addition F^ε may depend explicitly on t/ε^2 as well as t. All these possibilities will be discussed in the sequel and examples will be given.

The limit $\varepsilon \to 0$ with $0 \leqslant t \leqslant T$, $T < \infty$, in the above problems will lead to a diffusion approximation for $x^\varepsilon(t)$ so we may cal' it the diffusion limit or the white noise limit (because the scaling of $y^\varepsilon(s)$ makes it tend to white noise). However, both names are somewhat unsatisfactory because upon changing variables we find that we are considering the weak-coupling limit as follows. Suppose the primary problem is of the form

$$d\tilde{x}(\tau)/d\tau = F_0(\tilde{x}(\tau)) + \varepsilon\tilde{F}_1(\tilde{x}(\tau), y(\tau)), \qquad \tilde{x}(0) = x, \tag{3.3}$$

where $y(\tau)$ is a stationary process with correlation time equal to one, say.Here ε plays the role of coupling constant for the system (3.3); coupling the unperturbed problem ($\varepsilon = 0$) to the noise. For finite τ-intervals the limit $\varepsilon \to 0$ will simply produce the unperturbed problem. The effects of the coupling will not be felt unless τ is large, usually of order $1/\varepsilon^2$. Furthermore, for such cumulative effects to be meaningful, the unperturbed problem must have special behavior. For example, it may admit a change of dependent variables $\tilde{x} \to x$ and independent variable $t = \varepsilon^2\tau$, so that in the new representation it has the form (3.2) with suitable F^ε on the right-hand side.

All these points are well understood in the context of deterministic problems **[1]**, **[2]**, **[3]**. We shall see that the situation is quite similar for stochastic problems. Basic references for the analysis of stochastic problems from the point of view adopted here are **[3]**, **[4]**, **[5]** of the references listed in the introduction.

(ii) *Equations with Markov coefficients*. We shall begin with relatively familiar objects, stochastic differential equations whose coefficients form a stationary Markov process.

The following simple example has been analyzed by Goldstein **[4]**; see also the random evolution work in **[5]**. Let $\{y(t), t \geqslant 0\}$ be the random telegraph process that takes the values $\pm\alpha$ and with mean time between switching equal to $1/\beta$, $\beta > 0$. The transition function of the Markov process $y(t)$ is easily seen to be

$$(P\{y(t+s) = \pm\alpha | y(s) = \pm\alpha\}) = \frac{1}{2}\begin{pmatrix} 1 + e^{-2\beta t} & 1 - e^{-2\beta t} \\ 1 - e^{-2\beta t} & 1 + e^{-2\beta t} \end{pmatrix}, \tag{3.4}$$

so that

$$\frac{1}{2}\frac{d}{dt}\begin{pmatrix} 1+e^{-2\beta t} & 1-e^{-2\beta t} \\ 1-e^{-2\beta t} & 1+e^{-2\beta t}\end{pmatrix}\Bigg|_{t=0} = \begin{pmatrix} -\beta & \beta \\ \beta & -\beta \end{pmatrix} \equiv A. \tag{3.5}$$

The right-hand side of (3.5) is the infinitesimal generator (matrix) A of $\{y(t), t \geqslant 0\}$. Note that as $t \to \infty$ the transition function (matrix) tends to

$$\begin{pmatrix} \frac{1}{2} & \frac{1}{2} \\ \frac{1}{2} & \frac{1}{2} \end{pmatrix} \tag{3.6}$$

so that if $y(0)$ takes the values $\pm\alpha$ with probability $\frac{1}{2}$, then $\{y(t), t \geqslant 0\}$ is a stationary Markov process.

Consider now the stochastic differential equation

$$dx(\tau)/d\tau = y(\tau), \qquad x(0) = x, \tag{3.7}$$

which trivially integrates to

$$x(\tau) = x + \int_0^\tau y(s)\,ds.$$

We think of $y(\tau)$ as the velocity of a particle, taking values $\pm\alpha$, and $x(\tau)$ its position with x the initial position. Clearly $(x(\tau), y(\tau))$ jointly form a Markov process. Let us compute the infinitesimal generator of this process. Let $f(x, y)$ be a function on $R \times \{\alpha, -\alpha\}$, i.e., a 2-vector function of x, and let

$$u(\tau, x, y) = E_{x,y}\{f(x(\tau), y(\tau))\} \tag{3.8}$$

where (x, y) denotes the starting position x and velocity y $(= \pm\alpha)$ of the process $(x(\tau), y(\tau))$. Using (3.4) we deduce easily that

$$\partial u/\partial\tau = y\partial u/\partial x + Au, \qquad u(0, x, y) = f(x, y), \tag{3.9}$$

where A of (3.5) acts on $u(\tau, x, y)$ with the latter considered as a 2-vector $(u(\tau, x, +\alpha), u(\tau, x, -\alpha))$.

Let us now consider what happens if we take the weak coupling limit in (3.7), i.e., consider $dx(\tau)/d\tau = \varepsilon y(\tau)$ and let $\tau \sim 1/\varepsilon^2$. Let us define new variables

$$t = \varepsilon^2\tau, \quad y^\varepsilon(t) = y(t/\varepsilon^2), \quad x^\varepsilon(t) = x(t/\varepsilon^2). \tag{3.10}$$

Then we find that the equation for $x^\varepsilon(t)$ is

$$dx^\varepsilon(t)/dt = (1/\varepsilon)y^\varepsilon(t), \qquad x^\varepsilon(0) = x. \tag{3.11}$$

Thus, as mentioned in (i), the study of (3.11) in the limit $\varepsilon \to 0$ could be called the diffusion or white noise limit since the velocity $(1/\varepsilon)y^\varepsilon(t)$ tends to become white noise. The scaled version of (3.9) becomes

$$\begin{gathered} u^\varepsilon(t, x, y) = E_{x,y}\{f(x^\varepsilon(t), y^\varepsilon(t))\}, \\ \frac{\partial u^\varepsilon}{\partial t} = \frac{1}{\varepsilon}\frac{\partial u^\varepsilon}{\partial x} + \frac{1}{\varepsilon^2}Au^\varepsilon, \qquad u^\varepsilon(0, x, y) = f(x, y). \end{gathered} \tag{3.12}$$

Let us write (3.12) more explicitly with $u(t, x, +\alpha) = u_+(t, x)$ and $u(t, x, -\alpha) = u_-(t, x)$ as follows:

$$\frac{\partial}{\partial t}\begin{pmatrix} u_+(t,x) \\ u_-(t,x) \end{pmatrix} = \frac{1}{\varepsilon}\begin{pmatrix} +\alpha & 0 \\ 0 & -\alpha \end{pmatrix}\frac{\partial}{\partial x}\begin{pmatrix} u_+(t,x) \\ u_-(t,x) \end{pmatrix}$$
$$+ \frac{1}{\varepsilon^2}\begin{pmatrix} -\beta & \beta \\ \beta & -\beta \end{pmatrix}\begin{pmatrix} u_+(t,x) \\ u_-(t,x) \end{pmatrix}$$
$$u_\pm(0,x) = f_\pm(x).$$

In this form the problem becomes a singular perturbation of a second order linear hyperbolic system. If $f_+(x) = f_-(x)$, so that no initial layers arise, it is easy to verify that $w = u_+ + u_-$ satisfies the telegrapher's equation

$$\varepsilon^2 \frac{\partial^2 w}{\partial t^2} = \alpha^2 \frac{\partial^2 w}{\partial x^2} - 2\beta \frac{\partial w}{\partial t}, \qquad w(0,x) = 2f(x), \quad \frac{\partial w}{\partial t}(0,x) = 0. \tag{3.13}$$

The limit $\varepsilon \to 0$ in (3.12) will be studied formally in the next section for a large class of similar problems. From (3.13) it is clear however that we will end up with a diffusion equation which is no more than a statement of the central limit theorem for $\int_0^t y^\varepsilon(s)\,ds$ as $\varepsilon \to 0$.

The general case of stochastic equations with Markovian coefficients is similar. We shall present the problem in scaled form directly, the other case being only superficially different. Suppose $y(t)$ is an R^m-valued stationary Markov process, not necessarily a diffusion, with finite correlation time. We define $y^\varepsilon(t) \equiv y(t/\varepsilon^2)$ and let $x^\varepsilon(t)$ be the solution of

$$\begin{aligned} \frac{dx^\varepsilon(t)}{dt} &= \frac{1}{\varepsilon} F(x^\varepsilon(t), y^\varepsilon(t)) + G(x^\varepsilon(t), y^\varepsilon(t)), \qquad t > 0, \\ x^\varepsilon(0) &= x. \end{aligned} \tag{3.14}$$

We are interested in the asymptotic behavior of the statistics of $x^\varepsilon(t)$, as $\varepsilon \to 0$ and $0 \leqslant t \leqslant T$, $T < \infty$. The case where

$$E\{F(x, y(t))\} = 0 \tag{3.15}$$

is typically assumed in (3.14) because then both terms on the right side produce comparable effects. Without (3.15) the F/ε term in (3.14) dominates. Both cases are treated later on.

To study (3.14) we consider the process $(x^\varepsilon(t), y^\varepsilon(t))$ as a Markov process on $R^n \times R^m$ and determine its infinitesimal generator. It is not difficult to verify that if A is the infinitesimal generator of $y(t)$, i.e. if for suitable functions $f(y)$ on R^m

$$Af(y) = \lim_{h \downarrow 0} \frac{E\{f(y(h)) | y(0) = y\} - f(y)}{h},$$

then the infinitesimal generator of $(x^\varepsilon(t), y^\varepsilon(t))$, acting on suitable functions $f(x, y)$ on R^{n+m}, is given by

$$\begin{aligned} &\lim_{h \downarrow 0} \frac{E\{f(x^\varepsilon(h), y^\varepsilon(h))\} - f(x,y)}{h} \\ &\quad = \left(\frac{1}{\varepsilon} F(x,y) + G(x,y)\right)\frac{\partial f(x,y)}{\partial x} + \frac{1}{\varepsilon^2} Af(x,y). \end{aligned} \tag{3.16}$$

Here $G\,\partial f/\partial x$ denotes the dot product of the vector G with the x-gradient of x and A acts on f as a function of y only. Since we are specifically interested in the limit of $x^\varepsilon(t)$, $0 \leqslant t \leqslant T$, as $\varepsilon\downarrow 0$, we must analyze the asymptotic behavior of $u^\varepsilon(t, x, y) = E\{f(x^\varepsilon(t))\}$ satisfying

$$\begin{aligned}\frac{\partial u^\varepsilon(t, x, y)}{\partial t} &= \left(\frac{1}{\varepsilon} F(x, y) + G(x, y)\right)\frac{\partial u^\varepsilon(t, x, y)}{\partial x}\\ &\quad + \frac{1}{\varepsilon^2} A u^\varepsilon(t, x, y), \qquad t > 0, \qquad (3.17)\\ &\qquad\qquad u^\varepsilon(0, x, y) = f(x).\end{aligned}$$

Here $f(x)$ must belong to a sufficiently rich class of functions (compare with the results in §2(v)).

Example (3.14) leading to (3.17) is typical of the general situation that asymptotic problems for stochastic equations produce: singular perturbation problems of the form (3.17). As we shall see later (§3(vi)), non-Markovian problems give rise to problems formally similar to (3.17), and as mentioned in the introduction an even greater variety of problems also falls in the same category; see **[5]**–**[8]** of the introduction.

(iii) *Perturbation theory*. As we saw in the previous section, problems of the form

$$\frac{du^\varepsilon}{dt} = \frac{1}{\varepsilon^2}\mathcal{L}_1 u^\varepsilon + \frac{1}{\varepsilon}\mathcal{L}_2 u^\varepsilon + \mathcal{L}_3 u^\varepsilon, \qquad u^\varepsilon(0) = f, \qquad (3.18)$$

arise where $\mathcal{L}_1$, $\mathcal{L}_2$ and $\mathcal{L}_3$ are certain operators which are in general unbounded but for which (3.18) has solutions in a suitably generalized sense. The problem is to determine the asymptotic behavior of u^ε as $\varepsilon \to 0$, with $0 \leqslant t \leqslant T$, $T < \infty$.

There are many formal procedures for treating (3.18), but they all amount, ultimately, to perturbation theory (usually second order) for $\mathcal{L}_1 + \varepsilon\mathcal{L}_2 + \varepsilon^2\mathcal{L}_3$ [**6**, Chapter IV, for example]. The method given in **[2]** of the introduction (for nonlinear problems there, in fact) will be discussed first.

We shall assume that $\mathcal{L}_1$ has certain special properties corresponding to the fact that it is usually, but not always, the generator of a stationary (ergodic) Markov process. We assume specifically that the semigroup $e^{\mathcal{L}_1 t}$ generated by $\mathcal{L}_1$ converges to P as $t\uparrow\infty$

$$e^{\mathcal{L}_1 t} \to P, \qquad t\uparrow\infty, \qquad (3.19)$$

where P is the projection operator into the null space of $\mathcal{L}_1$. The operators are all defined on a Banach space B and the range of P, PB, is denoted by B_0. The details of the formalism that follows are in **[7]**, **[8]** so we shall be rather brief here. Let us also suppose at first that

$$P\mathcal{L}_2 P = 0, \qquad (3.20)$$

which corresponds to (3.15), since the expectation there is relative to the invariant distribution and P in (3.19) is, in case (3.14), expectation relative to the invariant distribution of the process $y(t)$.

Let u^ε be formally represented by a power series

$$u^\varepsilon = u_0 + \varepsilon u_1 + \varepsilon^2 u_2 + \cdots.$$

Inserting this into (3.18) and equating coefficients of equal powers of ε, we obtain the following sequence of problems:

$$\mathfrak{L}_1 u_0 = 0, \tag{3.21}$$

$$\mathfrak{L}_1 u_1 = -\mathfrak{L}_2 u_0, \tag{3.22}$$

$$\mathfrak{L}_1 u_2 = -\mathfrak{L}_2 u_1 - \mathfrak{L}_3 u_0 + du_0/dt, \tag{3.23}$$

$$\cdots.$$

We shall assume that $Pf = f$, which for (3.17) means that $f = f(x)$ and not a function of x and y. This typically is the interesting case in our problems. Let us first consider (3.21). It implies that

$$Pu_0 = u_0, \tag{3.24}$$

i.e., $u_0 = u_0(t)$ is in the range of P. Since $u_0(0) = f$ from (3.18), the condition $Pf = f$ avoids nonuniformities near $t = 0$, i.e., initial layers.

We insert (3.24) into (3.22) and consider the solvability of this equation for u_1. Clearly for the formal expansion to hold, u_1, u_2, etc., should be determinable in a recursive way; but so far u_0 is not determined except that it must satisfy (3.24). The solvability condition for u_1, u_2, etc., will determine u_0 and possibly other unknown functions, and this constitutes the essence of the method; see **[1]**, **[2]** of the introduction. The solvability condition for (3.22) is formally $P\mathfrak{L}_2 Pu_0 = 0$, and this is identically satisfied by (3.20). So we must continue with (3.23) to determine u_0 (second order theory). Inserting u_1 from (3.22) on the right-hand side of (3.23) and acting with P on the result yields the following equation for u_0.

$$P\mathfrak{L}_2\mathfrak{L}_1^{-1}\mathfrak{L}_2 Pu_0 - P\mathfrak{L}_3 Pu_0 + (d/dt)Pu_0 = 0, \qquad u_0(0) = f.$$

Remembering that (3.24) holds and $Pf = f$, we rewrite this as an evolution equation in $B_0 = PB$:

$$du_0/dt = \left(P\mathfrak{L}_3 P - P\mathfrak{L}_2\mathfrak{L}_1^{-1}\mathfrak{L}_2 P\right)u_0 \equiv \bar{\mathfrak{L}} u_0, \qquad u_0(0) = f. \tag{3.25}$$

We usually stop with u_0 in (3.20) because the solution of (3.25) is already quite complicated in most problems of interest. But the algorithm described above can continue without difficulty. In particular, up to terms in B_0, u_1 and u_2 are given by

$$u_1 = -\mathfrak{L}_1^{-1}\mathfrak{L}_2 u_0, \qquad u_2 = -\mathfrak{L}_1^{-1}\left[\mathfrak{L}_3 P - \mathfrak{L}_2\mathfrak{L}_1^{-1}\mathfrak{L}_2 P - \bar{\mathfrak{L}}\right]u_0, \tag{3.26}$$

assuming, of course, that the formal solvability conditions we have used (elimination of the component of the right-hand sides of (3.22), (3.23) in the null space of $\mathfrak{L}_1$) are truly correct. This actually turns out to be a delicate point with many problems which are formally of the form (3.18) (see §3(vi)).

Despite the fact that the above considerations are formal, we have all the elements of a rigorous proof at hand. The idea is to use (2.21) and then the Trotter-Kato theorem [7], **[8]**. We simplify the relevant steps here as follows.

Suppose that (3.25) has a nice solution u_0, and suppose u_1 and u_2 of (3.26) are well defined and bounded in the norm of our Banach space B. We wish to estimate $u^\varepsilon - u_0$; but, in view of what has just been assumed, it is enough to show that $u^\varepsilon - u_0 - \varepsilon u_1 - \varepsilon^2 u_2$ is small. This is a "bootstrap" approach familiar in singular perturbation. Now we have

$$\left(\frac{d}{dt} - \frac{1}{\varepsilon^2}\mathcal{L}_1 - \frac{1}{\varepsilon}\mathcal{L}_2 - \mathcal{L}_3\right)(u^\varepsilon - u_0 - \varepsilon u_1 - \varepsilon^2 u_2)$$
$$= \left(\frac{1}{\varepsilon^2}\mathcal{L}_1 + \frac{1}{\varepsilon}\mathcal{L}_2 + \mathcal{L}_3 - \frac{d}{dt}\right)(u_0 + \varepsilon u_1 + \varepsilon^2 u_2) \tag{3.27}$$
$$= \varepsilon\left[\left(\mathcal{L}_3 - \frac{d}{dt}\right)u_1 + \mathcal{L}_2 u_2\right] + \varepsilon^2\left(\mathcal{L}_3 - \frac{d}{dt}\right)u_2 = O(\varepsilon),$$

formally. Under favorable circumstances (3.27) implies that $u - u_0 - \varepsilon u_1 - \varepsilon^2 u_2$, and hence $u^\varepsilon - u_0$, is $O(\varepsilon)$ and the desired approximation theorem follows. Note that the bootstrap idea fits exactly into the scheme (2.21). Moreover, (2.19) cannot possibly be valid for problems such as (3.18).

In the event that (3.20) is not valid, then the first approximation u_0 will depend on ε and we will denote it by u_0^ε. The result is again that $u^\varepsilon - u_0^\varepsilon \to 0$ as $\varepsilon \to 0$, $0 \leqslant t \leqslant T$, where u_0^ε is in B_0 and satisfies the evolution equation

$$\frac{du_0^\varepsilon}{dt} = \left(\frac{1}{\varepsilon}P\mathcal{L}_2 P + P\mathcal{L}_3 P - P\mathcal{L}_2\mathcal{L}_1^{-1}(\mathcal{L}_2 - P\mathcal{L}_2)\right)u_0^\varepsilon = 0, \qquad u_0^\varepsilon(0) = f. \tag{3.28}$$

The details of this are given in [**8**].

The above analysis is quite satisfactory since it is simple and it gives a good clue for the actual estimation of the error. It is worthwhile, however, to give another formal procedure which is employed widely in the analysis of stochastic equations and other problems.

Let us assume for simplicity that $\mathcal{L}_3 \equiv 0$ in (3.18) and that (3.20) holds. Define

$$v^\varepsilon(t) = Pu^\varepsilon(t), \qquad w^\varepsilon(t) = (1 - P)u^\varepsilon(t), \tag{3.29}$$

where P is the projection operator of (3.19). Equation (3.18) now takes the form

$$\frac{dv^\varepsilon}{dt} = \frac{1}{\varepsilon}P\mathcal{L}_2 w^\varepsilon, \qquad v^\varepsilon(0) = f \quad (Pf = f),$$
$$\frac{dw^\varepsilon}{dt} = \frac{1}{\varepsilon^2}\mathcal{L}_1 w^\varepsilon + \frac{1}{\varepsilon}\mathcal{L}_2 v^\varepsilon + \frac{1}{\varepsilon}(1 - P)\mathcal{L}_2 w^\varepsilon, \qquad w^\varepsilon(0) = 0. \tag{3.30}$$

The second equation in (3.30) can be solved formally to yield

$$w^\varepsilon(t) = \frac{1}{\varepsilon}\int_0^t \exp\{(\mathcal{L}_1 + \varepsilon(1 - P)\mathcal{L}_2)(t - s)/\varepsilon^2\}\mathcal{L}_2 v^\varepsilon(s)\, ds.$$

Inserting this into the first equation in (3.30) we obtain a closed equation for $v^\varepsilon(t)$:

$$\frac{dv^\varepsilon(t)}{dt} = \frac{1}{\varepsilon^2}\int_0^t P\,\mathcal{L}_2 \exp\{(\mathcal{L}_1 + \varepsilon(1-P)\mathcal{L}_2)(t-s)/\varepsilon^2\}\,\mathcal{L}_2 v^\varepsilon(s)\,ds, \qquad (3.31)$$
$$v^\varepsilon(0) = f.$$

This equation is exact, of course. The first order smoothing approximation, so-called, consists in dropping the $\varepsilon(1-P)\mathcal{L}_2$ term in the exponent in (3.31) so that $v^\varepsilon_{\mathrm{FOS}}(t)$ satisfies

$$\frac{dv^\varepsilon_{\mathrm{FOS}}(t)}{dt} = \frac{1}{\varepsilon^2}\int_0^t P\,\mathcal{L}_2 \exp\{\mathcal{L}_1(t-s)/\varepsilon^2\}\,\mathcal{L}_2 v^\varepsilon_{\mathrm{FOS}}(s)\,ds, \qquad (3.32)$$
$$v^\varepsilon(0) = f.$$

Usually this equation is too difficult to solve, so we consider the long-time Markovian approximation $v_{\mathrm{LTM}}(t)$ which satisfies the equation

$$\frac{dv_{\mathrm{LTM}}(t)}{dt} = \int_0^\infty P\,\mathcal{L}_2 e^{\mathcal{L}_1 t}\mathcal{L}_2 P\,dt\; v_{\mathrm{LTM}}(t) = -P\,\mathcal{L}_2\mathcal{L}_1^{-1}\mathcal{L}_2 P v_{\mathrm{LTM}}(t), \qquad (3.33)$$
$$v_{\mathrm{LTM}}(0) = f.$$

Thus, the LTM approximation is the same approximation as the u_0 we obtained above. The name LTM comes from the fact that, in passing from (3.32) to (3.33), the unknown function $v^\varepsilon_{\mathrm{FOS}}(s)$ is pulled out of the integral and the integral is then stretched to ∞ so that "memory" effects are lost. A good reference for these ideas is [**5**] of the introduction, where additional references are given; but the terminology is not used consistently in the literature; see also [**9**].

It is believed that the FOS approximation is better than the LTM approximation, since one is a special case of the other, insofar as the former gives a more accurate description at very small t. Experience with applications so far shows that the LTM approximation is hard enough to implement so that the extra accuracy of the FOS does not have significant impact.

(iv) *Applications*. Let us first apply the approximation (3.25) to (3.12). Here $\mathcal{L}_1 = A$ and is given by (3.5) and $e^{\mathcal{L}_1 t}$ is given by (3.4) so that

$$P = \lim_{t\uparrow\infty} e^{\mathcal{L}_1 t} = \frac{1}{2}\begin{pmatrix} 1 & 1 \\ 1 & 1 \end{pmatrix}. \qquad (3.34)$$

The operator $\mathcal{L}_2$ is given by

$$\mathcal{L}_2 = \begin{pmatrix} \alpha & 0 \\ 0 & -\alpha \end{pmatrix}\frac{\partial}{\partial x},$$

and clearly $P\,\mathcal{L}_2 P = 0$. The operator $\mathcal{L}_3 \equiv 0$. An easy computation yields

$$-\mathcal{L}_1^{-1} = \int_0^\infty (e^{\mathcal{L}_1 t} - P)\,dt = \frac{1}{4\beta}\begin{pmatrix} 1 & -1 \\ -1 & 1 \end{pmatrix},$$

and hence

$$-P\,\mathcal{L}_2\mathcal{L}_1^{-1}\mathcal{L}_2 P = \frac{\alpha^2}{4\beta}\begin{pmatrix} 1 & 1 \\ 1 & 1 \end{pmatrix}\frac{\partial^2}{\partial x^2}.$$

Thus, if $f_+(x) = f_-(x) = f(x)$, then $u_\pm^\varepsilon(t, x)$ converges as $\varepsilon \to 0$ to $u_0(t, x)$ where

$$\frac{\partial u_0(t, x)}{\partial t} = \frac{\alpha^2}{2\beta} \frac{\partial^2 u_0(t, x)}{\partial x^2}, \qquad u_0(0, x) = f(x),$$

just as we expected from (3.13).

Naturally the example (3.12) is too simple to show the effectiveness of the approximation. Example (3.17), however, which corresponds to (3.14), is considerably more involved. The formal identification (3.17) with the notation of §3(iii) is as follows. $\mathfrak{L}_1 = A$, the infinitesimal generator of $\{y(t), t \geqslant 0\}$, $P = E\{\ \}$ where $E\{\ \}$ is expectation with respect to the invariant distribution of $y(t)$, $\mathfrak{L}_2 = F(x, y)\partial/\partial x$ so that, by (3.5), $P\mathfrak{L}_2 P = 0$ and $\mathfrak{L}_3 = G(x, y)\partial/\partial x$. The operator $\mathfrak{L}_2^{-1}$ is the only thing that requires some discussion. Let $P(t, y, dz)$ be the transition function of $\{y(t), t \geqslant 0\}$ and assume that

$$\chi(y, dz) = \int_0^\infty \left[P(t, y, dz) - \bar{P}(dz) \right] dt \tag{3.35}$$

is well defined. Here $\bar{P}(dz)$ is the invariant measure of $\{y(t), t \geqslant 0\}$ and (3.35) is called the recurrent potential kernel of the process; it is the kernel of the integral operator $-\mathfrak{L}_1^{-1}$, as can be easily verified. With these definitions we find that

$$\begin{aligned} \bar{\mathfrak{L}} f(x) = \int\int F(x, y) \frac{\partial}{\partial x} \left(F(x, z) \frac{\partial f(x)}{\partial x} \chi(y, dz) \bar{P}(dy) \right) \\ + \int G(x, y) \frac{\partial f(x)}{\partial x} \bar{P}(dy). \end{aligned} \tag{3.36}$$

Thus $x^\varepsilon(t)$, the solution of (3.14), converges to a diffusion process with generator $\bar{\mathfrak{L}}$ given by (3.36).

We rewrite $\bar{\mathfrak{L}}$ in another form that will be useful later.

$$\begin{aligned} \bar{\mathfrak{L}} f(x) = \int_0^\infty E\left\{ F(x, y(0)) \frac{\partial}{\partial x} \left(F(x, y(t)) \frac{\partial f(x)}{\partial x} \right) \right\} dt \\ + E\left\{ G(x, y(t)) \frac{\partial f(x)}{\partial x} \right\}. \end{aligned} \tag{3.37}$$

The identification of (3.36) and (3.37) is merely a matter of notation. We shall find in §3(vi) that the same result (3.37) holds for $x^\varepsilon(t)$ of (3.14) even if $y(t)$ is not Markovian but only stationary along with some additional assumptions discussed there.

As a final example consider the system

$$\begin{aligned} \frac{dx^\varepsilon(t)}{dt} &= \frac{1}{\varepsilon} y^\varepsilon(t), & x^\varepsilon(0) = x, \\ \frac{dy^\varepsilon(t)}{dt} &= \frac{1}{\varepsilon} F(x^\varepsilon(t)) + \frac{1}{\varepsilon}\left(\frac{dw(t)}{dt} - y^\varepsilon(t) \right), & y^\varepsilon(0) = y, \end{aligned} \tag{3.38}$$

where $w(t)$ is the Brownian motion processes and dw/dt is formal white noise. The backward equation for the diffusion Markov process $(x^\varepsilon(t), y^\varepsilon(t))$ is

$$\frac{\partial u^\varepsilon}{\partial t} = \frac{1}{\varepsilon} y \frac{\partial u^\varepsilon}{\partial x} + \frac{1}{\varepsilon} F(x) \frac{\partial u^\varepsilon}{\partial y} + \frac{1}{\varepsilon^2} \left(\frac{1}{2} \frac{\partial^2 u^\varepsilon}{\partial y^2} - y \frac{\partial u^\varepsilon}{\partial y} \right), \tag{3.39}$$

$$u^\varepsilon(0, x, y) = f(x),$$

with $u^\varepsilon(t, x, y) = E\{f(x^\varepsilon(t))\}$. Thus,

$$\mathcal{L}_1 = \frac{1}{2} \frac{\partial^2}{\partial y^2} - y \frac{\partial}{\partial y}, \quad \mathcal{L}_2 = y \frac{\partial}{\partial x} + F(x) \frac{\partial}{\partial y}, \quad \mathcal{L}_3 = 0.$$

The problem under consideration is called the Smoluchovski limit of the Ornstein-Uhlenbeck process in an external force field; see [**10**].

The perturbation analysis of §3(iii) yields the result that $x^\varepsilon(t)$ converges to a diffusion process, i.e., $u^\varepsilon(t, x, y) \to u_0(t, x)$ and

$$\frac{\partial u_0}{\partial t} = \frac{1}{2} \frac{\partial^2 u_0}{\partial x^2} + F(x) \frac{\partial u_0}{\partial x}.$$

We leave the details to the reader.

Let us remark that the Banach spaces in the applications to stochastic equations with Markovian coefficients are usually spaces of bounded continuous functions with sup norm so that convergence of u^ε to u_0 is in the sup norm. When it is difficult to ascertain the properties of $\mathcal{L}_1^{-1}$, the principal analytical and computational difficulty of our problem, other more appropriate spaces may be used. Some remarks on this are in [**8**].

(v) *Linear stochastic equations, averaging, master equations.* Linear equations are of particular interest so we discuss them separately.

Consider the system

$$\frac{d\tilde{x}_p(\tau)}{d\tau} = i\omega_p \tilde{x}_p(\tau) + i\varepsilon \sum_{q=1}^{n} y_{pq}(\tau)\tilde{x}_q(\tau),$$
$$p = 1, 2, \ldots, n, \; \tilde{x}_p(0) = x_{p_0}. \tag{3.40}$$

Here ω_p, $p = 1, 2, \ldots, n$, are positive real numbers and $(y_{pq}(t)) = (y_{qp}(t))$ is a real symmetric matrix-valued stochastic process which is Markovian with known infinitesimal generator. The Markovian assumption is removed in the next section.

We study (3.40) in the weak coupling limit, i.e., for $\varepsilon \to 0$ and τ of order $1/\varepsilon^2$. It is convenient therefore to introduce the scaled time and the adiabatic invariants

$$\tau = t/\varepsilon^2,$$

$$x_p^\varepsilon(t) = \exp\left[-i\omega_p t/\varepsilon^2\right]\tilde{x}_p(t/\varepsilon^2),$$

so that (3.40) yields

$$\frac{dx_p^\varepsilon(t)}{dt} = \frac{i}{\varepsilon} \sum_{q=1}^{n} y_{pq}(t/\varepsilon^2)\exp\left[i(\omega_q - \omega_p)t/\varepsilon^2\right]x_q^\varepsilon(t), \tag{3.41}$$

$$t > 0, p = 1, 2, \ldots, n, x_p^\varepsilon(0) = x_{po}.$$

The question of interest now is: what is the asymptotic distribution of $(x_1^\varepsilon(t), \ldots, x_n^\varepsilon(t))$ in the limit $\varepsilon \downarrow 0$, $0 \leqslant t \leqslant T$?

We shall assume that the matrix-valued process $(y_{pq}(t))$ is an ergodic Markov process with generator $\mathcal{L}_1$ whose exact form is not important. Instead of writing $P(t, y, A)$ for the transition function and $\bar{P}$ for the invariant measure, we shall use $E\{\cdot\}$ to express the answers. Implicit is the understanding that, in the final answers, we treat $(y_{pq}(t))$ as a stationary process, i.e., we assume that the initial values have $\bar{P}$ as their distribution. We also assume that $E\{y(t)\} = 0$.

To analyze our system, we proceed as follows. We consider $(x^\varepsilon(t), y^\varepsilon(t))$, $y^\varepsilon(t) \equiv y(t/\varepsilon^2)$, jointly as a Markov process. Now, however, the right-hand side of (3.41) depends on t/ε^2 explicitly and x^ε is complex valued. So we must change the format of §3(ii) somewhat to cope with this complication. The explicit t/ε^2 dependence in (3.41) is essential for the results we want.

Let $\tau^\varepsilon(t) = t/\varepsilon^2$ and consider $y(t) = (y_{pq}(t))$ jointly with $\tau(t) = t$. The generator of $(y(t), \tau(t))$ is $\mathcal{L}_1 + \partial/\partial\tau$, and it acts on functions of real symmetric matrices and $\tau \in (-\infty, \infty)$. Note that $\mathcal{L}_1$ and $\partial/\partial\tau$ commute. Since $x^\varepsilon(t)$ is complex valued, we could treat its real and imaginary parts separately. It is more convenient, however, to deal with $x^\varepsilon(t)$ and $x^{\varepsilon^*}(t)$, its complex conjugate. If $x = x^R + ix^i$, we shall employ the notation

$$\frac{\partial}{\partial x} = \frac{1}{2}\left(\frac{\partial}{\partial x^R} - i\frac{\partial}{\partial x^I}\right), \qquad \frac{\partial}{\partial x^*} = \frac{1}{2}\left(\frac{\partial}{\partial x^R} + i\frac{\partial}{\partial x^I}\right).$$

With this notation, the infinitesimal generator of the Markov process $(x^\varepsilon(t), y^\varepsilon(t), \tau^\varepsilon(t))$ is

$$\frac{1}{\varepsilon^2}\left(\mathcal{L}_1 + \frac{\partial}{\partial\tau}\right) + \frac{1}{\varepsilon}(\mathcal{L}_2 + \mathcal{L}_2^*), \tag{3.42}$$

where

$$\mathcal{L}_2 = i \sum_{p,q=1}^{n} y_{pq}\exp\left[i(\omega_q - \omega_p)\tau\right]x_q \frac{\partial}{\partial x_p},$$

$$\mathcal{L}_2^* = -i \sum_{p,q=1}^{n} y_{pq}\exp\left[-i(\omega_q - \omega_p)\tau\right]x_q^* \frac{\partial}{\partial x_p^*},$$

and (3.42) acts on functions of (x, y, τ), $x \in C^n$, y a real symmetric matrix, $\tau \in (-\infty, \infty)$.

Let $f(x)$ be smooth bounded real-valued function of $x \in C^n$ and let

$$u^\varepsilon(t, x, y, \tau) = E\{f(x^\varepsilon(t))\} \tag{3.43}$$

with $x^\varepsilon(0) = x$, $y^\varepsilon(0) = y$, $\tau^\varepsilon(0) = \tau$. We wish to find the limit of u^ε, in the appropriate sense, as $\varepsilon \to 0$. The equation satisfied by u^ε, the backward equation, is

$$\frac{\partial u^{\varepsilon}}{\partial t} = \frac{1}{\varepsilon^2}\left(\mathcal{L}_1 + \frac{\partial}{\partial \tau}\right)u^{\varepsilon} + \frac{1}{\varepsilon}(\mathcal{L}_2 + \mathcal{L}_2^*)u^{\varepsilon}, \qquad u^{\varepsilon}(0, x, y, \tau) = f(x). \quad (3.44)$$

This problem is of the form (3.18) and so the perturbation theory applies. The operator $\mathcal{L}_1$ of (3.18) is $\mathcal{L}_1 + \partial/\partial\tau$ in (3.44) and the net effect of this is that we will perform an averaging [**1**] simultaneously with the other approximations.

If we assume that $e^{\mathcal{L}_1 t} \to E\{\cdot\}$ (expectation with respect to the invariant measure $\overline{P}$) strongly and, say, exponentially fast, then, with $\tilde{\mathcal{L}}_1 = \partial/\partial\tau$,

$$e^{(\mathcal{L}_1 + \tilde{\mathcal{L}}_1)t} \to P = \lim_{T\uparrow\infty} \frac{1}{T}\int_0^T E\{\cdot\}\, ds,$$

which corresponds to (3.19). Thus the projection operator is averaging with respect to the invariant measure and time averaging; see [**7**], [**13**]. Since $E\{y(t)\} = 0$, it follows that $P\mathcal{L}_2 P = P\mathcal{L}_2^* P = 0$; and hence, from (3.25), $u^{\varepsilon}(t, x, y, \tau)$ of (3.43) converges as $\varepsilon \to 0$, $0 \leqslant t \leqslant T$, to $u_0(t, x)$ where

$$du_0/dt = -P(\mathcal{L}_2 + \mathcal{L}_2^*)(\mathcal{L}_1 + \tilde{\mathcal{L}}_1)^{-1}(\mathcal{L}_2 + \mathcal{L}_2^*)u_0,$$
$$t > 0,\ u_0(0, x) = f(x),\ \tilde{\mathcal{L}}_1 = \partial/\partial\tau. \quad (3.45)$$

We must now find the explicit form of the operator on the right-hand side of (3.45).

Using again the fact that $e^{\mathcal{L}_1 t} \to E\{\cdot\}$ rapidly, it is easy to see that

$$-(\mathcal{L}_1 + \tilde{\mathcal{L}}_1)^{-1}(\mathcal{L}_2 + \mathcal{L}_2^*)f = \int_0^\infty e^{\mathcal{L}_1 s}[e^{\mathcal{L}_1 s} - P]\, ds(\mathcal{L}_2 + \mathcal{L}_2^*)f$$

$$= i\sum_{p,q=1}^{n}\int_0^\infty E\{y_{pq}(s)|y(0)\}\exp[i(\omega_q - \omega_p)(\tau + s)]\, ds x_q \frac{\partial f}{\partial x_p}$$

$$-i\sum_{p,q=1}^{n}\int_0^\infty E\{y_{pq}(s)|y(0)\}\exp[-i(\omega_q - \omega_p)(\tau + s)]\, ds x_q^* \frac{\partial f}{\partial x_p^*}.$$

Assuming now that $\omega_1, \omega_2, \ldots, \omega_n$ are distinct along with their sums and differences, equation (3.45) takes the following form.

$$\frac{\partial u_0(t, x)}{\partial t} = \Bigg[(\delta_{pq'}\delta_{qp'} + \delta_{q'p'}\delta_{pq})$$
$$\cdot\int_0^\infty E\{y_{pq}(s)y_{p'q'}(0)\}\exp[i(\omega_q - \omega_p)s]\, ds x_{q'}\frac{\partial}{\partial x_{p'}}\left(x_q \frac{\partial}{\partial x_p}\right)$$
$$-(\delta_{q'q}\delta_{p'p} + \delta_{q'p'}\delta_{qp})$$
$$\cdot\int_0^\infty E\{y_{pq}(s)y_{p'q'}(0)\}\exp[i(\omega_q - \omega_p)s]\, ds \quad (3.46)$$
$$\cdot x_{q'}\frac{\partial}{\partial x_{p'}}\left(x_q^* \frac{\partial}{\partial x_p^*}\right) + \text{c.c.}\Bigg]u_0(t, x), \quad u_0(0, x) = f(x).$$

Here we have employed the summation convention and we have used the abbreviation c.c. for complex conjugate, i.e., the formal complex conjugate of the operator preceding c.c. in (3.46). Note also that u_0 and f are thought of as functions of x and x^*; they are real and the diffusion operator on the right side of (3.46) takes real functions into real functions.

One can deduce many interesting facts from the result that $u^\varepsilon \to u_0$, the solution of (3.46). One result is as follows. Suppose $f = f(x, x^*) = x_p x_p^*$ successively for $p = 1, 2, \ldots, n$. Let

$$w_p(t) = \lim_{\varepsilon \downarrow 0} E\left\{|x_p(t)|^2\right\}.$$

Then (3.46) yields the following equation for $w_p(t)$.

$$\frac{dw_p(t)}{dt} = \sum_{q=1}^{n} \left[A_{pq} w_q(t) - A_{qp} w_p(t)\right], \qquad t > 0,\ w_p(0) = |x_{p_0}|^2, \tag{3.47}$$

where

$$\begin{aligned} A_{pq} = A_{qp} &= \int_0^\infty E\left\{y_{pq}(s) y_{pq}(0)\right\} \cos(\omega_q - \omega_p)s\, ds \\ &= \text{power spectrum of } \{y(t), t \geq 0\} \text{ at} \\ &\quad\ \text{the difference frequency } \omega_q - \omega_p. \end{aligned} \tag{3.48}$$

These equations have the form of master equations or transport equations. They control the way in which energy is exchanged among the oscillators (3.40) in the weak coupling limit. Further results of this kind, with Markovian or non-Markovian $\{y(t), t \geq 0\}$ can be found in **[11]**, **[12]**, **[13]** and references cited therein. In **[11]** the role of the condition that $\omega_1, \omega_2, \ldots, \omega_n$ be distinct is examined closely, and in **[12]** a more complicated system of the form (3.40) is studied in connection with underwater sound propagation.

(vi) *Non-Markovian problems*. The case where the coefficients $y^\varepsilon(t) \equiv y(t/\varepsilon^2)$ in (3.14) constitute a Markov process is typical of more general problems and it was treated first for illustrative purposes. We shall now consider more general situations as follows.

Consider the stochastic differential equation

$$\frac{dx^\varepsilon(t)}{dt} = \frac{1}{\varepsilon} F(x^\varepsilon(t), \omega^\varepsilon(t)) + G(x^\varepsilon(t), \omega^\varepsilon(t)), \qquad t > 0,\ x^\varepsilon(0) = x, \tag{3.49}$$

where, to distinguish it from the Markovian case (3.14), we denote the random "coefficients" by $\omega^\varepsilon(t) \equiv \omega(t/\varepsilon^2)$. The "coefficients" $\omega(t)$ are defined on a probability space $(\Omega, \mathfrak{F}, P)$ so that $\omega \to \omega(t)$ $(\omega(0) = \omega)$, $-\infty < t < \infty$, is a one-parameter group of measure-preserving transformations on Ω. This means that if $A \in \mathfrak{F}$ and if $A_t = \{\omega \in \Omega | \omega(-t) \in A\}$, then

$$P(A_t) = P(A), \qquad -\infty < t < \infty.$$

The functions $F(x, \omega)$ and $G(x, \omega)$ on $R^n \times \Omega \to R^m$ are appropriately restricted so that (3.49) has solutions for all $t \geqslant 0$ when $\varepsilon > 0$.

We shall refer to the motion $\omega \to \omega(t)$ as the bath and to $x^\varepsilon(t)$ as the system. This terminology is used to emphasize the point that (3.49) represents a classical dynamical system driven by interaction with the motion $\omega(t)$. The state of the system $x^\varepsilon(t) = x^\varepsilon(t, x, \omega)$, i.e., it depends on the initial state of the system and the initial state of the bath. The initial state of the bath is distributed randomly with measure P. Thus, the only way randomness enters into (3.49) is via the initial distribution of the bath.

Let

$$F(x, t, \omega) = F(x, \omega(t)), \qquad G(x, t, \omega) = G(x, \omega(t)). \tag{3.50}$$

For each x, $F(x, t, \omega)$ and $G(x, t, \omega)$ are stationary random functions, often abbreviated $F(x, t)$, $G(x, t)$, and we may write (3.49) in the form

$$\frac{dx^\varepsilon(t)}{dt} = \frac{1}{\varepsilon} F\left(x^\varepsilon(t), t/\varepsilon^2\right) + G\left(x^\varepsilon(t), t/\varepsilon^2\right), \qquad t > 0,\ x^\varepsilon(0) = x. \tag{3.51}$$

This is the familiar form in stochastic differential equations [**14**], [**15**]. We shall assume that

$$E\{F(x, t, \cdot)\} = \int_\Omega F(x, \omega) P(d\omega) = 0 \tag{3.52}$$

which corresponds to the case $P\mathfrak{L}_2 P = 0$ in (3.20). The extension to the other case is not difficult.

How do we proceed with the asymptotic analysis of (3.51) as $\varepsilon \downarrow 0$? We attempt to follow the pattern of the Markovian case and establish an equation of the form (3.17), i.e., a backward equation. However, $(x^\varepsilon(t), \omega^\varepsilon(t))$ do not constitute a Markov process, so we must approach the problem in a more general way. This is done in [**7**] in the manner of [**16**]. Direct but more involved procedures are [**14**], [**15**]. We shall not give details here but we shall summarize the facts as follows.

First of all the final results are the same as in the Markovian case. That is, if $f(x)$ is a smooth bounded function on R^n, then $E\{f(x^\varepsilon(t))\}$, with $x^\varepsilon(0) = x$, converges as $\varepsilon \to 0$, $0 \leqslant t \leqslant T < \infty$, to the solution $u(t, x)$ of the diffusion equation

$$\partial u(t, x)/\partial t = \bar{\mathfrak{L}} u(t, x), \qquad u(0, x) = f(x),$$

with $\bar{\mathfrak{L}}$ given by (3.37). In (3.37) we replace $y(t)$ by $\omega(t)$, i.e., the expectations are taken relative to the stationary processes defined by (3.50). This identity of the results in the Markovian and non-Markovian cases does not, of course, tell us what the necessary assumptions on F and G of (3.50) are for its validity. In the Markovian case, ergodicity of $\{y(t), t \geqslant 0\}$ in a sufficiently strong sense (a Fredholm alternative for $\mathfrak{L}_1$) was sufficient. In the non-Markovian case sufficient conditions are given in [**7**]. A typical one expressed in words says: The best estimate of the state of the bath at time $t + s$ as observed, i.e., $F(x, \omega(t + s))$, given all observable information up to some t, tends to zero (because of (3.52)) sufficiently rapidly as $s \uparrow \infty$. This "unpredictability" of the bath is what makes the system go to a diffusion process in the weak coupling limit.

We conclude this section by stating again that, under appropriate conditions, the asymptotics for Markovian and non-Markovian coefficients are the same. Thus, all of our above examples, suitably reinterpreted, serve as non-Markovian examples as well. In particular, the results of §3(v) on the randomly coupled oscillators carry over immediately.

(vii) *Case* $P \mathfrak{L}_2 P \neq 0$. *Small diffusion*. When condition (3.15) or (3.52) does not hold then we set

$$\overline{F}(x) = E\{F(x, t, \cdot)\}, \qquad \overline{G}(x) = E\{G(x, t, \cdot)\} \tag{3.53}$$

and the result is that $E\{f(x^\varepsilon(t))\}$ with $x^\varepsilon(0) = x$ is approximated well when $\varepsilon \ll 1$ by the solution $\overline{u}^\varepsilon(t, x)$ of the diffusion equation

$$\partial \overline{u}^\varepsilon(t, x)/\partial t = \overline{\mathfrak{L}}^\varepsilon \overline{u}^\varepsilon(t, x), \qquad \overline{u}^\varepsilon(0, x) = f(x), \tag{3.54}$$

where, from (3.28),

$$\begin{aligned} \mathfrak{L}^\epsilon f(x) = \frac{1}{\epsilon} \overline{F}(x) \frac{\partial f(x)}{\partial x} &+ \overline{G}(x) \frac{\partial f(x)}{\partial x} \\ &+ \int_0^\infty E\left\{(F(x, 0, \cdot) - \overline{F}(x)) \frac{\partial}{\partial x}\left((F(x, t, \cdot) - \overline{F}(x)) \frac{\partial f(x)}{\partial x}\right\}\right\} dt. \end{aligned} \tag{3.55}$$

The approximation of $E\{f(x^\varepsilon(t))\}$ by $\overline{u}^\varepsilon(t, x)$ is valid in an interval $0 \leqslant t \leqslant T$ with $T < \infty$ but arbitrary.

The analysis of the diffusion equation (3.54), a small diffusion problem, is of special interest independently of the way it was obtained here. Specifically, on rescaling the time we may write it as

$$\begin{gathered} \frac{\partial \overline{u}^\varepsilon(t, x)}{\partial t} = \varepsilon \sum_{i,j=1}^n a_{ij}(x) \frac{\partial^2 \overline{u}^\varepsilon(t, x)}{\partial x_i \partial x_j} + \sum_{i=1}^n \left(\overline{F}_i(x) + \varepsilon b_i(x)\right) \frac{\partial \overline{u}^\varepsilon(t, x)}{\partial x_i}, \\ \overline{u}^\varepsilon(0, x) = f(x), \end{gathered} \tag{3.56}$$

with the coefficients identified from (3.55). We may now ask several questions about (3.56) or boundary value problems associated with it. If the solutions of

$$d\overline{x}(t)/dt = \overline{F}(\overline{x}(t)), \qquad \overline{x}(0) = x, \tag{3.57}$$

are periodic orbits for all $x \in R^n$, then one may employ the method of averaging on (3.56). First we introduce new coordinates in space, corresponding to "action-angle" variables, and then do the averaging. The analysis of this step is carried out in [**18**].

If the structure of the solutions of (3.57) is more complex, for example if there are stable or unstable equilibria, limit cycles or other invariant manifolds, etc., then approximations to (3.56) lead to a very rich and potentially extremely useful theory. The simplest form of that theory is the Gauss-Markov theory discussed in the next section. There we study the solution of (3.56) near the orbits of (3.57). The behavior of the diffusion process associated with (3.56) away from the orbits of (3.57) is a problem in the theory of large deviations. It has been studied in detail in the fundamental work of Ventsel and Freidlin ([**12**] of §2). Interesting applications and many illuminating remarks can be found in [**19**].

(viii) *Gauss-Markov limit*. As we mentioned above this is a relatively crude theory which is used when (3.15) or (3.52) do not hold and when the orbits of (3.57) are complicated but results are needed only near these orbits. Because of the importance of the results we shall start the analysis afresh and we shall give a self-contained treatment following [**20**] and [**21**]. The analysis extends to stochastic delay differential equations [**21**] but the limiting process is not Markovian in this case.

Consider agian the stochastic equation

$$dx^\varepsilon(t)/dt = F(x^\varepsilon(t), \omega^\varepsilon(t)), \qquad x^\varepsilon(0) = x, \tag{3.58}$$

where $\omega^\varepsilon(t) \equiv \omega(t/\varepsilon)$ and $\omega(t)$ is the motion defined on $(\Omega, \mathfrak{F}, P)$ in §3(vi). Note, however, that the scaling of (3.58) is different from that of (3.49) or (3.51). We may also write (3.58) in the form

$$dx^\varepsilon(t)/dt = F(x^\varepsilon(t), t/\varepsilon), \qquad x^\varepsilon(0) = x, \tag{3.59}$$

with $F(x, t, \omega) = F(x, \omega(t))$. We define $\overline{F}(x)$ by

$$\overline{F}(x) = E\{F(x, \cdot)\} = E\{F(x, t, \cdot)\} \tag{3.60}$$

and $\overline{x}(t)$ as the solution

$$d\overline{x}(t)/dt = \overline{F}(\overline{x}(t)), \qquad \overline{x}(0) = x. \tag{3.61}$$

By an elementary application of the perturbation procedure of §3(iii), or better by direct analysis, it is easy to conclude that

$$\lim_{\varepsilon \downarrow 0} \sup_{0 \leqslant t \leqslant T} E\{|x^\varepsilon(t) - \overline{x}(t)|\} = 0. \tag{3.62}$$

What happens on a longer time interval ($t \sim 1/\varepsilon$) corresponds to the analysis of §3(vii) (with $G \equiv 0$ and $t \to \varepsilon t$ there). Let us ask, however, for more modest information. How does the fluctuation process, defined by

$$(x^\varepsilon(t) - \overline{x}(t))/\sqrt{\varepsilon} = z^\varepsilon(t), \tag{3.63}$$

behave in $0 \leqslant t \leqslant T$ as $\varepsilon \downarrow 0$? First we find a differential equation for $z^\varepsilon(t)$. From (3.63), (3.59) and (3.61) it follows that

$$\frac{dz^\varepsilon(t)}{dt} = \frac{1}{\sqrt{\varepsilon}}\left[F\big(\overline{x}(t) + \sqrt{\varepsilon}\, z^\varepsilon(t), t/\varepsilon\big) - \overline{F}(\overline{x}(t))\right], \quad z^\varepsilon(0) = 0. \tag{3.64}$$

Next we expand the right-hand side of (3.64) in powers of $\sqrt{\varepsilon}$.

$$dz^\varepsilon(t)/dt = (1/\sqrt{\varepsilon})\big[F(\bar{x}(t), t/\varepsilon) - \bar{F}(\bar{x}(t))\big] + \frac{\partial F(\bar{x}(t), t/\varepsilon)}{\partial x} z^\varepsilon(t) + O(\sqrt{\varepsilon}), \qquad z^\varepsilon(0) = 0. \tag{3.65}$$

Now the result is clear. Under appropriate conditions [20], [21] $z^\varepsilon(t)$ will converge to a Gaussian process because

$$\frac{1}{\sqrt{\varepsilon}} \int_0^t \big[F(\bar{x}(s), s/\varepsilon) - \bar{F}(\bar{x}(s))\big]\, ds \tag{3.66}$$

will converge as $\varepsilon \to 0$ to a Gaussian process $w(t)$, say, and hence $z^\varepsilon(t) \to \bar{z}(t)$ where

$$\bar{z}(t) = w(t) + \int_0^t \frac{\partial \bar{F}(\bar{x}(s))}{\partial x} \bar{z}(s)\, ds. \tag{3.67}$$

Actually (3.66) converges to a Gaussian process with independent increments, i.e., a time-inhomogeneous Brownian motion with $E\{w(t)\} = 0$; and if ($*$ = transpose)

$$A(x) = \lim_{T\uparrow\infty} \frac{1}{T} \int_0^T \int_0^T E\big\{\big(F(x, s) - \bar{F}(x)\big) \cdot \big(F(x, \sigma) - \bar{F}(x)\big)^*\big\}\, d\sigma ds, \tag{3.68}$$

then $E\{w(t)w^*(t)\} = A(\bar{x}(t))$ is the covariance matrix of $w(t)$. Thus $\bar{z}(t)$, the limiting fluctuation process defined by (3.67), is a time-inhomogeneous Ornstein-Uhlenbeck process. We may call the approximation $\bar{x}(t) + \sqrt{\varepsilon}\, \bar{z}(t)$, $0 \leqslant t \leqslant T$, of $x^\varepsilon(t)$ the guided O.U. approximation because the mean follows the orbit $\bar{x}(t)$ and the fluctuation is an O.U. process centered at the orbit $\bar{x}(t)$ and with covariance matrix $C(t)$ computed as follows.

Let $U(t, s)$ be the fundamental matrix solution of the linear equations

$$\frac{dU(t, s)}{dt} = \frac{\partial \bar{F}(\bar{x}(t))}{\partial x} U(t, s), \qquad t > s,\ U(s, s) = I.$$

Then (3.67) yields $\bar{z}(t) = \int_0^t U(t, s)\, dw(s)$, and hence

$$E\{\bar{z}(t)\bar{z}^*(t)\} = \int_0^t U(t, s) A(\bar{x}(s)) U^*(t, s)\, ds = C(t). \tag{3.69}$$

Thus, $C(t)$ can be computed explicitly if we have access to the fundamental matrix U.

As a simple application we consider the symmetric competition two-species model of interacting populations [21] in a noisy environment.

Let $x(t)$ and $y(t)$ denote the population of species one and two respectively and assume that they satisfy the evolution equations

$$\begin{aligned} \frac{dx^\varepsilon(t)}{dt} &= x^\varepsilon(t)(k + \gamma_1(t/\varepsilon) - x^\varepsilon(t) - \alpha y^\varepsilon(t)), \qquad x^\varepsilon(0) = x, \\ \frac{dy^\varepsilon(t)}{dt} &= y^\varepsilon(t)(k + \gamma_2(t/\varepsilon) - y^\varepsilon(t) - \alpha x^\varepsilon(t)), \qquad y^\varepsilon(0) = y. \end{aligned} \tag{3.70}$$

Here $\alpha \in [0, 1]$ is the competition coefficient and measures the degree of interaction between the species, k is the deterministic carrying capacity of the environment and $\gamma_1(t/\varepsilon)$, $\gamma_2(t/\varepsilon)$ are zero-mean stochastic processes which fluctuate rapidly since $\varepsilon \ll 1$. We assume they are stationary and denote the covariance by $R_{ij}(t) = E\{\gamma_i(t+s)\gamma_j(s)\}$. If $\mathbf{x} = (x, y)$ and

$$F(\mathbf{x}, t/\varepsilon) = \begin{bmatrix} x(k + \gamma_1(t/\varepsilon) - x - \alpha y) \\ y(k + \gamma_2(t/\varepsilon) - y - \alpha x) \end{bmatrix},$$

then

$$\bar{F}(\mathbf{x}) = \begin{bmatrix} x(k - x - \alpha y) \\ y(k - y - \alpha x) \end{bmatrix}.$$

We note that $\bar{F}(\mathbf{x})$ has two equilibrium points, $(0, 0)$ and the point $(k/(1+\alpha), k/(1+\alpha))$, the latter being stable.

Let us assume that (3.70) is solved with initial conditions $(x, y) =$

$$(k/(1+\alpha), k/(1+\alpha)) = \mathbf{x}^0.$$

Then, $\bar{\mathbf{x}}(t) = \mathbf{x}^0$ for all $t > 0$ and

$$\left.\frac{\partial \bar{F}(\mathbf{x})}{\partial \mathbf{x}}\right|_{\mathbf{x}=\mathbf{x}^0} = \frac{-k}{1+\alpha}\begin{bmatrix} 1 & \alpha \\ \alpha & 1 \end{bmatrix} \equiv B. \tag{3.71}$$

Furthermore, from (3.68) we find that (assuming finite integrals)

$$A(\mathbf{x}^0) = \left(\frac{k}{1+\alpha}\right)^2 \int_0^\infty dt \begin{bmatrix} R_{11}(t) & R_{12}(t) \\ R_{12}(t) & R_{22}(t) \end{bmatrix}. \tag{3.72}$$

We are ready now to compute the covariance of the limit fluctuation process. We have chosen the initial conditions $(x, y) = \mathbf{x}^0 =$ the stable equilibrium point in order to make the mean $\bar{\mathbf{x}}(t)$ as simple as possible. The fluctuation process $\bar{\mathbf{z}}(t)$ will be a Gaussian process with independent increments, mean zero and covariance given by (3.69). To simplify the calculations we assume further that the noise intensities for the two species are the same so we may write $A(\mathbf{x}^0)$ in the form

$$A = \left(\frac{k}{1+\alpha}\right)^2 \sigma^2 \begin{bmatrix} 1 & \rho \\ \rho & 1 \end{bmatrix}.$$

Now however A and B of (3.71) commute so $C(t)$ of (3.69) is given by

$$C(t) = \int_0^t e^{2Bs}\, ds\, A. \tag{3.73}$$

From this formula one finds easily that the ellipses of constant probability for the Gaussian process $\bar{\mathbf{z}}(t)$ are rotated by 45° and that in this 45° frame of reference the covariance matrix is diagonal and is given by

$$\begin{bmatrix} \dfrac{\sigma^2 k}{2} \dfrac{1+\rho}{(1+\alpha)^2}(1-e^{-2kt}) & 0 \\ 0 & \dfrac{\sigma^2 k}{2}\dfrac{1-\rho}{1-\alpha^2}(1-e^{-2k((1-\alpha)/(1+\alpha))t}) \end{bmatrix}. \tag{3.74}$$

When $t = \infty$ this simplifies to

$$\begin{bmatrix} \dfrac{\sigma^2 k}{2} \dfrac{1+\rho}{(1+\alpha)^2} & 0 \\ 0 & \dfrac{\sigma^2 k}{2}\dfrac{1-\rho}{1-\alpha^2} \end{bmatrix}, \tag{3.75}$$

and a number of interesting conclusions can be drawn from this simple result. We note however, since the results hold for $0 \leqslant t \leqslant T < \infty$, it is not, strictly speaking, proper to set $t = \infty$ in (3.74). The significance of (3.75) is, however, relative; it is simply an approximation to (3.74) which is appropriate and useful in certain ranges of the parameters.

(ix) *Higher order terms, larger intervals of validity, etc*. Throughout the preceding discussion, approximations were valid as $\varepsilon \downarrow 0$, $0 \leqslant t \leqslant T < \infty$. It is natural to ask if the results are valid in $0 \leqslant t < \infty$. This has considerable practical significance since solving diffusion equations to obtain information about the limiting process is frequently a hopeless task. On the other hand, finding time-independent solutions is simpler. The question is, then, what is the significance of such equilibrium solutions of the limiting diffusion equation relative to the original problem? (See [17].)

An interesting answer to this question is given by the following result of Norman [**22**], which we state in the context of diffusion approximations for equations with Markov coefficients.

Let $u^\varepsilon(t, x, y)$ be the solution of (3.17), and suppose that

$$\lim_{\varepsilon\downarrow 0} \sup_{0\leqslant t\leqslant T} \sup_{x,y} |u^\varepsilon(t,x,y) - \bar{u}(t,x)| = 0, \tag{3.76}$$

where $\bar{u}(t, x)$ satisfies the equation

$$\partial \bar{u}(t,x)/\partial t = \bar{\mathfrak{L}}\bar{u}(t,x), \qquad \bar{u}(0,x) = f(x), \tag{3.77}$$

with $\bar{\mathfrak{L}}$ defined by (3.36). Suppose that there is an $\bar{f}(x)$ such that

$$\lim_{t\uparrow\infty} \sup_{x} |\bar{u}(t,x) - \bar{f}(x)| = 0. \tag{3.78}$$

Let $\bar{u}^\varepsilon(t, x, y)$ be the solution of (3.17) with $\bar{u}^\varepsilon(0, x, y) = \bar{f}(x)$ and suppose that

$$\lim_{\varepsilon\downarrow 0} \sup_{t\geqslant 0} \sup_{x,y} |\bar{u}^\varepsilon(t,x,y) - \bar{f}(x)| = 0. \tag{3.79}$$

Then (3.76) holds with $T = \infty$. The proof of this result is quite elementary [**22**].

Clearly (3.79) is the hypothesis that is hardest to verify since it looks about as difficult as the original problem. It may happen, however, that the limit of $\bar{u}(t, x)$ as $t \uparrow \infty$ is very simple, for example, $\bar{f}(x) =$ constant. In this case (3.79) is trivially true and hence we have the result (3.76) with $T = \infty$.

As an example, consider (3.41), the linear randomly coupled oscillators, and take $f(x) = x_p x_p^*$, $p = 1, 2, \ldots, n$. Assuming nondegeneracy, (3.74) yields $\bar{u}(t, x) \to 1/n$ as $t \to \infty$ if $\Sigma_{p=1}^n |x_{p0}|^2 = 1$, say. Thus,

$$E\left\{|x_p^\varepsilon(t)|^2\right\} \underset{\varepsilon \downarrow 0}{\to} w_p(t),$$

the solution of (3.47), and the result holds for all $t \geq 0$. This example is typical of what one may expect in general. It should be contrasted with the situation described in §3(vii) where it is expected that the above argument will fail.

Concerning higher order terms in the expansions, valid in the same time interval, the procedure of §3(iii) is perfectly adequate. It goes without saying that calculations become prohibitively complex very rapidly.

(x) *Boundary value problems and shooting*; *one-dimensional waves*. Linear two-point (or multi-point) boundary value problems for stochastic equations can be treated by the simple device of obtaining the statistical properties of the fundamental solution matrix. From this one can obtain all relevant information about the statistics of the solution to the boundary value problem. This is the method of shooting.

We shall describe an example, one-dimensional wave propagation, that illustrates the general ideas. The example is also of independent interest, and many other remarks and references to the literature can be found in **[23]**, **[24]**.

Consider a one-dimensional random medium occupying the interval $0 \leq x \leq l$. Let $u(x)$ and $n(x)$ be the wave field (with the time factor $e^{-i\omega t}$ omitted) and the index of refraction at location x, respectively. We assume that $u(x)$, $-\infty < x < \infty$, satisfies the reduced wave equation

$$d^2u(x)/dx^2 + k^2 n^2(x) u(x) = 0, \qquad -\infty < x < \infty, \tag{3.80}$$

$$n^2(x) = \begin{cases} n_1^2, & x < 0, \\ 1 + \varepsilon\mu(x), & 0 \leq x \leq l, \\ n_2^2, & x > l. \end{cases}$$

$$u(x) \text{ and } du(x)/dx \text{ continuous, } x \in (-\infty, \infty).$$

Here $\mu(x)$ denotes a real zero-mean stationary random process with x the "time" variable and $\varepsilon \ll 1$ characterizing the size of the fluctuations of the refractive index in $0 \leq x \leq l$. If we denote by R and T the complex-valued reflection and transmission coefficients, then we have

$$\begin{aligned} u(x) &= e^{ikn_1x} + Re^{-ikn_1x}, \qquad x < 0, \\ u(x) &= Te^{ikn_2x}, \qquad x > l. \end{aligned} \tag{3.81}$$

From (3.80) and (3.81) it follows that $u(x)$ satisfies the following stochastic two-point boundary value problem.

$$\frac{d^2u(x)}{dx^2} + k^2[1 + \varepsilon\mu(x)]u(x) = 0, \qquad 0 \leqslant x \leqslant l,$$

$$\frac{1}{2}\left[u(0) + \frac{1}{ikn_1}\frac{du(0)}{dx}\right] = 1, \qquad \frac{1}{2}\left[u(l) - \frac{1}{ikn_2}\frac{du(l)}{dx}\right] = 0. \quad (3.82)$$

To reduce (3.82) to a first order system we define $A(x)$ and $B(x)$ by

$$u(x) = e^{ikx}A(x) + e^{-ikx}B(x),$$
$$du(x)/dx = ik\left[e^{ikx}A(x) - e^{-ikx}B(x)\right], \qquad 0 \leqslant x \leqslant l.$$

It follows that A and B satisfy the stochastic two-point boundary value problem

$$\frac{dA(x)}{dx} = \varepsilon\frac{ik\mu(x)}{2}\left[A(x) + B(x)e^{-2ikx}\right],$$
$$\frac{dB(x)}{dx} = -\varepsilon\frac{ik\mu(x)}{2}\left[A(x)e^{2ikx} + B(x)\right], \qquad 0 \leqslant x \leqslant l, \quad (3.83)$$
$$A(0) = E_g + \Gamma_g B(0), \qquad B(l) = \Gamma_l A(l)$$

with

$$E_g = \frac{2n_1}{1+n_1}, \quad \Gamma_g = \frac{1-n_1}{1+n_1}, \quad \Gamma_l = e^{2ikl}\frac{1-n_2}{1+n_2}.$$

Since $E\{\mu(x)\} = 0$ in (3.83) we are naturally interested in using the methods of previous sections to study the statistics of A and B when $\varepsilon \ll 1$ and $l \sim 1/\varepsilon^2$, i.e., the weak coupling limit.

Let us express the solution of (3.83) in terms of fundamental matrix solutions. Let us assume for simplicity that $n_1 = n_2 = 1$ so that $\Gamma_g = \Gamma_l = 0$ in (3.83). Let $Y(x)$ be the 2×2 matrix solution of

$$\frac{dY(x)}{dx} = \varepsilon\frac{ik\mu(x)}{2}\begin{bmatrix} 1 & e^{-2ikx} \\ -e^{2ikx} & -1 \end{bmatrix}Y(x), \qquad Y(0) = I. \quad (3.84)$$

It is easily verified that $Y(x)$ is of the form

$$Y = \begin{bmatrix} a & b \\ \bar{b} & \bar{a} \end{bmatrix}, \qquad |a|^2 - |b|^2 = 1, \quad (3.85)$$

where $a = a(x)$ and $b = b(x)$ are complex random functions. Note that $Y = Y(x, 0)$, i.e., it is the fundamental matrix or propagator from location 0 to location x.

Let

$$Y_1 = \begin{pmatrix} a_1 & b_1 \\ \bar{b}_1 & \bar{a}_1 \end{pmatrix} = Y(x, 0) \quad \text{and} \quad Y_2 = \begin{pmatrix} a_2 & b_2 \\ \bar{b}_2 & \bar{a}_2 \end{pmatrix} = Y(l, x).$$

Then, it can be verified by direct computation that the solution $A(l, x)$, $B(l, x)$ of (3.83) (with $\Gamma_g = \Gamma_l = 0$) is given by

$$A(l, x) = \frac{\bar{a}_2}{\bar{b}_2 b_1 + \bar{a}_2 \bar{a}_1}, \quad B(l, x) = \frac{-\bar{b}_2}{\bar{b}_2 b_1 + \bar{a}_2 \bar{a}_1}, \qquad 0 \leqslant x \leqslant l. \quad (3.86)$$

Formulas (3.86) give the desired expression of A and B in terms of fundamental matrices.

The stochastic analysis can now begin with the matrix initial value problem (3.84). The procedures of the previous sections apply directly and we are led quickly to a diffusion equation for the transition probabilities of the limiting (matrix valued) Markov process.

The solution of this diffusion equation turns out to be obtainable in closed form in terms of known functions. Using this explicit solution we can then return to (3.86) and compute statistical averages of A, B and higher powers. The calculations are straightforward but somewhat lengthy, so we refer to [**23**], [**24**] for details and discussion of the results.

REFERENCES

1. N. N. Bogoljubov and Ju. A. Mitropolskiĭ, *Asymptotic methods in the theory of nonlinear oscillators*, 2nd rev. ed., Fizmatgiz, Moscow, 1958; English transl., Internat. Monographs on Advanced Math. and Phys., Hindustan, Delhi, 1961; Gordon and Breach, New York, 1962. MR **20** #6812; **25** #5242.

2. V. M. Volosov, *Averaging in systems of ordinary differential equations*, Uspehi Mat. Nauk **17** (1962), no. 6 (108), 3–126 = Russian Math. Surveys **17** (1962), no. 6, 1–126. MR **26** #3976.

3. J. Cole, *Perturbation methods in applied mathematics*, Blaisdell, Waltham, Mass., 1968. MR **39** #7841.

4. S. Goldstein, *On diffusion by discontinuous movements, and on the telegraph equation*, Quart. J. Mech. Appl. Math. **4** (1951), 129–156. MR **13**, 960.

5. R. Hersh, *Random evolutions: a survey of results and problems*, Rocky Mountain Math. J. **14** (1974), 443–496.

6. W. Heitler, *The quantum theory of radiation*, Oxford Univ. Press, Oxford, 1954.

7. G. C. Papanicolaou, *Some probabilistic problems and methods in singular perturbations*, Rocky Mountain J. Math. **6** (1976), 653–674.

8. T. G. Kurtz, *A limit theorem for perturbed operator semigroups with applications to random evolution*, J. Functional Analysis **12** (1973), 55–67.

9. U. Frisch, *Wave propagation in random media*, Probabilistic Methods in Appl. Math., vol. 1, Academic Press, New York, 1968, pp. 75–198. MR **42** #4088.

10. H. Kramers, *Brownian motion in a field of force and the diffusion model of chemical reactions*, Physica **7** (1940), 284–304. MR **2**, 140.

11. G. C. Papanicolaou and R. Burridge, *Transport equations for the Stokes parameters starting from Maxwell's equations in a random medium*, J. Mathematical Phys. **16** (1975), 2074–2085.

12. W. Kohler and G. C. Papanicolaou, *Sound propagation in a randomly inhomogeneous ocean* (to appear).

13. G. C. Papanicolaou, *A kinetic theory for power transfer in stochastic systems*, J. Mathematical Phys. **13** (1972), 1912–1918. MR **47** #4577.

14. R. Z. Has′minskiĭ, *A limit theorem for solutions of differential equations with a random right hand part*, Teor. Verojatnost. i Primenen. **11** (1966), 444–462 = Teor. Probability Appl. **11** (1966), 390–406. MR **34** #3637.

15. G. C. Papanicolaou and W. Kohler, *Asymptotic theory of mixing stochastic ordinary differential equations*, Comm. Pure Appl. Math. **27** (1974), 641–668.

16. T. G. Kurtz, *Semigroups of conditioned shifts and approximations of Markov processes*, Ann. of Prob. **3** (1975), 618–642.

17. G. Blankenship and G. C. Papanicolaou, *Stability and control of stochastic systems with wide band noise disturbances*, SIAM J. Appl. Math. (to appear).

18. R. Z. Has'minskiĭ, *The averaging principle for parabolic and elliptic differential equations and Markov processes with small diffusion*, Teor. Verojatnost. i Primenen. **8** (1963), 3–25 = Theor. Probability Appl. **8** (1963), 1–21. MR **28** #4253.

19. D. Ludwig, *Persistence in dynamical systems under random perturbations*, SIAM Rev. **17** (1975), 605–640.

20. R. Z. Has'minskii, *On stochastic processes defined by differential equations with a small parameter*, Teor. Verojatnost. i Primenen. **11** (1966), 240–259 = Theor. Probability Appl. **11** (1966), 211–228. MR **34** #3636.

21. B. White, Dissertation, N.Y.U., 1974.

22. M. F. Norman, *Ergodicity of diffusion and temporal uniformity of diffusion approximations*, Ann. Probability (to appear).

23. W. Kohler and G. C. Papanicolaou, *Power statistics for wave propagation in one dimension and comparison with radiative transport theory*, J. Mathematical Phys. **14** (1973), 1733–1745. MR **49** #4442.

24. ——— , *Power statistics for wave propagation in one dimension and comparison with radiative transport theory*. II, J. Mathematical Phys. **15** (1974), 2186–2197.

COURANT INSTITUTE OF MATHEMATICAL SCIENCES, NEW YORK UNIVERSITY, NEW YORK, NEW YORK 10012

Lectures in Applied Mathematics
Volume 16, 1977

Lectures in Population Dynamics

G. Oster

1. Introduction. When I first started thinking about ecology several years ago I asked a friend who had been trained as an ecologist: "What is it that ecologists want to know? What are the central issues, the great questions?" He replied that that was simple: "Ecologists want to understand the laws governing the distribution and abundance of species in the world." At the time I remember being bothered by the way he spoke of "laws" in ecology in just the way I spoke about "laws" in physics.

After being in the field for several years now I realize that my unease about his casual use of the word "laws" was well founded. The disturbing fact is that a great deal of the organization and regularity we see in the ecosphere may well be a consequence of historical accident and not the operation of a set of mathematical laws. For example: how are we to explain the dissimilarities between kangaroos in Australia and buffalo in North America, despite the fact that they play almost identical roles in their respective ecosystems? (That is, they are both "primary consumers" on a grasslands biome.) How could one possibly formulate mathematical laws that could deduce such phenomena from a set of "fundamental principles"?

Fortunately, in their day-to-day fieldwork most ecologists do not think in mathematical terms. They know that the right "language" to use for a subject as complicated as ecology is not mathematics, but English! Mathematics can only deal with very simple systems–and even there we need computers often as not.

Nevertheless, mathematics can provide some service to ecology by delineating the logical consequences of simple models which, while admittedly oversimplified caricatures of reality, do capture some aspect of the population's behavior. However, one must never take these mathematical models seriously in the absence of confirming experimental data, for the distance between theory and experiment is far greater in ecology than in most other fields. Indeed, the literature abounds with speculative models which, although clever and plausible, are little more than educated guesses.

Our approach in this set of notes will be to investigate the consequences of the conservation laws for populations in the setting of a particular set of laboratory conditions. It must be borne in mind, however, that the extrapolation from the laboratory to the real world is far riskier in population ecology than in engineering and physics. Therefore, we shall have to exercise restraint

AMS (MOS) subject classifications (1970). Primary 92-05, 92A15.

in attempting to generalize our conclusions. Nevertheless, the deductions we shall draw from our analyses will have analogs in the real world.

2. The balance equations.

2.1. Whatever the subtleties and complexities of ecological systems, we can be certain of one thing: populations surely obey a conservation law of the form

$$dN/dt = \text{Births} - \text{Deaths} \pm \text{Migration}. \tag{2-1}$$

Therefore, it behooves us to commence our study of biological populations by understanding the consequences of this basic balance law. This is not so easy as it seems. Indeed, the science of hydrodynamics boils down to deducing the consequences of the balance laws for mass, momentum and energy under various conditions–a nontrivial task!

2.2. The earliest population model was constructed by Thomas Malthus in 1798. He assumed that birth and death rates were proportional to population size: $B = bN$, $D = dN$. Since his "system" was the entire human race, he could ignore migration and write: $\dot{N} = (b - d)N \equiv rN$. (While the consequences of this growth law were appalling to Malthus, his writings on the subject inspired Darwin to infer the role that competition for limited resources must play as the instrument of natural selection.) Paul Verhulst, in 1838, incorporated the notion of a negative feedback on population growth-rate due to competition for and depletion of resources. He assumed that, whatever the details of the biology, birthrates should decrease and deathrates increase as the population density increases. A linear law is the simplest: $B = b_0 - b_1N$, $D = d_0 + d_1N$. Thus the balance equation becomes:

$$\dot{N} = r_0N(1 - N/K) \tag{2-2}$$

where $r_0 = b_0 - d_0$ is the initial rate of exponential growth and $K = (b_0 - d_0)/(b_1 + d_1)$ is the equilibrium population. Substituting $u = (K - N)/N$ yields the familiar sigmoid, or logistic, growth curve.

A great deal of mathematical ecology has been elaborations on this theme: deriving various birth and death laws from assumptions about resource consumption, including time lags to account for maturation delays, adding stochastic terms to model the fluctuating environment and combining logistic-like equations to model competition between species. A lucid account of this work can be found in Krebs [**1972**] and May [**1975**].

2.3. A major deficiency in these types of models becomes apparent when one inquires as to what is an adequate state description of a population. Clearly a mere numerical census, $N(t)$, is insufficient to predict the future course of population growth. What if all members of the population were males? Or past their reproductive years? Or too starved to bear viable offspring? More information about the population is required than numbers, and so as a next step we must expand our description to include the age structure and other phenotypic characteristics which might influence population growth such as mass, size, state of health, etc. That is, we must write our balance law (equation (2-1)) for a density function, $n(t, a, \mathbf{x})$, where a is the chronological age and $\mathbf{x}$ is a vector of the relevant phenotypic traits.

Therefore, we take as our state space $\mathbf{R}^{k+1}$ with coordinates $(a, x_1, x_2, \ldots, x_k)$. A conservation equation for the density function $n(\cdot)$: $\mathbf{R}^{n+1} \to \mathbf{R}$ can be written as:

$$\frac{\partial n}{\partial t} = -\nabla \cdot \mathbf{J} + B - D + M. \tag{2-3}$$

The birth, death and migration terms may all be functions of n, a and $\mathbf{x}$. For the flux, $\mathbf{J}$, we shall take

$$\mathbf{J} = \mathbf{v}n, \tag{2-4}$$

i.e. a convective flow with velocity $\mathbf{v} = (1, g_1, \ldots, g_k)^T$ where $g_i = dx_i/dt$ is the growthrate of trait i and $da/dt = 1$. Let us restrict ourselves to a single trait, x, and rewrite (2-3) as

$$\frac{\partial n}{\partial t} + \frac{\partial n}{\partial a} + \frac{\partial}{\partial x}(gn) = -\mu n + M, \tag{2-5a}$$

$$n(t, 0, x) = B, \tag{2-5b}$$

where $\mu(t, a, x)$ is the deathrate per individual. In (2-5b) we have split the birthrate off into a boundary condition; it is this boundary condition that gives the population equations their unique character.

2.4. To write the boundary condition (2-5b) explicitly we are forced to take cognizance of a fundamental characteristic of biological populations: the mechanism of Mendelian inheritance whereby parents pass on their characteristics to their offspring. Let us put this question aside for the time being, however, since it greatly complicates the mathematics and focus our attention on characteristics, x, for which we need not take into account the mechanics of sexual reproduction (e.g., nutritional state). Then we can write (2-5b) in the following form:

$$n(t, 0, x) = \int_0^\infty \int_\alpha^{\alpha+\gamma} b(t, a, x', x) n(t, a, x')\, da\, dx'. \tag{2-6}$$

The birthrate kernel $b(\cdot)$ describes how parents of age a and trait x' give rise to offspring at age 0 and trait x. The age interval $(\alpha, \alpha + \gamma)$ is the reproductive window.[1]

2.5. Remark. The equations (2-5a) and (2-6) constitute a positive feedback system for which we can devise a "mechanical analog" which may prove helpful in understanding the dynamics. If we neglect the term in x, equation (2-5a) is formally identical to the equation for convective transport of particles of density $n(t, a)$ carried by a stream moving at unit velocity. If x be interpreted as the size of the particles then $g(\cdot)$ gives the rate at which the particles grow in size as they move down the stream (i.e. age) (cf. Figure 1).

[1]Throughout our discussion we shall deal with females only and assume that either the sex ratio is 50 : 50, or that there are sufficient males so that fertilization rates are not limiting. For a discussion of two-sex models see Frederickson [**1971**].

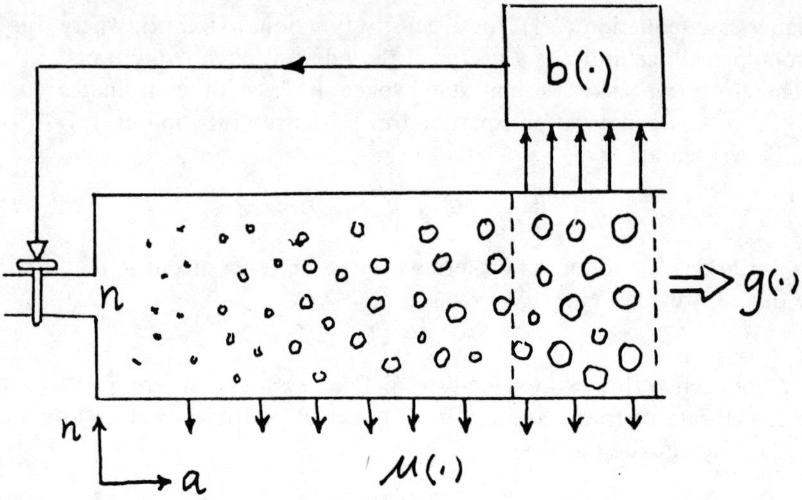

FIGURE 1

Death is represented as a loss of particles (individuals) as they move down the stream. The birth feedback can be viewed as a device which measures the number of individuals at age a, multiplies by $b(\cdot)$ and opens a valve at $a = 0$, injecting $b(\cdot)n(\cdot)$ "newborn" into the age stream.

2.6. REMARK. Equation (2-5a) can be regarded as an approximation to a more general model (Weiss [**1968**]). If we define $\Psi(t, x, x'|a)$ as the transition rate at time t from x to the interval $(x + x', x + x' + dx')$, conditional on a chronological age a, then $\Phi(t, a, x) = \int_0^\infty \Psi\, dx'$ is the total transition rate from x. The balance law for $n(t, a, x)$ can then be written as:

$$\frac{\partial n}{\partial t} + \frac{\partial n}{\partial a} = -(\mu + \Phi)n + \int_0^x \Psi(t, a, x - x', x')n(t, a, x - x')\, dx'. \quad (2\text{-}7)$$

If transitions $x \mapsto x'$ are such that $|x - x'| \ll 1$, then the integrand on the right can be expanded in a Taylor's series about x in much the same way as one proceeds from the Chapman-Kolmogorov equation to the Fokker-Planck equation. If second order expansion terms are retained, the resulting equation includes a dispersion term, $\partial^2(\sigma^2 n)/\partial x^2$, to account for the fact that all individuals do not grow synchronously. Except for one instance where its effects are important, we shall neglect dispersion in our treatment here.

2.7. The model equations we shall deal with are (2-5a) and (2-6):

$$\partial n/\partial t + \partial n/\partial a + \partial(gn)/\partial x = -\mu n + M,$$

$$n(t, 0, x) \overset{\Delta}{=} B(t, x) = \int_0^\infty \int_\alpha^{\alpha+\gamma} b(t, a, x', x)n(t, a, x')\, da\, dx'.$$

They contain unknown functions (constitutive relations) for birth, $b(\cdot)$, death, $\mu(\cdot)$, and growth, $g(\cdot)$. These are analogous to the parameters governing viscosity, heat conduction, etc. in the hydrodynamic equations, and must be supplied either by experiment or by more detailed submodels.

2.8. Equation (2-5a) is a first order, quasi-linear hyperbolic partial differential equation. Its solution can be obtained by the method of characteristics (Courant and Hilbert [**1966**]). Recall that equation (2-5a) can be interpreted geometrically as follows (Mac Lane [**1968**]); cf. Figure 2:

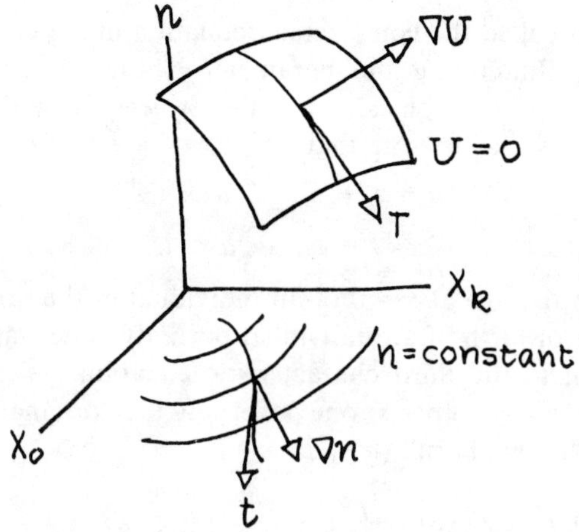

FIGURE 2

The graph of $n(t, a, \mathbf{x}) \equiv n(x_0, x_1, x_2, \ldots, x_k)$ as a level set in $\mathbf{R}^{k+1}$ is $U(\mathbf{x}, n) = z - n(\mathbf{x})$, where $z(\cdot)$ is the projection onto the n-axis. The normal to this surface is given by the gradient

$$\nabla U(\mathbf{x}, n) = (\partial n/\partial x_0, \partial n/\partial x_1, \ldots, \partial n/\partial x_k, -1).$$

Then if we define the vector field $T(\mathbf{x}, n)$ on $\mathbf{R}^{k+1}$ by $T(\mathbf{x}, n) = (1, \mathbf{g}(\cdot), -\mu(\cdot))^T(\mathbf{x}, n)$, equation (2-5a) can be written:

$$\mathbf{T} \cdot \nabla U = 0. \tag{2-8}$$

Thus the solution surface $n(t, a, x)$, whose graph is specified by the gradient field ∇U, is swept out by the integral curves of the vector field $\mathbf{T}$, which generates the tangents to the (graph of the) solution surface, $U = 0$. Thus we must find the integral curves of T, $s \mapsto \gamma(s)$,[2] by solving the ordinary differential equations

$$d\gamma(s)/ds = \mathbf{T}(\gamma(s)), \tag{2-8a}$$

i.e.,

$$\frac{dt}{ds} = 1, \quad \frac{da}{ds} = 1, \quad \frac{dx}{ds} = g(\cdot), \quad \frac{dn}{ds} = -\mu n + M. \tag{2-8b}$$

2.9. The characteristic equations (2-8) can be solved analytically only in restricted cases, some of which we shall leave as exercises. The relevant features of the solution can be appreciated by examining the special case

[2]Note: we are using sloppy notation by identifying the coordinate functions x_i: $\mathbf{R}^{n+1} \to \mathbf{R}$ with the coordinate of the curve $x_i \circ \gamma$: $\mathbf{R}^1 \to \mathbf{R}^1$ (cf. Mac Lane [**1968**]).

when we neglect the dependence on the phenotypic variable, x:

$$\partial n/\partial t + \partial n/\partial a = -\mu n, \tag{2-9a}$$

$$B(t) = \int_{\alpha}^{\alpha+\gamma} b(t, a)n(t, a)\, da. \tag{2-9b}$$

Equation (2-9a) is called the von Foerster equation in the ecological literature (Trucco [**1965**]). Eliminating the parameter s, the characteristic equation $da/dt = 1$ yields $a = t +$ constant, i.e., the projection of the characteristics on the base plane is linear. Note that:

$$a = t + a_0, \qquad a > t, \tag{2-10a}$$

$$= t - t_0, \qquad a < t, \tag{2-10b}$$

where a_0 is the initial age at $t = 0$ of an individual in the original population and t_0 is the time of birth of an individual born after the start of the process. Thus the solution to the third characteristic equation, $dn/da = -\mu n$, must therefore be written in two parts: one governing the "original settlers" ($a > t$) and one for the "native born" ($a < t$):

$$n(t, a) = n(0, t - a)\exp\left[-\int_0^t \mu(\xi, a - t + \xi)\, d\xi\right], \qquad a > t, \tag{2-11a}$$

$$= n(t - a, 0)\exp\left[-\int_0^a \mu(t - a + \zeta, \zeta)\, d\zeta\right], \qquad a < t. \tag{2-11b}$$

The parametrization of the base characteristics is shown in Figure 3. (In the demographic literature the a-t plane is called a Lexis diagram (Keyfitz [**1968**]).)

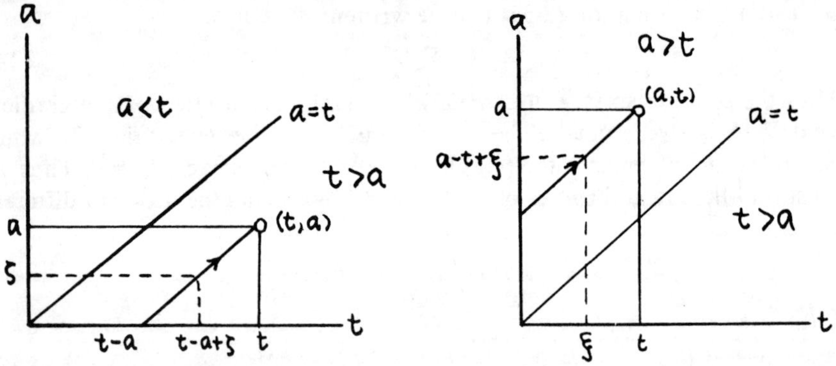

FIGURE 3

For the most part we shall be concerned with the long term behavior of the population, so we can write the equation for the "native", (2-11b), as

$$n(t, a) = B(t - a)l(t, a) \tag{2-12}$$

where $l(t, a) = \exp[-\int_0^t \mu(\xi, a - t + \xi)\, d\xi]$ is the fraction of individuals born at $t - a$ which survive to age a (cf. Figure 3).

2.10. EXERCISE. If g is a function of a and t only show that the solution of equation (2-5a) is

$$n(t, a, x) = n\left(0, a - t, x - \int_0^t g(t', a - t + t')\, dt'\right)$$

$$\cdot \exp\left[-\int_0^t \mu\left(\hat{t}, a - t + \hat{t}, x - \int_0^t g(\hat{t}, \hat{t} + a - t)\, d\hat{t}\right) dt' \right], \qquad a > t,$$

$$n(t, a, x) = n\left(t - a, 0, x - \int_{t-a}^t g(t', t' - t + a)\, dt'\right)$$

$$\cdot \exp\left[-\int_{t-a}^t \mu\left(t', t' - t + a, x - \int_{t'}^t g(\hat{t}, \hat{t} - t + a)\, d\hat{t}\right) dt \right], \qquad a < t.$$

2.11. EXERCISE. Equation (2-5a) can be approximated by a set of five coupled ordinary differential equations for the first two moments of the distribution $n(t, a, x)$. This is worked out in Streiffer and Istock [**1973**].

2.12. EXERCISE. The von Foerster model in equation (2-9) is related to the theory of branching processes. Assume that $\mu = 0$ and that reproduction is by binary fission: $B(t) = 2\dot{N}(t)$, where $N(t) = \int_0^\infty n(t, a)\, da$. Then show that the equation for the total population, $N(t)$, is

$$N(t) = 2\int_0^t N(t - a) f(a)\, da + N(0)[1 - F(t)]$$

where $f(a)$ is the generation time distribution and $F(a) = \int_0^a f(a')\, da'$. Show that the solution is

$$N(t) = \sum_{k=1}^{\infty} 2^k \{f_k * \phi\}(t) + \phi(t)$$

where $\phi(t) = N(0)(1 - F(t))$ and $*$ denotes convolution. Repeat the calculation for $\mu > 0$.

2.13. EXERCISE. If the migration term $M(t, a)$ is included on the right side of equation (2-9a) show that the solution can be written as:

$$n(t, a) = \underbrace{B(t - a) l(t, a)}_{\text{free response}} + \underbrace{J(t, a)}_{\text{forced response}} \qquad \text{(2-13)}$$

where $J(t, a) = l(t, a) \int_0^a (M(u, t - a + u)/l(t - a + u, u))\, du$ (Langhaar [**1972**]).

2.14. Equation (2-12) does not constitute a solution to the von Foerster model (2-11a, b) since the boundary condition (2-9b) contains the unknown function $n(\cdot)$ as well. Therefore we must eliminate one of the two unknown functions, $B(t)$ or $n(t, a)$. If (2-12) is substituted into (2-9b) the problem is reduced to solving the integral equation

$$B(t) = \int_\alpha^{\alpha+\gamma} b(t, a) l(t, a) B(t - a)\, da. \qquad \text{(2-14a)}$$

If migration is included then we use (2-13) in (2-9b):

$$B(t) = \psi(t) + \int_{\alpha}^{\alpha+\gamma} b(\cdot)l(\cdot)B(t-a)\,da \tag{2-14b}$$

where $\psi(t) = \int b(a, t)J(a, t)\,da$ is the contribution to the birthrate from first generation immigrants.

Since the per capita birthrate function $b(\cdot)$ vanishes outside the breeding window $(\alpha, \alpha + \gamma)$(e.g., approximately age 14–44 for human females) we can write (2-14b) as

$$B(t) = f(t) + \int_0^t \phi(t, a)B(t-a)\,da \tag{2-15a}$$

where $f(t) = \psi(t) + \int_t^\infty \phi(t, a)B(t-a)\,da$ is the effect of the original population to the birthrate and $\phi(t, a) = b(t, a)l(t, a)$. Letting $\tau = t - a$, (2-15a) can be written as a Volterra equation

$$B(t) = f(t) + \int_0^t B(\tau)\phi(t, t-\tau)\,d\tau. \tag{2-15b}$$

In this form the von Foester model is called the Lotka equation. If $f(\cdot)$ and $\phi(\cdot)$ are continuous and bounded then by using a contraction mapping argument one can show that there exists a unique solution on the finite interval $[0, T]$.

2.15. Most work has been done on the case where $\phi(\cdot)$ is a function of age only. In that case (2-15b) can be solved by Laplace transform:[3]

$$\tilde{B}(s) = \tilde{F}(s)/(1 - \tilde{\phi}(s)). \tag{2-16}$$

The solution can be written as a sum of exponentials

$$B(t) = \sum Q_k e^{r_k t}$$

where r_k are the roots of $\tilde{\phi}(s) = 1$, i.e.

$$\int_0^\infty \phi(a)e^{-ra}\,da = 1. \tag{2-17}$$

The important fact about the characteristic equation (2-17) is that it has one real root, r_0, larger than the amplitude of any of the other roots, r_i, $i = 1, 2, \ldots, \infty$. Thus the birthrate is asymptotically exponential

$$B(t) \sim Ae^{r_0 t} \quad \text{as } t \to \infty. \tag{2-18}$$

Moreover, the age distribution asymptotically approaches a stationary shape, i.e.,

$$c(a, t) \stackrel{\Delta}{=} \frac{n(a, t)}{N(t)} \to \eta(a), \qquad t \to \infty. \tag{2-19}$$

A discussion of these properties can be found in Lopez [**1967**] and Pollard

---

[3]Note that if equation (2-16) is written in block diagram form then we can see explicitly that the population model is a positive feedback system as mentioned in §2.5

[**1973**]. If $\phi(\cdot)$ is time dependent then these results may not hold; the effect of the initial conditions may not decay.

A more general result has been proven by Gurtin and MacCamy [**1974**]. They allow the birthrates and deathrates to be functionals of the total population: $b(a, N)$, $\mu(a, N)$. The resulting model consists of two coupled integral equations: (2-15) and the definition of $N(t)$ which, using a contraction mapping argument again, they show have a unique solution $(B(t), N(t))$ on $[0, \infty)$. Moreover, if

$$1 = \int_0^\infty b(a, N)l(a, N)\, da \overset{\Delta}{=} \text{net reproductive rate (NRR)} \qquad (2\text{-}20)$$

then there is a unique stationary age distribution given by $n(a) = Nl(a, N)/\int_0^\infty l(a, N)da$. Moreover, this distribution is locally stable if the eigenvalues r_i of the characteristic equation (obtained by linearizing of equations of motion and substituting $n(a, t) = \bar{n}(a)e^{rt}$) have no roots in the right half-plane.

2.16. The finite dimensional version of the balance equations is frequently employed in the ecological and demographic literature. To derive it we take the difference scheme shown in Figure 4. We obtain

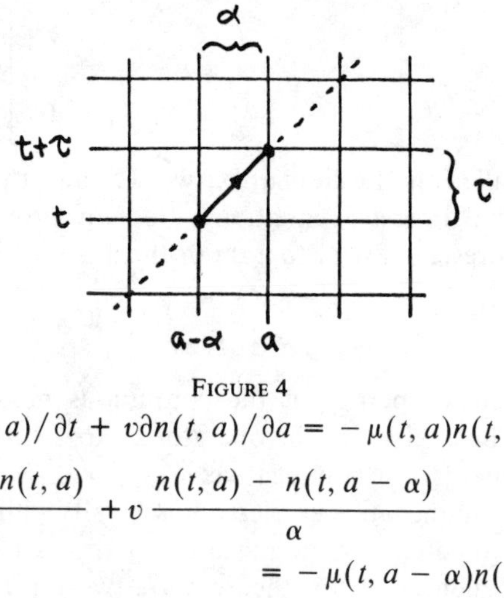

FIGURE 4

$$\partial n(t, a)/\partial t + v\partial n(t, a)/\partial a = -\mu(t, a)n(t, a), \qquad (2\text{-}21a)$$

$$\frac{n(t+\tau, a) - n(t, a)}{\tau} + v\frac{n(t, a) - n(t, a-\alpha)}{\alpha} = -\mu(t, a-\alpha)n(t, a-\alpha). \qquad (2\text{-}21b)$$

In (2-21) we have included an aging velocity, v, so as to include the general case where the age step α is different from the time step τ. This occurs frequently in the entomological literature where insect development rates are nearly proportional to ambient temperature and so it is necessary to use "physiological age" which may proceed at a different rate than chronological age, i.e., $da/dt = v[T(t)]$. Equation (2-21b) may be rewritten as

$$n(t+\tau, a) = n(t, a)\left[1 - (\tau/\alpha)v\right] + n(t, a-\alpha)\left[(\tau/\alpha)v - \mu(t, a-\alpha)\right]. \qquad (2\text{-}22)$$

If $\tau = \alpha \stackrel{\Delta}{=} 1$, $v = 1$, the difference equation (2-22) can be written:

$$n_{t+1}(a) = (1 - \mu(t, a-1))n(t, a-1). \tag{2-23}$$

If we define a column vector, $\mathbf{n}_t(a)$, whose elements are the number of individuals in age class a, we can write (2-23) as a matrix equation

$$\mathbf{n}_{t+1} = \mathbf{S}\mathbf{n}_t, \qquad S_{i,i+1} = 1 - \mu(t, a-1).$$

The matrix $\mathbf{S}$ has only subdiagonal elements consisting of the fraction of individuals in age class $a - 1$ surviving to age a. (Clearly the subdiagonal entries in $\mathbf{S}$ just multiply the kth element in $\mathbf{n}_t$ and promote it to the $(k+1)$st entry in $\mathbf{n}_{t+1}$.) If $v \neq 1$ and/or $\tau \neq \alpha$ there would be diagonal entries as well.

From the boundary condition we can add a top row to account for births into the first age class:

$$n(t, 0) = \int b(t, a)n(t, a)\, da \to \sum_{a=1}^{A} \beta(t-1, a)n(t-1, a), \tag{2-24}$$

so that the matrix operator corresponding to the von Foerster model is:

$$\mathbf{L} = \begin{bmatrix} \beta_0 & \beta_1 & \beta_2 & \cdots & \beta_A \\ s_0 & 0 & 0 & & \\ 0 & s_1 & 0 & 0 & \\ & & \ddots & & \\ & 0 & & \ddots & \\ & & & s_{A-1} & 0 \end{bmatrix}. \tag{2-25}$$

The matrix $\mathbf{L}$ is called the Leslie matrix; we see that it is the finite difference version of the von Foerster equation. We can project out the orbit of $\mathbf{n}_t \in \mathbf{R}^A$–which represents the histogram of the age profile–by simply iterating $\mathbf{L}$:

$$\mathbf{n}_t = \mathbf{L}^t\mathbf{n}_0. \tag{2-26}$$

2.17. Most of the properties of the continuous model carry over to the discrete version. The analysis of the operator $\mathbf{L}$: $\mathbf{R}^A \to \mathbf{R}^A$ depends on the fact that all of its elements are nonnegative: $S_{i,i+1} \geqslant 0$, $\beta_{1i} \geqslant 0$, and that every iterate $\mathbf{L}^t$ has only nonnegative entries. Thus $\mathbf{L}$ falls within the purview of the Perron-Frobenius theorem for nonnegative matrices (Seneta [**1973**], Parlett [**1970**]). The consequences of this theorem are two-fold:

(a) $\mathbf{L}$ has only one eigenvalue, λ_0, such that its eigenvector has all positive components (the only one, therefore, meaningful as a population vector). Moreover, λ_0 is larger in amplitude than all the other eigenvalues. Clearly, this eigenvalue gives the same asymptotic geometric growth rate as the dominant root of equation (2-17) did for the Lotka model: $\mathbf{n}_t = \mathbf{L}^t\mathbf{n}_0 = \lambda_0^t(\mathbf{y}^T\mathbf{n}_0)\mathbf{x} + \mathcal{O}(\lambda_0^t)$, where $\mathbf{x}$ and $\mathbf{y}$ are the right and left eigenvectors, respectively, of $\mathbf{L}$ ($\mathbf{Lx} = \lambda\mathbf{x}$, $\mathbf{y}^T\mathbf{L} = r\mathbf{y}$).

(b) The rows of $\mathbf{L}^t$ tend to proportionality as $t \to \infty$, i.e., there is a stable age distribution:

$$\lim_{t\to\infty} \frac{\mathbf{x}^T\mathbf{L}^t}{\mathbf{x}^T\mathbf{L}^t\mathbf{1}} = \mathbf{y}^T.$$

The right and left eigenvectors of $\mathbf{L}$ can be written explicitly:

$$x_j = \prod_{k=0}^{j-1} S_k \cdot \lambda_0^{-j}, \qquad y_j = \sum_{n=j}^{k} \left[\prod_{k=j}^{n-1} S_k \right] \beta_n \lambda_0^{j-n-1}.$$

The expression for the left eigenvector has an interesting biological interpretation; it can be more easily appreciated in the continuous version:

$$\frac{v(a)}{v(0)} = \int_a^\infty e^{-r\tau} l(\tau) b(\tau)\, d\tau / e^{-ra} l(a).$$

$v(a)$ is called the "reproductive value" of a female of age a (relative to that at age 0). $v(a)$ measures the expected number of future offspring borne by a female from age a onward. $v(0)$ is just the net reproductive rate (equation (2-20)). $v(a)$ is an important functional of the population from an evolutionary standpoint when studying the optimization of reproductive strategies by natural selection (Pianka [**1974**]).

3. An application of the balance equations: Nicholson's blowflies.

3.1. In this section we shall use the population balance equations to construct a model of a laboratory "ecosystem". The predictions of our model will fall into two categories. On the one hand, by employing all of the available experimental information we shall be able to achieve a reasonable quantitative agreement between the model and the data. However, ecological experiments can rarely (if ever) supply data of sufficient resolution to completely characterize all of the constitutive relations required to implement the balance equations. We will inevitably be left with some degrees of freedom with which to empirically "fit" the model to the data. Therefore, we must also look for qualitative features of the data which, if they turn up in the model, may lead to an understanding of their underlying mechanism. Thus our modelling strategy will be as follows. First we will construct a detailed model which includes all of the biological features that we imagine play a role in the dynamics. Such a model will necessarily be too complex to do much with except simulate on a computer. We do this only to satisfy ourselves that we have gotten the biology right; our only defense against our model becoming a "high powered curve fit" is to adhere to the data as closely as possible. We cannot afford to investigate the model's characteristics in every region of the parameter space; only the behavior in the "biological region" need concern us. The next step is to simplify the model in several directions; the aim of each simplication is to elucidate the mathematical mechanism underlying a particular qualitative feature of the larger model.

3.2. The Australian entomologist A. J. Nicholson maintained a population of sheep blowflies under carefully controlled conditions for several years. Despite the formidable experimental difficulties such an experiment entailed he was able to maintain an accurate census of the various life stages of the flies. One section of his data is shown in Figure 5 (Nicholson [**1957**]), where a plot of the number of adult flies over a 700-day period is presented. In this experiment the flies were maintained under constant conditions of temperature and food supply. The major feature of the data is the regular periodic oscillations of approximately 35–40 days which persist for more than 400 days. In addition to this basic limit cycle there are at least three other phenomena discernable in Figure 5. (i) The population peaks appear to have a regular "fine structure" consisting of a double or occasional triple peak. (ii) There appears to be a "subharmonic" superimposed on and spanning three periods of the basic limit cycle. (iii) After about 425 days (10–11 cycles) the cycles become more irregular, appearing almost chaotic till near the 600th day whereupon there is a large excursion; thereafter the basic limit cycle appears to reassert itself till the end of the data record. (Below the experimental curve we have included a Fourier spectrum of the data for those readers unwilling to trust their eyes.)

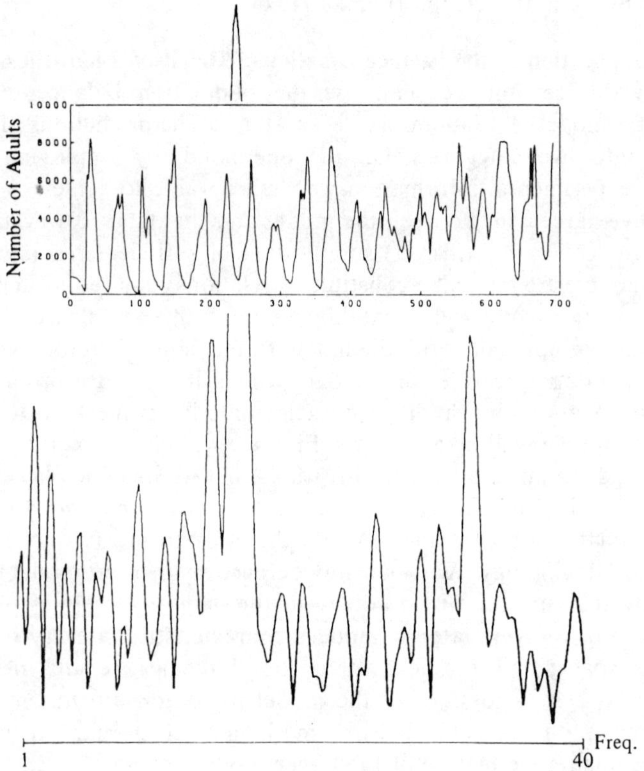

FIGURE 5[4]

[4]Inset adapted from a figure appearing in an article by A. J. Nicholson, Symposium on Quantitative Biology (1957), Vol. 22, pp. 153–173, with permission of Cold Spring Harbor Laboratory.

Thus there are four phenomena occurring on four disparate time scales to be explained by our models. The biological explanation for the major limit cycle is easy to express in words. It turns out that the capacity of the adult flies to manufacture and lay eggs is strongly dependent on their ability to procure sufficient protein. Nicholson deliberately maintained the experimental conditions such that there was insufficient protein to supply all of the adult flies at the population peaks. Thus there was strong competition for protein and consequently few eggs were laid at the peaks. The subsequent generation therefore was much smaller and thus could obtain enough protein to realize their maximum fecundity. The histogram superimposed on the data record is a daily egg count; it is easy to see the inverse relation between the population size and the number of eggs laid. Consequently, there is an alternation of large and small generations which produces a population cycle whose duration is about two "generation times" (if we define a generation time to be the average age of reproduction: $\bar{a} = (1/N)\int_0^\infty an(t, a)\, da$).

The explanations for the other three phenomena are less obvious and we defer their discussion until the model has been constructed.

3.3. We commence by writing a balance equation for the density of flies, $n(t, a, x)$. From the discussion above we can see what the phenotypic variable must be: x is a measure of the amount of protein present in a fly which is available for manufacturing eggs. (Recall that x is to be interpreted as an average for all individuals in a cohort.) Thus we can write

$$\frac{\partial n}{\partial t} + \frac{\partial n}{\partial a} + \frac{\partial}{\partial x}\left[g(\cdot)n\right] = -\mu(\cdot)n, \tag{3-1a}$$

$$n(0, a, x) = E_0\delta(a)\delta(x), \text{ i.e., start with } E_0 \text{ eggs}, \tag{3-1b}$$

$$dx/dt = g(t, a, x, A(t)), \tag{3-2}$$

where

$$A(t) = \int_0^\infty \int_\alpha^\infty n(t, a, x)\, da\, dx \tag{3-3}$$

is the total number of adults (α is the age of eclosion). The birthrate boundary condition is, as usual:

$$n(t, 0, x) = \int_0^\infty \int_\alpha^\infty b(t, a, x', x)n(t, a, x')\, dx'. \tag{3-4}$$

The growthrate, $g(\cdot)$, is the difference between consumption and metabolism of protein

$$g(\cdot) = C(\cdot) - M(\cdot). \tag{3-5}$$

Finally we must include the material balance for the food (protein) supplied by the experimenter.

$$\begin{aligned}\frac{df}{dt} &= \text{supply} - \text{consumption}\\ &= u(t) - C(\cdot), \qquad f(0) = f_0.\end{aligned} \tag{3-6}$$

In the experiment shown in Figure 5 the protein supplied to the adults, $u(t)$, was constant.[5] In another series of experiments Nicholson imposed a periodic food supply. However, to analyze these experiments it is first necessary to understand the autonomous system. The structure of the model is shown schematically in Figure 6.

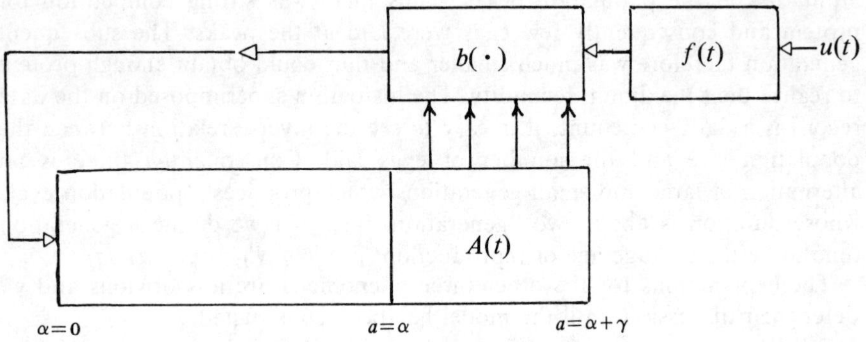

FIGURE 6

3.4. The balance equations (3-1)–(3-6) contain the following constitutive functions which must be determined experimentally or obtained from a submodel.

(1) *Deathrate*, $\mu(t, a)$. In these experiments the preadult mortality was negligible; the adult deathrate appeared to be approximately constant with age (i.e., a Poisson process with exponential survivorship).

Therefore:

$$\mu(a) = \mu_0 - H(a - \alpha) \tag{3-7}$$

where $H(\cdot)$ is a step function at the age of adult emergence, α.

(2) *Growthrate*, $g(\cdot) = C(\cdot) - M(\cdot)$. We shall assume (for want of better information) that the rate of protein metabolism per individual is a first order process:

$$M/A = kx. \tag{3-8}$$

The rate constant k can be measured experimentally. In general, we expect that the per capita consumption rate should be a saturating function of the food available, as shown in Figure 7a.

A function of the following form has been selected:

$$C(f)/A = D(1 - e^{-f/D}) \tag{3-9a}$$

where

$$D(t) = \int\int \mathfrak{D}(a)n\, da\, dx \tag{3-9b}$$

[5]While protein was necessary for reproduction, both adult and immature stages survived quite well on sugar-water with which they were abundantly supplied.

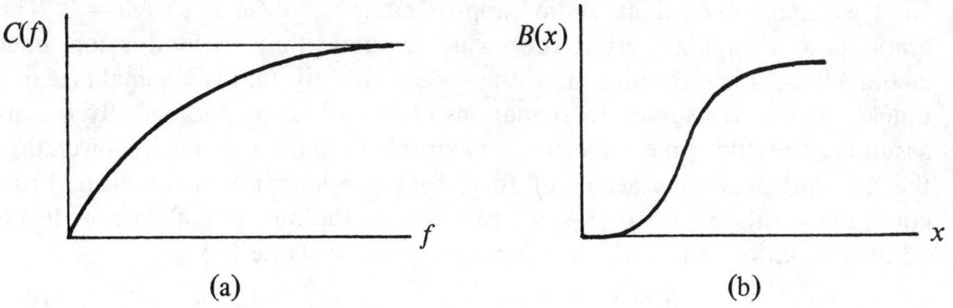

FIGURE 7

is the "demand" per individual, i.e. the amount a fly would consume if given access to unlimited protein.

(3) *Birthrate*, $b(\cdot)$. The dependence of fecundity on protein is of the general form shown in Figure 7b. Therefore, we take as the birthrate:[6]

$$b(a, x) = b_{\max} H(a - \alpha) \beta(x). \tag{3-10}$$

Given the parameters in the above constitutive functions, and the initial conditions, the model is completely specified. A full account of the numerical simulations is presented elsewhere (Oster, Auslander and Allen [**1975**]). For our purposes here it suffices to present a short segment of the results (Figure 8) to demonstrate the kind of accuracy one can expect in fitting the frequency and wavelength of the major limit cycle. Of more concern to us will be the various approximations to the complete model and the conclusions to be drawn from them.

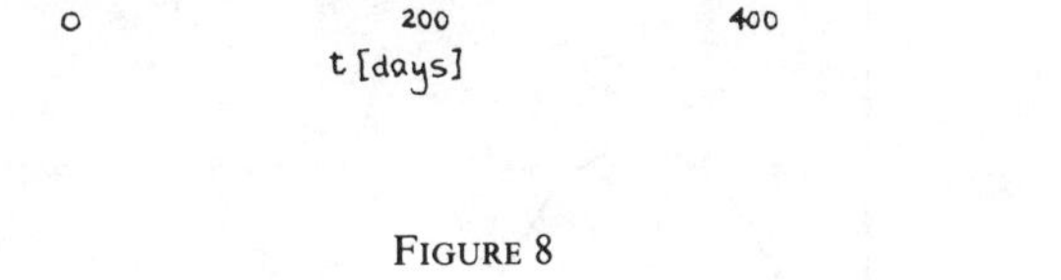

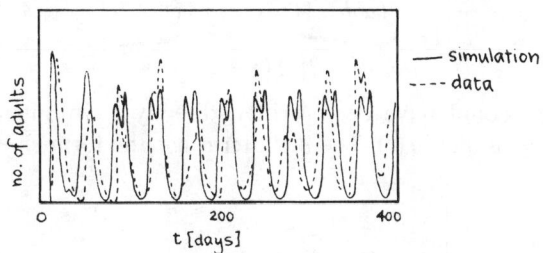

FIGURE 8

4. Approximations.

4.1. Since the experimental record consists of a running census of the adult blowfly population it seems reasonable to see how far we can get by reducing the model to an equation for the adult flies only. We can do this in two steps.

[6] There is a problem concerning the nutritional state of the eggs; the fecundity of offspring can be correlated with that of the mother. Moreover, new adult flies can occasionally lay eggs by using protein stores accumulated earlier. We shall neglect these effects here.

First we make the "quasi-static" approximations: $df/dt = dx/dt = 0$. This amounts to a singular perturbation which assumes that: (i) food is consumed as quickly as it is introduced into the system, and (ii) that is is metabolized as quickly as it is consumed (cf. equations (3-5) and (3-6)). Biologically, we are assuming that the time scale for demographic changes is much slower than the physiological time scales of food consumption and metabolism. From equations (3-5), (3-6) and (3-8) we can express the nutritional state, x, to the adult population, $A(t)$, which compete for the available food.

$$u = C = M = kxA, \qquad x = u/kA. \tag{4-1}$$

Second, we average out the dependence on age and nutritional state by integrating over a and x. The resulting equation is:

$$\frac{dA}{dt} = n(t, \alpha) - \int_\alpha^\infty \mu n \, da = B(t - \alpha) l(\alpha) - \mu_0 A. \tag{4-2}$$

From the boundary condition (3-4),

$$B(t - \alpha) = \int b[a, A(t - \alpha)] n(t - \alpha, a) \, da.$$

Defining

$$\bar{b}(A) \equiv \int b[a, A] n \, da / \int n \, da \tag{4-3}$$

the rate of egg laying is

$$B(t - \alpha) = \bar{b}[A(t - \alpha)] A(t - \alpha). \tag{4-4}$$

Thus the equation of motion for the adult population is:

$$\frac{dA(t)}{dt} = \underbrace{l(\alpha)\bar{b}[A(t - \alpha)]A(t - \alpha)}_{B[A(t - \tau)]} - \mu_0 A(t). \tag{4-5}$$

This equation could have been written directly from Figure 6 by doing a balance on the adults, $A(t)$. The birth term has the form shown in Figure 9a:

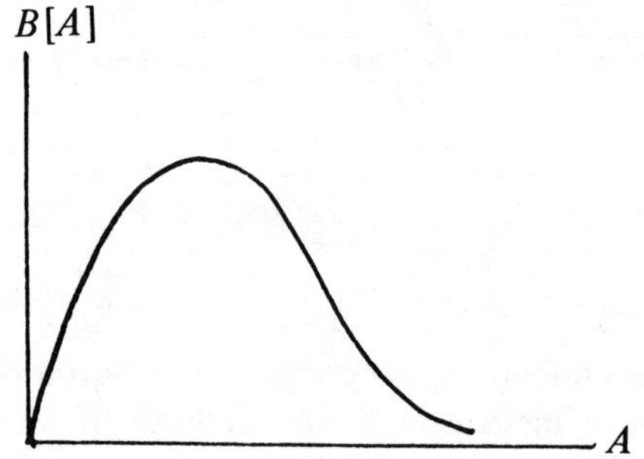

FIGURE 9a

4.2. There are a number of ways of investigating the consequences of equation (4-5). The most obvious is numerical simulation, and so in Figure 9b a plot of equation (4-5) is shown. Two features are plain: (1) the basic limit

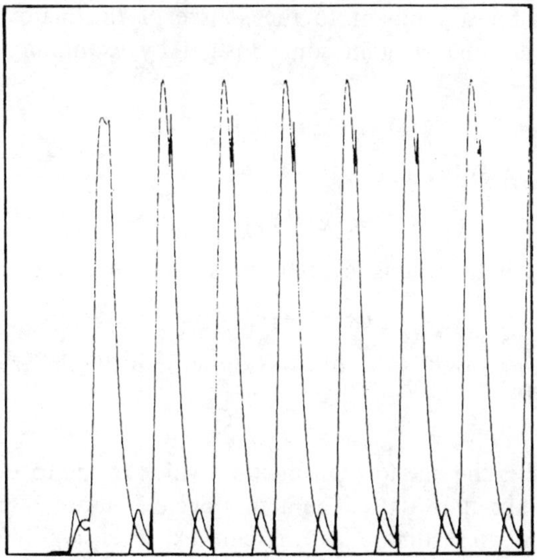

FIGURE 9b

cycle has survived the approximations, and (2) there is a clearly discernible high frequency component, manifested as a "double peak".

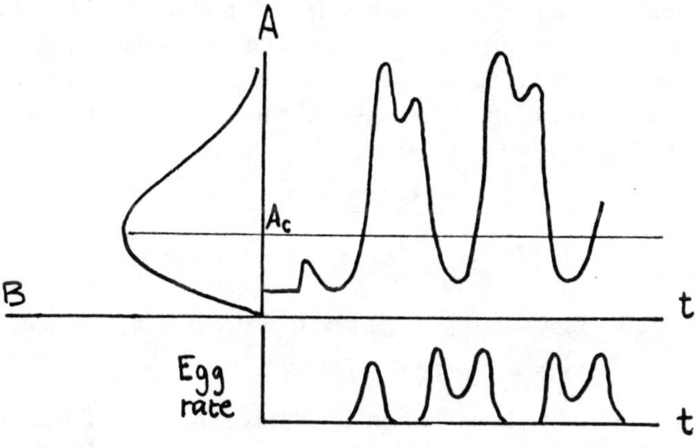

FIGURE 10

If we accept for the moment the existence of the major limit cycle it is easy to see how this double peak arises. In Figure 10 we have plotted the egg production term (Figure 9a) on the vertical axis along side of the limit cycle. The important feature of the birthrate curve is that it has a critical point, $dB[A_c]/dA = 0$. Any such function acts as a "frequency multiplier": if a periodic signal, $A(t)$, whose amplitude is greater than A_c is fed into $B[\cdot]$ the output has twice the frequency. Thus, if the amplitude of the limit cycle is greater than A_c the egg production curve will be double-peaked as shown below the limit cycle. Therefore, the succeeding peak in adult numbers will also be double-peaked providing that the preadult mortality is sufficient.

4.3. Returning for a moment to the source of the major limit cycle in equation (4-5) we can also gain some insight by examining the linearized version.

$$\dot{x}(t) = \kappa x(t - \tau) - \delta x(t). \tag{4-6}$$

Letting $y = xe^{\delta t}$, (4-6) becomes

$$\dot{y} = (\kappa e^{d\tau})y(t - \tau). \tag{4-7}$$

Substituting $e^{\omega t}$ the characteristic equation for the linearized frequencies is

$$\omega\tau = -(\tau\kappa e^{\delta\tau}). \tag{4-8}$$

A plot of the left and right sides of (4-8) shows that the linear estimate of the limit cycle period is

$$\tau < T < 2\tau \tag{4-9}$$

where τ we recall is the age to reproduction which in the case of Nicholson's experiments was about 18 days. The actual cycle is about 38 days long. This is about as close an estimate as we can expect using linear analysis–a nonlinear analysis we will get to presently will do better.

4.4. REMARK. A possible source of "subharmonics" superimposed on the limit cycle is the "overlapping generations effect". Referring to Figure 6 we see that if the width of the reproductive window, γ, is larger than the age to reproduction, α, then it is possible for a female to live to see her grandchildren, i.e., a mother and her daughter reproducing simultaneously. That this can introduce harmonics onto the limit cycle can be seen by examining the linearization of the original model. Letting $n(t, a) = \bar{n}(a) + x(t, a)$, where $\|x\| \ll 1$, we can write

$$n_t + n_a = -\mu_0 n, \tag{4-10a}$$

$$n(t, 0) = \int_{\alpha}^{\alpha+\gamma} b(x)n\,da \approx c_1 x + bA, \tag{4-10b}$$

$$0 = \dot{f} = u - c(f)A \approx u - c_2 f - c_3 A, \tag{4-10c}$$

$$0 = \dot{x} = c(f) - kx \approx c_2 f - kx, \tag{4-10d}$$

where the c_i's and b are linearization constants. To obtain the characteristic equation for this system we take the Laplace transform of (4-10a) substitute from (4-10b):

$$\tilde{n}_a = -(\mu + s)\tilde{n}, \qquad \tilde{n}(s, a) = e^{-(\mu+s)a}[c_1 x + b\tilde{A}].$$

Using (4-10c, d), x may be eliminated and both sides integrated over $[\alpha, \alpha + \gamma]$ to yield an equation in $\tilde{A}(s)$ only of the form $\tilde{A}(s) = N(s)/D(s)$ where

$$D(s) = \mu + s + c_1 e^{-(\mu+s)\alpha}(1 - e^{-(\mu+s)\gamma}). \tag{4-11}$$

The roots of $D(s) = 0$ yield the resonant frequencies of the system. In particular, if we look for the roots on the imaginary axis, $s = i\omega$, the characteristic equation takes the form

$$K_1 \sin \omega\alpha + K_2\omega \cos \omega\alpha = \sin \omega\gamma. \tag{4-12}$$

An investigation of equation (4-12) shows that the number of roots changes suddenly near $\gamma = \alpha$ indicating a change in the frequency spectrum.[7]

4.5. EXERCISE. For the nonlinear model (4-5) to possess a limit cycle the equilibrium $\bar{A}$ at $\bar{b}[\bar{A}] = \mu_0/l$ must be unstable. Show that the stability conditions for equation (4-6) are (i) $\delta\tau > 1$, and (ii) $\delta\tau > \kappa\tau > -[\rho^2 + \delta^2\tau^2]^{1/2}$, where ρ is the root of $\rho = \delta\tau \tan \rho$.

4.6. EXERCISE. Linearize the functional equation $\dot{x}(t) = bx(t-\tau)^2 - \delta x(t)$ about its equilibria. Then expand the delay about t to second order, thus obtaining a harmonic oscillator equation, and estimate the frequency of the oscillation.

4.7. REMARK. A convenient "rule of thumb" for estimating the stability of a delay differential equation is: the system is likely to be (linearly) *un*stable if the delay, τ, is greater than the "natural response time" of the system (e.g., $r\tau \gtrsim \mathcal{O}(1)$ where r is the exponent in the linearized solution–or the slowest eigenvalue in the vector case).

4.8. EXERCISE. A better model than the discrete delay equation (4-5) might be a distributed delay, e.g.,

$$\dot{A}(t) = rA(t)\left[1 - \int_{-\infty}^{t} A(t')Q(t-t')\,dt'\right]$$

where $Q(\cdot)$ weights the past history of the population. (i) If the recent past is most important (e.g., starvation) then the weighting kernel might be

$$Q(t) = \begin{cases} \tau^{-1}e^{(t-t')/\tau}, & t' < t, \\ 0, & t' > t \end{cases}.$$

Show that the condition for the onset of oscillations is $r\tau > \frac{1}{4}$. (Hint. Linearize and differentiate to obtain the harmonic oscillator equation.)

(ii) If the distant past is most important,

$$Q(t) = (1/K\tau)t/\tau^{-t/\tau},$$

show that the equilibrium $A = K$ is locally stable if $r\tau < 2$.

(iii) If $Q(t)$ is a delta function at $t/\tau = 1$, show that the stability condition becomes $r\tau < \pi/2$. Compare these results with the "rule of thumb" given above (cf. R. May [**1973**]).

4.9. REMARK. There is an interesting analogy between the limit cycle oscillations in Nicholson's blowflies and a phenomenon in human population dynamics called the "Easterlin effect" (Frauenthal [**1975**]). Individuals born in a large cohort will grow up under conditions of severe economic competition and will consequently tend to have fewer children. Conversely, small cohorts can afford to have more children than the average. Economic competition thus mimics the competition for food amongst the blowflies. A simple model of this situation is:

[7] For investigating the frequency response of delay systems a root locus plotting program for exponential polynomials is very handy (Krall [**1970**], Oster and Takahashi [**1974**]).

$$n_t + n_a = -\mu n, \qquad B(t) = \int_\alpha^\beta b(t,a)n(t,a)\,da,$$

where the birthrate kernel is given by

$$b(t,a) = m(a)[1 + \lambda(B_0 - B(t-a))/(B(t-a))].$$

That is, fertility is proportional to cohort size; B_0 is the birthrate corresponding to a stable age distribution and λ measures the strength of the competition. Substituting an exponential solution into the linear model yields the characteristic equation:

$$\int_0^\infty e^{-rx}m(x)e^{-\mu x}\,dx = \frac{1}{1-\lambda}, \qquad \lambda \neq 1.$$

Letting $r = i\omega$, multiplying the characteristic equation by $e^{-i\omega\bar{x}}$ and expanding about $\bar{x} = \int xl(x)m(x)\,dx$ (= average age of reproduction yields) for the period of oscillation $T \approx 2\bar{x}$, which is close to the cycle length in Nicholson's data.

4.10. Now is the time to introduce another biological fact about the blowflies. So far, we have assumed that each female (or each cohort of females) lays eggs continuously throughout her reproductive period. However, it turns out that a biological clock plays a crucial role in their reproductive physiology.

Females do not lay eggs continuously but do so in "bursts" every few days, depending on their protein consumption. Moreover, the egg-laying takes place within a few hours during the morning. Therefore, we are certainly justified in approximating the birthrate kernel by a string of delta functions:

$$b(a,A) = \sum_{i=1}^{K} b_i(A)\delta(a - \alpha_i). \tag{4-13}$$

The $b_i(A)$ are monotonically decreasing as before and the spacing of the reproductive bursts can depend on the food ingested: $\delta(a - \alpha_i(x))$. Here we shall assume they are fixed. Note that while each cohort reproduces discontinuously, the population can reproduce continuously if all reproductive age classes are represented.

The equations of motion now become:

$$\partial n/\partial t + \partial n/\partial a = -\mu n, \tag{4-14}$$

$$\begin{aligned} n(t,0) &= \int \sum b_i(A)\delta(a-\alpha_i)n(t,a)\,da \\ &= \sum b_i(A)n(t,\alpha_i), \end{aligned} \tag{4-15}$$

$$A(t) = \int_0^\infty n(t,a)\,da. \tag{4-16}$$

So the model reduces to the following coupled difference-integral equations (Figure 11).

$$B(t) = \sum b_i(A) l(\alpha_i) B(t - \alpha_i), \tag{4-17}$$

$$A(t) = \int_0^\infty l(a) B(t - a)\, da. \tag{4-18}$$

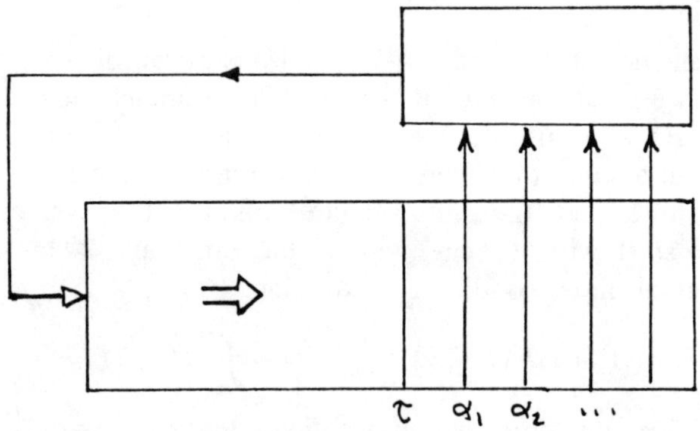

FIGURE 11

This is equivalent to a "density dependent" Leslie model

$$\mathbf{n}_{t+1} = \mathbf{L}(\mathbf{n})\mathbf{n}_t \tag{4-19}$$

where the top row depends on the population vector $\mathbf{n}_t$ (i.e., (4-17) is the equation for the top row of $\mathbf{L}(\mathbf{n})$ and $\mathbf{A} = \sum_{k=\tau}^{M} n_k$).

To get an idea of how this new fact affects the model behavior a simulation of equations (4-17), (4-18) is shown in Figure 12 together with the experimental data (one-half day age classes were employed in the simulation shown). The fundamental limit cycle is still evident. In addition, two other features are prominent: (i) there is more "fine structure" on the major peaks than before (which might at first glance be taken for "noise", but which we see is quite deterministic). (ii) There is clearly evident a long term "subharmonic" which appears as the envelope of alternate peak heights. It is remarkable that the individual egg-laying "clocks"–a phenomenon occurring on the physiological time scale–should manifest itself at the population level on a time scale much longer than the lifetime of any individual. (In the simulation shown in Figure 12 the width of the reproductive window, γ, was slightly greater than twice the age to reproduction, α. In the light of the discussion in §4.4, we see how this long term "beat" might arise.)

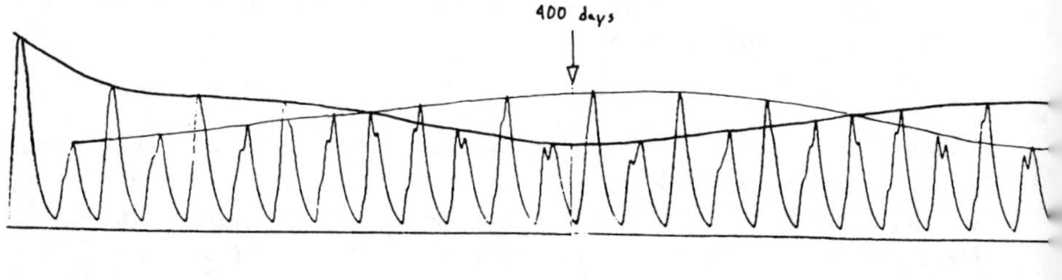

FIGURE 12

This is an intrinsically nonlinear phenomena, and not amenable to analysis by linearization. Before we can understand how this comes about we must make a short excursion into some poorly charted waters in modern dynamical systems theory. We shall do this by making one final approximation to our model.

4.11. The behavior of the model (4-17), (4-18) is extraordinarily complex; to understand it we must understand the orbits of nonlinear maps on $\mathbf{R}^N$. To gain some insight we shall examine the special case of a 1-dimensional map. That is, we shall assume that there is but one reproductive burst: $i = 1$, and moreover, assume that the initial conditions for the system are $B(0) = E_0\delta(t)$, i.e. start with E_0 eggs, all of the same age. (This is close to Nicholson's actual initial conditions.) Equations (4-17), (4-18) then reduce to

$$B(t) = \bar{b}(A)l(\alpha)B(t-\alpha), \qquad A(t) = \int_\alpha^\infty l(a)B(t-a)\,da,$$

or, taking $\alpha \equiv 1$ as our time step, and noticing that there is only one cohort moving through the age structure:

$$B_{t+1} = \phi(B_t)B_t = F(B_t) \tag{4-20}$$

where $\phi(B_t) = \phi\{\int_\alpha^\infty l(a)B(t-a)\,da\}$ is a monotone decreasing function of B_t. Equation (4-20) is a nonlinear first order difference equation; in Figure 13 we have plotted the graph of the map F: $\mathbf{R} \to \mathbf{R}$

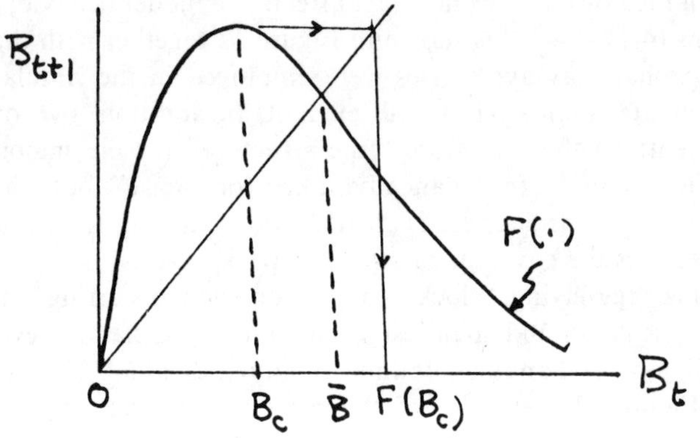

FIGURE 13

Depending on the shape of $\beta(x)$ in Figure 7 the map $F(\cdot)$ can be fitted by a number of functional forms, e.g.

$$F_1(x) = xe^{a(1-x)}, \quad F_2(x) = ax(1-x), \quad F_3(x) = ax/(1+x^b)\ldots$$

(May and Oster [**1975**]). The generic features of these maps are: (1) If $x_t = 0$, then $x_{t'} = 0\ \forall t' > t$. (2) They possess a single critical point, $F'(x_c) = 0$, i.e., "density dependence". (3) They have a single fixed point in the interior of the interval $(0, F(x_c))$ where $F(\cdot)$ intersects the identity map $(1x = x)$: $F(\bar{x}) = \bar{x}$.

It is easy to project out the trajectory of the map $F(\cdot)$ by simply alternating between the graph and the 45 degree line as shown in Figure 14a. The fixed point, $\bar{x}$, is locally attracting (stable) if the eigenvalue of $F'(\bar{x})$ (i.e., the slope of $F(\cdot)$ at $\bar{x}$) has absolute value less than unity (cf. Figure 14).

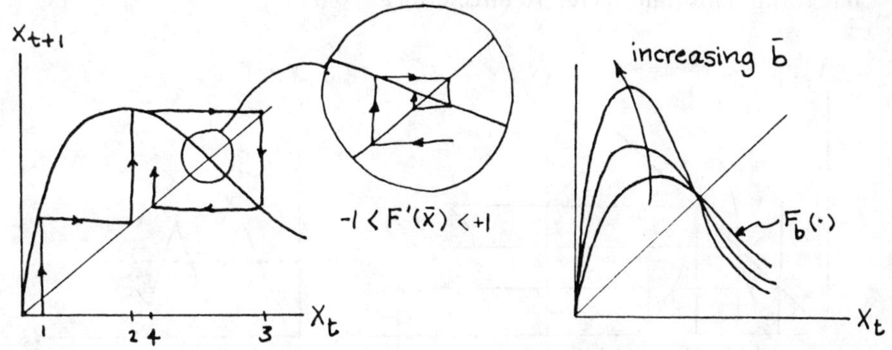

(a)

FIGURE 14

4.12. REMARK. It is worth bearing in mind that there is no simple relationship between a particular difference equation and a differential equation. That is, there is no unique way to "pass to the limit". Indeed, if the difference equation $x_{t+1} = F(x_t)$ has a critical point, $F'(x_c) = 0$, then there is no (explicit) "corresponding" differential equation, $\dot{x} = f(x)$. (There may be an implicit equation, $g(x, \dot{x}) = 0$.) This is easy to see since the approach to equilibrium is oscillatory in the former case (cf. Figure 14) but monotone for the latter. Indeed, difference equations such as (4-20) are more like delay equations than ordinary differential equations as the following exercise demonstrates.

4.13. EXERCISE. (a) Show that the differential equation "corresponding" to the difference equation $x_{t+1} = x_t(\lambda x_t^{-b})$ is the Gompertz equation $\dot{x} = x(-\beta \ln x/K)$. (b) Show that the Gompertz equation with a delay in the log term is locally stable if $0 < \beta\tau < \pi/2$ (cf. §4.7). (c) We know that the Malthus equation $\dot{x} = rx$ has only monotone solutions while the delayed Malthus model $\dot{x} = rx(t-\tau)$ has an infinite number of frequencies (roots to the characteristic equation $i\omega = re^{-\omega\tau}$). Can you give a (geometrical) argument to show that a time dependent O.D.E. $\dot{x} = f(x, t)$ cannot exhibit subharmonics while the corresponding delay differential equation can? (Hint. The phase space is the cylinder and a subharmonic wraps at least twice around.) First order difference equations can have subharmonics.

4.14. Let us turn now to the question of what happens when the fixed point of the difference equation (4-20) becomes unstable. This comes about when the birthrate b increases so that the curve $F(\cdot)$ becomes more "nonlinear" as shown in Figure 14b. At some value of the parameter, b, the slope at the fixed point, $F'(\bar{x})$, becomes less than -1 and the trajectory begins to wind out from $\bar{x}$. Where does it go? To see where we need only examine the second iterate of the map, $F^2(\cdot) \stackrel{\Delta}{=} F \circ F(\cdot)$, obtained by plotting x_t vs. x_{t+2}. This composition can be performed graphically (how?) and the results are shown in Figure 15. Just as the map $F(\cdot)$ had one hump (one critical point), $F^2(\cdot)$ has two humps (and therefore three critical points).

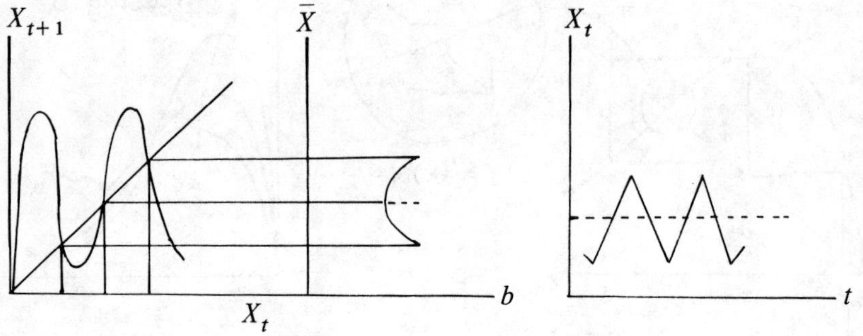

FIGURE 15

As the birthrate is increased the F^2 map eventually cuts the 45 degree line at three fixed points. This happens just at the point where $F'(\bar{x}) = -1$. Thus the fixed point $\bar{x}$ bifurcates into two new fixed points as shown on the branching diagram to the right in Figure 15. The trajectory then alternates between these new "period 2" points. This period 2 cycle is stable, and remains so as b is increased until the slope of $F^2(\cdot)$ at the period 2 points passes through -1, whereupon each point bifurcates into a "pitchfork" giving rise to a period 4 oscillation as the old fixed points become unstable and the trajectory alternates amongst the new period 4 points (cf. Figure 16).

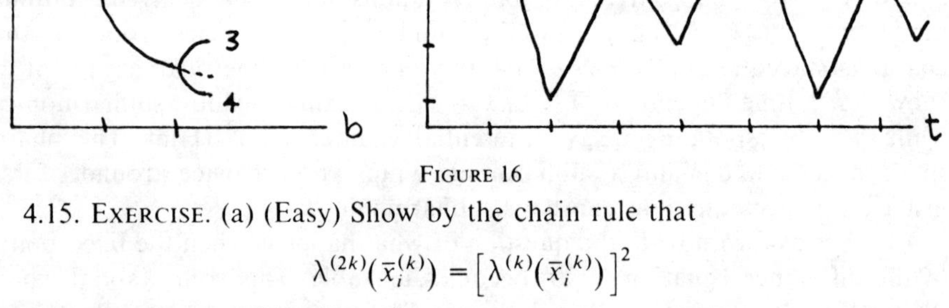

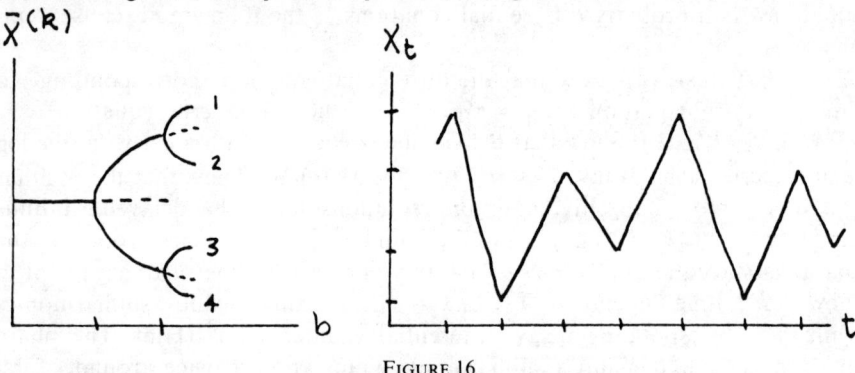

FIGURE 16

4.15. EXERCISE. (a) (Easy) Show by the chain rule that

$$\lambda^{(2k)}(\bar{x}_i^{(k)}) = \left[\lambda^{(k)}(\bar{x}_i^{(k)})\right]^2$$

where $\lambda^{(k)}(\bar{x}_i^{(k)}) = F^{(k)'}(\bar{x}_i^{(k)})$ is the eigenvalue at a period k point. (b) (Hard) Using this fact show the sequence of points followed by the stable orbit is 1-4-2-3 as shown in Figure 16.

4.16. This process of successive bifurcation continues as b is increased, each period k cycle branching to a period $2k$ cycle as $F^{(k)'}(x_i^{(k)}) < -1$. Thus cycles of arbitrary length 2^n are excited successively by increasing birthrates. However, as the order of the bifurcation increases, the distance between the branch points along the b axis decreases so that the domain of stability for higher cycles becomes progressively smaller. Heuristically, this is clear since every iterate of the map, $F(\cdot)$, doubles the number of humps and so crowds more fixed points into the same interval. Finally, at a critical value of b, a completely new phenomenon occurs. Some of the wrinkles in a high iterate of F which had not been deep enough to cross the 45 degree line now do so; the ensuing bifurcation creates a fixed point where previously there had been none. This new fixed point promptly splits into two branches as shown in Figure 17. These "tangent" bifurcations occur when $F^{(k)'}(\bar{x}_i^{(k)}) = +1$, and by the time they do occur, all of the 2^n cycles have become unstable.

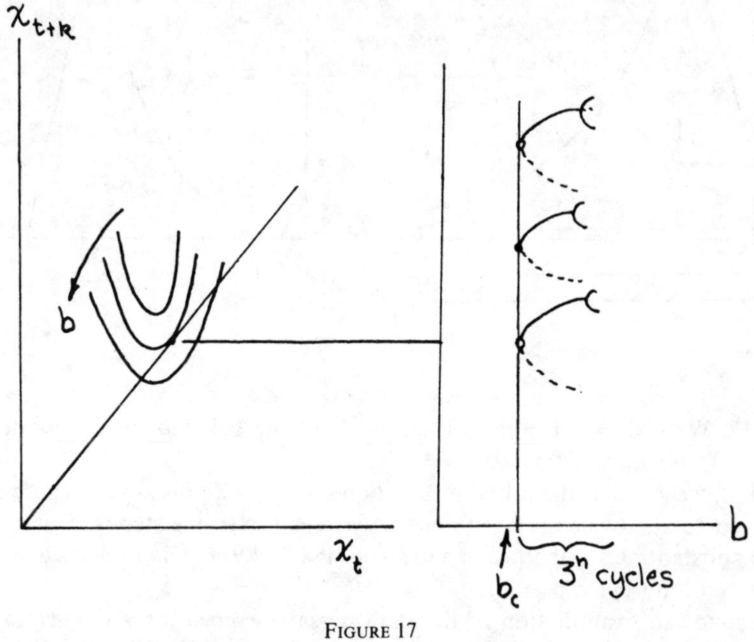

FIGURE 17

The remarkable fact is that once a 3-point cycle has appeared beyond the critical value of b_c then there coexist cycles of *arbitrary* length, although for maps on $\mathbf{R}^1$ such as ours with a single critical point there can be no more than one stable (nontrivial) cycle at a time. Furthermore, once the 3-point cycle has appeared there are initial conditions for which the orbit is completely *aperiodic*. A proof of these statements can be found in Guckenheimer and Oster [**1975**]. Fortunately, for 1-dimensional maps these aperiodic points have measure zero. But for higher-dimensional nonlinear Leslie models this need not be so: many stable periodic orbits can coexist and the "chaotic" attracting sets may be large. Moreover, in general, as the dimensionality of the system increases, the "amount" of nonlinearity required to produce a "chaotic" orbit is less. On $\mathbf{R}^1$ we require that $F(\cdot)$ have a critical point; this is not necessary in higher dimensions. We can see that the presence of a critical point in $F(\cdot)$ means that $F(\cdot)$ "folds" the interval $(0, F(x_c))$ onto itself (Figure 18). At the 3-point cycle this folding is sufficient so that there is an interval, ac, which completely covers itself under the action of $F(\cdot)$. Under iterates of $F(\cdot)$, there are subintervals of $(0c)$ which fold on themselves an infinite number of times, and thus, by the fixed point property, there are an infinite number of periodic points.

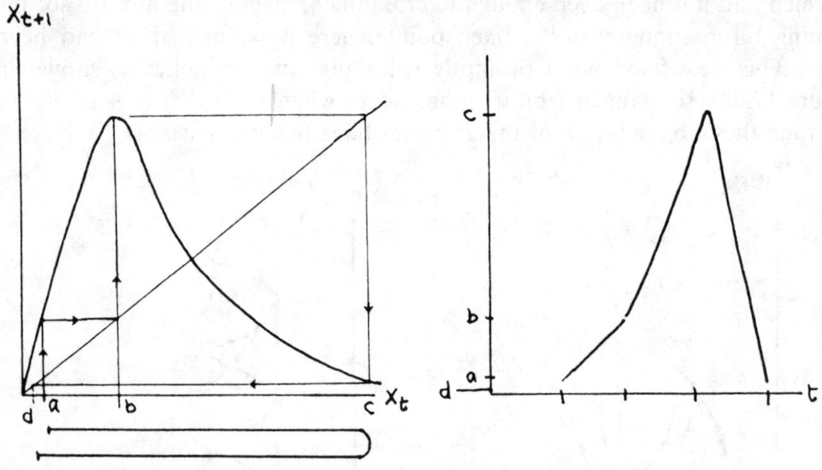

FIGURE 18

4.17. What does this analysis tell us about the behavior of our model as it relates to Nicholson's data?

First, we see that the initial bifurcation creates a 2-point cycle whose length is perforce 2α: twice the age to maturity (recall our time step is in units of α). This substantiates our linear analysis in §§4.3, 4.9 and is in accord with the observed limit cycle period.

Second, an examination of the data reveals a repeating sequence of three peak heights over the first 400 days (i.e., the first peak is the highest, thence a rising sequence of three, which repeats). The Fourier spectrum in Figure 5 does indeed show a component with a period of about three cycles. Thus we are led to suspect that the system's periodic behavior is a reflection of a higher order bifurcation than the first. It is important to note several things in drawing this conclusion:

(i) The difference equation (4-19) has many age classes while our analysis is based on equation (4-20) which has but one age class. The higher-dimensional system is capable of a much richer spectrum of periodic orbits. For example, comparing Figure 1 with the 1-dimensional model one might suspect that the system is in an 8-point cycle (after three bifurcations). However, using the results of §4.15 one can verify that the sequence of peak heights for the stable

8-point cycle is (numbering the branches from highest amplitude to lowest as in Figure 16): 1-8-4-5-2-7-3-6. (Sketch this for yourself and verify the pattern: "the higher the peak, the lower the crash".) This amplitude sequence is not the same as the 8-point cycle in the data. Equation (4-19), however, can produce such a cycle under multiple bifurcations.

(ii) As each successive bifurcation occurs a new component appears in the frequency spectrum. However, the old component does not disappear, but attenuates gradually. For example, the 4-point cycle in Figure 16 shows strong traces of the 2-point cycle (since the orbit is forced to alternate between the upper branch pair and the lowest). Thus while the data is dominated by the basic cycle one should expect to see lower frequencies superimposed indicating multiple bifurcations.

4.18. If the multiple bifurcation explanation of the major limit cycle and the superimposed subharmonics is true then we have in hand also a ready explantion for the apparent descent into chaos of the data after the 400th day. Recall that the domains of stability along the parameter axis of the cycles decreased as the cycle lengths increased. Thus, for higher order cycles small changes of the parameter can force the system through a large number of higher branches initiating, quite rapidly, cycles of much greater length, or even chaotic orbits. Moreover, upon closer examination the data between 400 and 700 days appears to retain many of the periodic features of the preceding cycles and is probably not yet into the chaotic regime.

4.19. What could cause the transition to higher order bifurcations–and the concommitant irregular appearing cycles? That is, how was the effective birthrate being increased as the experiment proceeded? The answer was provided by Nicholson himself and involves a characteristic of populations we have heretofore omitted from our model: genetics. He realized that his experiments placed a high premium on a fly's ability to lay eggs under conditions of protein starvation. Thus during the population peaks, when competition for protein is fiercest, those individuals capable of laying more eggs project their genes into the next generation more effectively. As these periodic bouts of intense protein competition are repeated there is a selection for individuals capable of maintaining their fecundity at low protein levels. Since the protein supply was maintained constant throughout the experiment the average birthrate increased slowly. Indeed, when Nicholson measured the protein dependence of fecundity for flies that had survived the course of the experiment against a group of control flies he found that the former group did much better. His experiments, as much as anything else, were demonstrations of the process of natural selection! Before showing how to include this phenomenon in our model there are some other aspects of the bifurcation process we should discuss.

4.20. From a practical standpoint it is probably irrelevant whether an orbit is aperiodic or is in a long stable cycle. Seldom will there be a data record of sufficient length or resolution to differentiate the two. This raises some unsettling issues about the usefulness of deterministic models in population ecology, a subject discussed more fully in May and Oster [**1975**] (where one can also find a more extensive account of the bifurcations of 1-dimensional difference equations). Let us only note in passing that we have in hand a possible explanation for a puzzling and ubiquitous phenomenon in the insect world. Frequently, an insect population will persist at quite low levels for many seasons punctuated by occasional outbreaks whereupon the population levels explode to levels four or five orders of magnitude higher than before. These epidemics may occur over spans of 11 or 12 years and appear uncorrelated with any known meteorological cycle. Simulations of equations (4-19) and (4-20) with high density dependence show this same phenomenon: long periods of near zero population interspersed with high amplitude excursions of short duration. The reason for this is clear from Figure 14. If $F(\cdot)$ is very "humped" so that the critical point lies close to the origin, then a small initial condition may take many iterates to work its way up to x_c. Near x_c, however, the trajectory rapidly rises to near its maximum amplitude, whereupon it crashes to near its minimum amplitude and the processes repeats.

4.21. Consider the difference equation generated by the triangle function

$$\begin{aligned} F(x) &= 2x \qquad (0 \leq x \leq \tfrac{1}{2}), \\ &= -2x \qquad (\tfrac{1}{2} \leq x \leq 1). \end{aligned}$$

This could be viewed as an approximation to the quadratic function $x(1 - x)$ analyzed by Li and Yorke [**1975**]. If we agree to count as a "0" every time the orbit lands in the left half of the unit interval and as a "1"

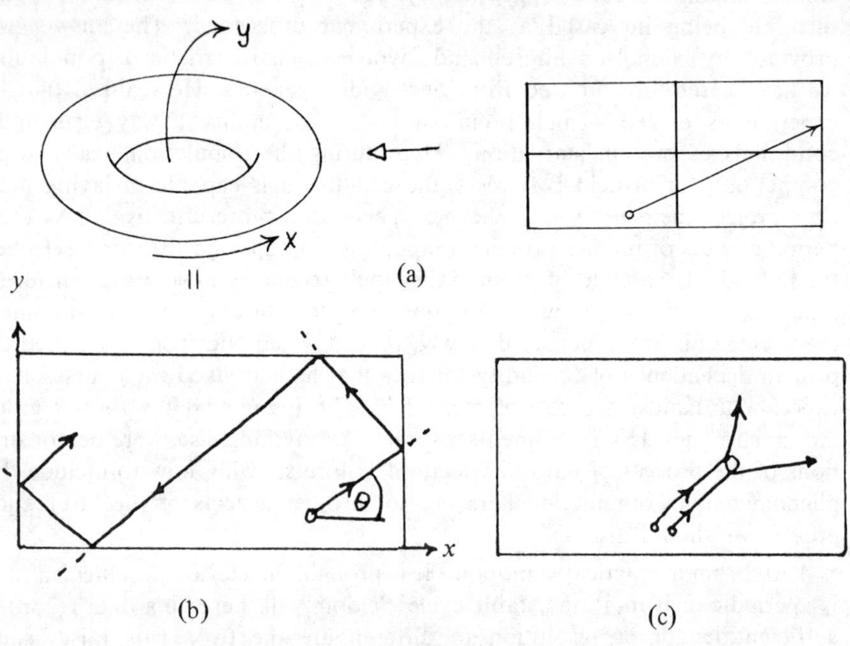

(a)

(b)

(c)

FIGURE 19

when it lands in the right half, then the trajectory generates a sequence of 0's and 1's. This sequence can be shown to be indistinguishable from a Bernoulli process (e.g., flips of a coin)! Thus the orbits of nonlinear difference equations can be as "random as possible". One is therefore left with the question: is this sort of behavior "typical" of nonlinear systems? And if so, can we hope to "understand" these complex behaviors? The answer to both is a complicated "yes".

4.22. A heuristic appreciation of how nonlinearities can generate chaotic motion can be gained by considering the trajectory of a ball on a pool table, Figure 19a.

By the symmetric nature of the reflections off the sides, it is clear that the trajectory of the ball is a flow on the torus, $T = S' \times S'$ (Figure 19a). If the initial angle, θ, is rational the ball will trace out a periodic trajectory on T; if θ is irrational a dense orbit is generated. In both cases, however, nearby initial conditions generate orbits which stay close. If, however, we place a small pin (or several) in the middle of the table then nearby trajectories which collide with the pin diverge far apart (Figure 19c).The high radius of curvature of the pin spreads the orbits so that very quickly it becomes impossible to predict the future trajectory. (The identical argument about molecular collisions is employed in statistical mechanics to justify the assumption of "molecular chaos".)

The large curvature around the critical point of the map $F(\cdot)$ in Figure 13 acts in the same way: nearby orbits diverge so that correlations between initial conditions are rapidly lost. As we have mentioned previously, in higher dimensions the "amount" of nonlinearity required for chaotic behavior is less.

4.23. The next question we must address is how to extend our conclusions about the dynamics of difference equations to the continuous time models we started with. The generalization is provided by the following theorem whose proof can be found in Marsden and McCracken [**1976**].

THE HOPF BIFURCATION THEOREM. *Given a vector field* $\mathbf{X}(\mathbf{x}, \mu)$ *on* $\mathbf{R}^2$ *depending on parameter,* μ, *we view the flow of* $\mathbf{X}(\cdot)$ *on* $\mathbf{R}^2 \times \mathbf{R}^1$ *as shown in Figure* 20. *Let the following conditions hold*:

1. $\mathbf{X}(\mathbf{0}\mu) = \mathbf{0}$, $\mathbf{x} = \mathbf{0}$ (*i.e., the* μ*-axis*) *is a fixed point for* $\mathbf{X}(\cdot, \mu)$ *and the origin is an attractor for* $\tilde{\mathbf{X}}(\cdot, \cdot)$ *for* $\mu \leqslant 0$.

2. *The Jacobian of* $\mathbf{X}(\cdot)$ *at* $(\mathbf{0}, 0)$, $D\mathbf{X}(\mathbf{0}, 0)$ *and has a conjugate pair of simple eigenvalues* $\lambda_1(\mu) \pm i\lambda_2(\mu)$ *such that*: (i) $\lambda_1(0) = 0$, $\lambda_2(0) \neq 0$; (ii) $\partial\lambda_1/\partial\mu|_{\mu=0} \neq 0$; *i.e., the complex digenvalues move smoothly across the imaginary axis–of the real axis–as* μ *is increased through* $\mu = 0$.

Then there exists a 1*-parameter family of closed orbits for* $\mathbf{X}(\cdot)$ *near* $\mathbf{0}$. *These orbits are stable under the "vague attractor" hypothesis* (1), *i.e., that the origin is still attracting when* $\lambda_1(0) = 0$.

We have stated the theorem for vector fields in the plane; its generalization to $\mathbf{R}^n$ employs the center manifold theorem which reduces the problem back to $\mathbf{R}^2$. The generalization to infinite dimensions can be found in Marsden [**1974**].

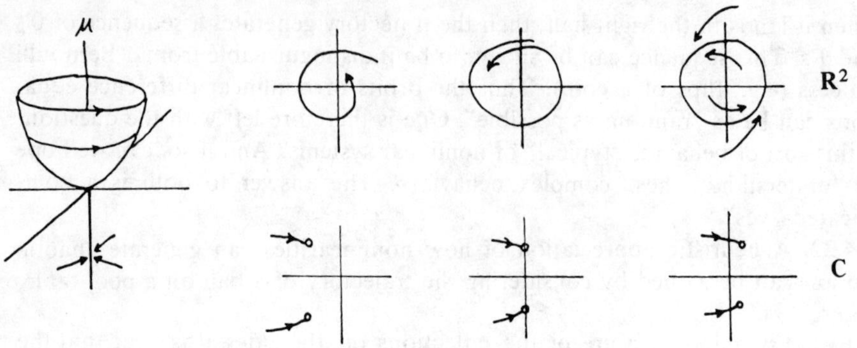

FIGURE 20

Thus the search for limit cycles is facilitated by a root locus–in the complex plane of the linearized system–as the parameters are varied. Unfortunately, the vague attractor conditions are quite tedious to compute except in simple cases. However, it is fairly easy to verify the conditions of the theorem by testing numerically that near $\lambda_1(0) = 0$ the system is still stable. (Easy, that is, if most of the system parameters are tied down by experimental data: an unrestricted search of parameter space is clearly not practicable in general.)

4.24. REMARK. When more than one parameter at a time may be varying the limit cycles generated may appear suddenly at finite amplitude rather than growing continuously from zero amplitude. This is what happens in certain 2-population systems (Auslander, Oster and Huffaker [**1974**]). A discussion of the theory of 2-parameter Hopf bifurcations can be found in Takens [**1973**]. Higher dimensional bifurcations can be treated by the methods of "catastrophe theory" (Thom [**1975**], Jones [**1975**]) (which, unfortunately, is capable of handling only gradient flows).

4.25. EXAMPLE. The following system has been used as a model for predator-prey dynamics (May [**1974**]):

$$\dot{x} = x\left[r(1 - x/K) - ky(1 - e^{-cx})\right], \qquad \dot{y} = y\left[-b + \beta(1 - e^{-fx})\right].$$

Computing the Jacobian at the equilibrium it is easy to check that the signs of the determinant and trace in the characteristic equation, $\lambda^2 - \mathrm{Tr} \cdot \lambda + \det = 0$, depend on the parameter set $\{r, K, k, c, b, \beta, f\}$. Moreover, one can find, numerically, a family of curves parametrized by some members of this set which carry the eigenvalues into the RHP (cf. Figure 21).

Since we can be sure that large radius orbits move inward, the limit cycles which arise from this model are generated by the Hopf mechanism (Oster and Guckenheimer [**1976**]).

4.26. Higher order bifurcations than the first produce more complicated periodic orbits. They are analyzed by reducing the problem back to the difference equation case. This is accomplished by constructing a hyperplane, M, transverse to the limit cycle as shown in Figure 22a. Then the flow of the vector field **X** induces a map **P**: $M \to M$ (the "return map" or Poincaré

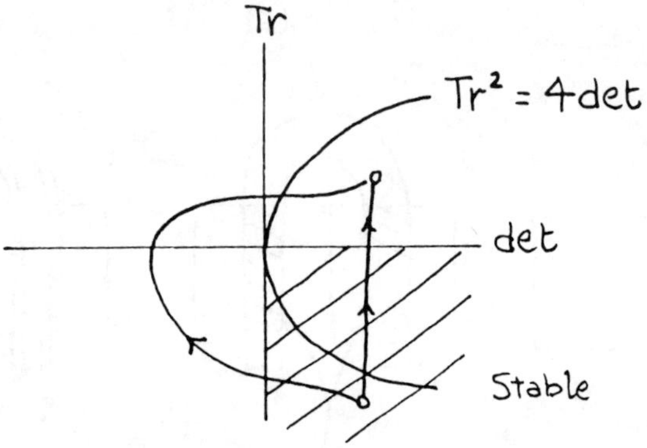

FIGURE 21

map). Thus, on M, $\mathbf{P}$ is a difference equation; it bifurcates to a 2-point cycle when the eigenvalues of $D\mathbf{P}(\mathbf{x}_0)$ leave the unit circle. The resulting 2-point cycle for P corresponds to a limit cycle which "goes around twice", as shown in Figure 22b. As $\mathbf{P}$ bifurcates to higher order cycles the orbit of $\mathbf{X}$ becomes more and more convoluted, so that the record of any observable (i.e., f: $\mathbf{R}^n \to \mathbf{R}$, e.g., one coordinate $x_i(\cdot)$) becomes progressively harder to interpret deterministically. [*Note*. By construction $\mathbf{P}$ is a diffeomorphism. It is also possible to go the other way: given a diffeomorphism one can regard it as the return map of a flow embedded in a higher dimensional manifold. By this process it is possible–in principle–to "suspend" a difference equation so as to obtain a differential equation. Unfortunately, this is not always easy to do in practice, and the embedding space may have a more complicated topology than $\mathbf{R}^n$. However, given this "theoretical equivalence", most work in this area has been done on maps rather than flows (Nitecki [**1971**]).]

4.27. In fact, the situation can get even worse. The attracting set of a dynamical system can be a very complicated object–in contrast to the familiar simple attracting sets: equilibria and limit cycles. These limit sets are so complicated, in fact, they can only be characterized statistically! One of the earliest examples of such a "strange limit set" was the "horseshoe map" (Smale [**1965**]). This is shown in Figure 23.

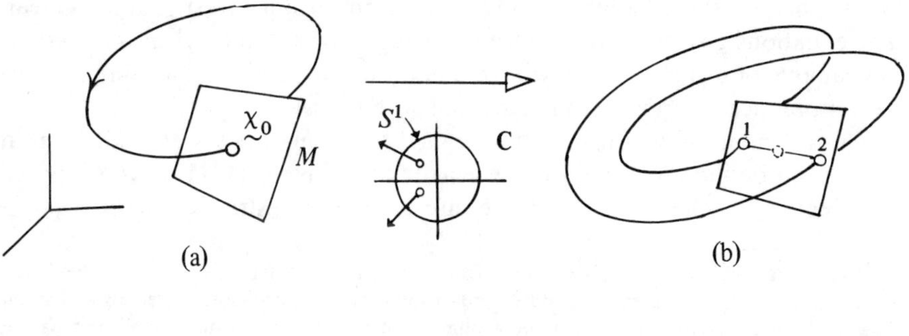

FIGURE 22

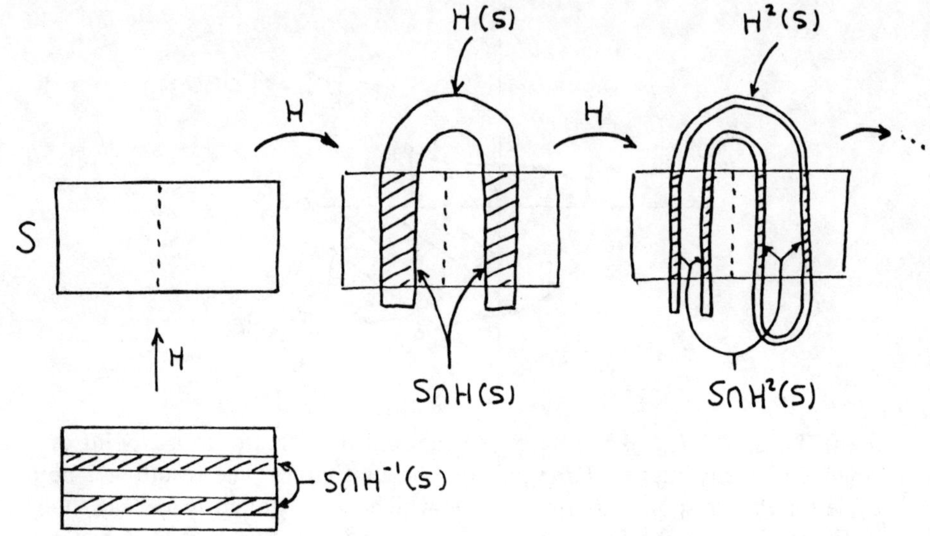

FIGURE 23

Clearly H is very nonlinear: it consists of stretching the square, S, bending it into a horseshoe and then placing it atop the domain S so that $S \cap H(S)$ consists of two vertical strips (shown shaded). Iterating the map once more creates four thinner vertical strips, and so on: $S \cap H^n(S)$ is 2^n thin strips. Iterating H backwards n times produces a set $S \cap H^{-n}(S)$ consisting of 2^n horizontal strips. The limiting set $\Omega = S \cap \prod_{k=-\infty}^{\infty} H^k(S)$ consists of a countable number of points whose structure resembles a Cantor set.[8] Just as for the difference equation in §4.20 we can classify the points of Ω in a binary code according to whether they fall in the left strip ("0") or the right strip ("1") of $S \cap H(S)$. The behavior of the map H is quite "random" in a well defined sense, and the attracting set of the dynamical system obtained by suspending H in a flow is an example of a strange attractor.

4.28. We shall not delve deeper into these matters here since we would quickly become enmeshed in technicalities beyond the scope of this paper. (A readable introduction to this aspect of dynamical systems by J. Guckenheimer should appear soon.) The lesson for our investigations is clear. The equations governing the deterministic dynamics of biological populations are capable of exhibiting behavior so complex as to render it indistinguishable for all practical purposes from a stochastic process!

This dilemma is, of course, not restricted to population systems, although it appears to be more common there. Ruelle and Takens [**1971**] have proposed the identical mechanism of multiple bifurcations to a strange attractor as the

[8]Recall that the Cantor set is obtained from the unit interval by deleting first the (open) middle third (1/3, 2/3), then the middle third of the remaining two closed intervals and so on. The limiting set (consisting of an infinite number of points) is compact, closed and has no isolated points; yet it has measure zero. Marsden [**1974**, pp. 77, 295].

mechanism for generating turbulence in the Navier-Stokes equation (the parameter being tuned is, of course, the Reynolds number). Recent numerical studies appear to confirm this. Indeed, Poincaré himself lamented the dynamical complexity of mechanical systems possessing "homoclinic points" (e.g., 3-body problems). A homoclinic point, h, arises when the unstable manifold intersects the stable manifold as shown in Figure 24.

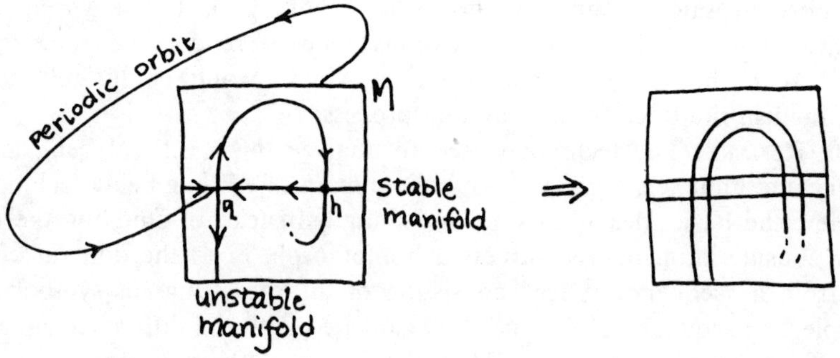

FIGURE 24

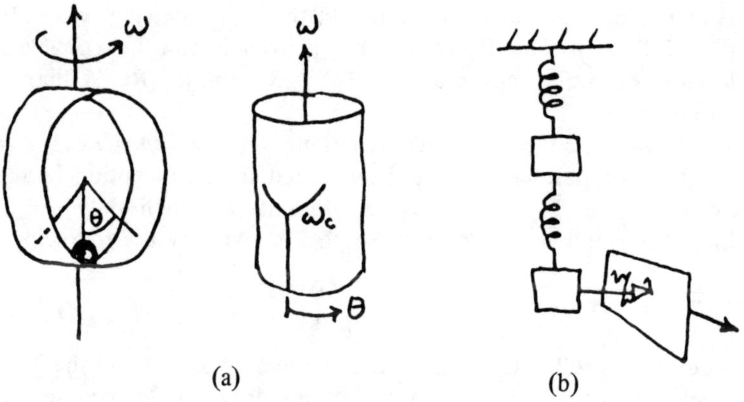

FIGURE 25

By examining a pair of strips about the stable and unstable manifolds under forward and reverse iterates of P we see that M contains a horseshoe–so it is easy to appreciate why the flow is so complex.[9]

4.29. REMARK. The idea of "statistical" behavior can turn up even in rather simple systems. Consider the mechanical system shown in Figure 25 (due to Marsden et al. [**1972**]) consisting of a hoop on a vertical axis with a ball at the bottom. When the hoop is rotated with increasing angular velocity ω the ball, initially at rest at $\theta = 0$, moves to a new equilibrium at either $\pm\theta(\omega)$ when $\omega > \omega_c$. The phase space is the cylinder $(\theta, \omega) \in S' \times \mathbf{R}$ (Figure 25). (This

[9]Notice that the homoclinic point h is not an equilibrium point. Rather, h is characterized by the property that it approaches the equilibrium q under both forward and reverse iterations of the map $\mathbf{P}$.

bifurcation is initiated when the first eigenvalue crosses the imaginary axis on the real axis.) The point is that after the bifurcation at $\omega = \omega_c$, which branch the ball chooses to follow is virtually a random event. If we imagine the system passing through a sequence of such bifurcations then the track of the ball is perfectly equivalent to a sequence of Bernoulli trials. The location of the ball after a bifurcation is no more predictable than the flip of a coin.

A more prosaic example is shown in Figure 25b. If the springs are nonlinear then frequency multiplications quickly render the trace quite chaotic. After the system has been running awhile it would be difficult indeed to distinguish the trace from a random process.

4.30. REMARK. The technique used to analyze the statistical behavior of complicated limit sets is called "symbolic dynamics". While being technically complex, the basic idea is easy to understand. Instead of studying the map under consideration one constructs a homomorphism to the domain of the map from a "sequence space" consisting of infinite strings of symbols, for example, the sequence of 0's and 1's associated with the difference equation in §4.20 or the horseshoe map. Then the action of the map on the original space is equivalent to the unit "shift map" on the sequence space, i.e., the shift map simply moves the sequence to the right one step: $\sigma(\ldots 0110 \ldots) = (\ldots 110 \ldots)$. The flavor of this approach can be conveyed by the following example (communicated by S. Smale, R. Williams and J. Guckenheimer).

Recall that, for the difference equation $x_{t+1} = F(x_t) = rx_t(1 - x_t)$, $0 \leqslant r \leqslant 4$, when the parameter r has been tuned to the inception of the 3-point cycle, the cycles of all periods are present (although only the 3-point is stable). The number of cycles of length i is given by the following formula:

$$N_i = \mathrm{Tr}\left[\begin{pmatrix} 0 & 1 \\ 1 & 1 \end{pmatrix}^i\right].$$

It is derived as follows. In Figure 26a the map is drawn and the 3-point cycle is labeled 1-2-3. Since it is stable we can draw small open sets about the periodic points as shown. The complement of these intervals consists of four closed intervals which we denote by $\{\alpha, \beta, \gamma, \delta\}$. Now notice that under the action of $F(\cdot)$ the interval α gets stretched till it covers both α and β. Similarly, one can verify that β is mapped onto γ, γ onto $\gamma + \beta$ and δ onto α. We can represent this transition scheme by the "Markov chain"-like graph shown in Figure 26b. In turn, this transition scheme can be represented by the "connection matrix" $\mathbf{T}$ for the graph shown in Figure 26c. ($T_{kj} = 1$ if there is a path from state (row) k to state (column) j, and 0 otherwise.) Thus the trajectories of points within the interval set $I = \{\alpha, \beta, \gamma, \delta\}$ are completely accounted for by the linear map $\mathbf{T}$ which permutes the elements of I exactly like the map $F(\cdot)$. Moreover, an examination of $\mathbf{T}$ (or the transition graph) shows that only the states β and γ can sustain a periodic orbit. (In the language of Markov chains $\{\beta, \gamma\}$ is "absorbing".) Therefore, only the middle block $\mathbf{B} = \left[\begin{smallmatrix} 0 & 1 \\ 1 & 1 \end{smallmatrix}\right]$ is relevant for periodic orbits. Now we would like to

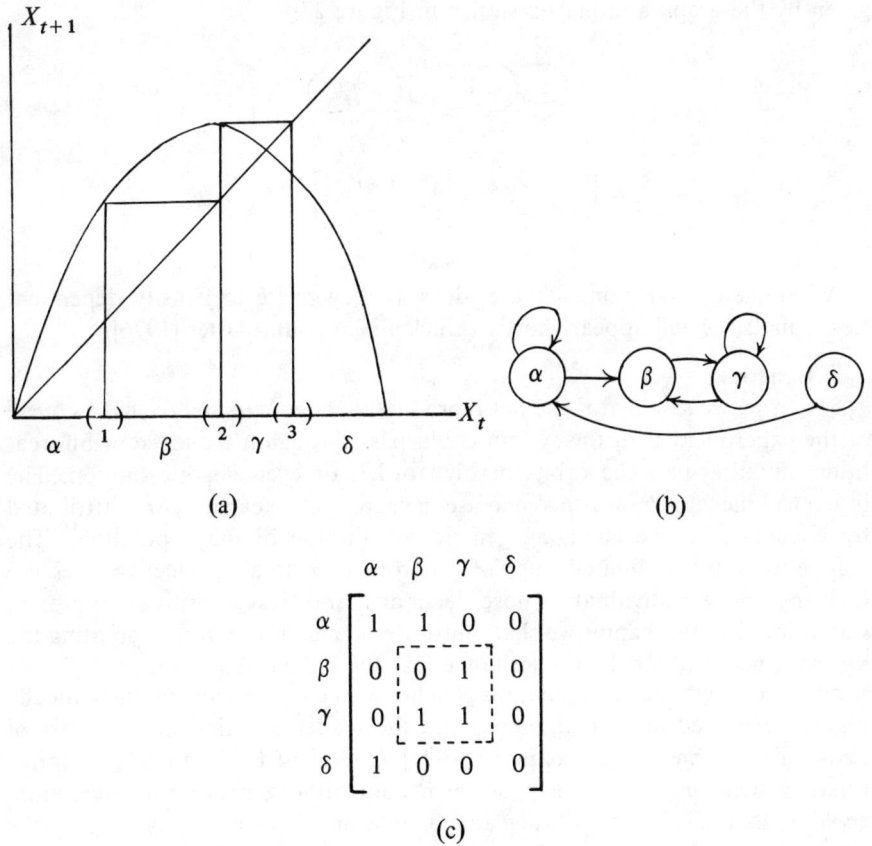

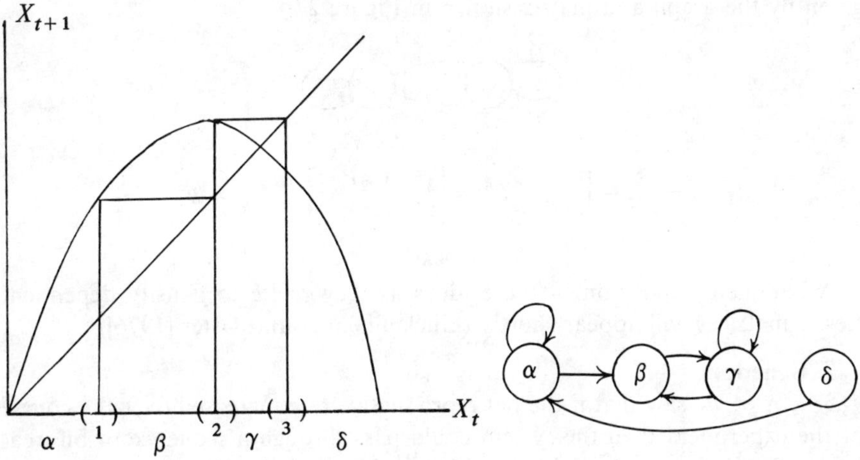

(a) (b)

$$
\begin{array}{c} \\ \alpha \\ \beta \\ \gamma \\ \delta \end{array}
\begin{array}{c} \begin{array}{cccc} \alpha & \beta & \gamma & \delta \end{array} \\ \left[\begin{array}{c|cc|c} 1 & 1 & 0 & 0 \\ \hline 0 & 0 & 1 & 0 \\ 0 & 1 & 1 & 0 \\ \hline 1 & 0 & 0 & 0 \end{array}\right] \end{array}
$$

(c)

FIGURE 26

compute the number of periodic paths of length i between the intervals β and γ. We can do this by simply counting the number of paths through the graph starting and ending in state i. It is easy to verify the following result from graph theory (or Markov chain theory):

$$(\mathbf{B}^k)_{ij} = \text{number of paths of length } k+1 \text{ beginning in state } i \text{ and ending in state } j.$$

In particular, we are interested in $(\mathbf{B}^i)_{kk}$ giving the number of cycles of length i, so we can add up all the periodic orbits by computing $\operatorname{Tr} \mathbf{B}^i$. This can be simplified by diagonalizing $\mathbf{B}$ by a matrix $\mathbf{A}$ (or putting it in Jordan form and noting that $(\mathbf{ABA}^{-1})^i = \mathbf{AB}^i\mathbf{A}^{-1}$. Since the trace is an invariant for $\mathbf{B}$, $\operatorname{Tr} \mathbf{B}^i = \lambda_1^i + \lambda_2^i$ where λ_1, λ_2 are the eigenvalues of $\mathbf{B}$. (In fact, it is easy to see that, because of the structure of $\mathbf{B}$, the λ_i can be expressed as Fibonacci numbers, a result which would surely appear magical without the above analysis!) Thus $N_i = \{1, 3, 4, 7, \ldots\}$.

A nearly identical calculation can be carried out for the "triangle" difference equation and the horseshoe map. There the transition structure is given by the graph and matrix shown in Figure 27.

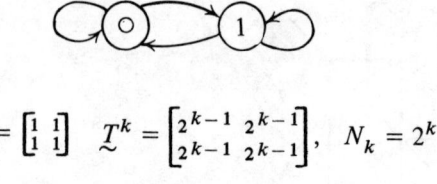

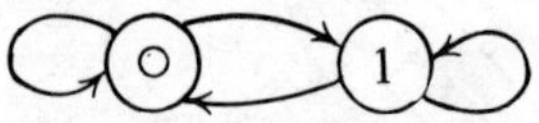

$$\underset{\sim}{T} = \begin{bmatrix} 1 & 1 \\ 1 & 1 \end{bmatrix} \quad \underset{\sim}{T}^k = \begin{bmatrix} 2^{k-1} & 2^{k-1} \\ 2^{k-1} & 2^{k-1} \end{bmatrix}, \quad N_k = 2^k$$

FIGURE 27

A complete exposition of these ideas as they relate to density dependent Leslie matrices will appear shortly (Guckenheimer and Oster [**1976**]).

5. Genetics.

5.1. In §4 we saw that if the net reproductive rate increased over the course of the experiment then the system could pass through a sequence of bifurcations initiating periodic orbits of high order, or even chaotic motion. The biological mechanism for this increase in reproductive capacity was attributed by Nicholson to the changing genetic constitution of the population.[10] The competition for a limited supply of protein created a selective pressure favoring those individuals whose fecundity was less sensitive to protein starvation. In this chapter we shall illustrate one method for incorporating the mechanisms of Mendelian inheritance into the balance equations.

5.2. The discipline of population genetics is one of the most mathematically highly developed areas of biology, and the reader can find any number of texts on the subject (e.g., Jacquard [**1974**], Crow and Kimura [**1970**]). Unfortunately there has been a gap between population genetics and population ecology that has been unbridgeable heretofore. The difficulty is that the mechanics of Mendelian inheritance are quite complicated when expressed in mathematical terms. Thus very little is known about the properties of genetic equations except in special cases and these cases generally bear little relevance to ecological situations. In particular, traits relevant to complex behavioral and/or physiological performance are controlled by many genes, while the mathematical apparatus of population genetics is confined to treating one or a few loci (with one or a few alleles at each locus). Indeed, the mechanics of polygenic systems is only beginning to be worked out (e.g., L. Hood et al. [**1975**]). Therefore, in treating genetic dynamics it behooves us to emulate physics once again and attempt a "thermodynamic" formulation which averages out all of the "statistical mechanical" details of the underlying genetic processes. Such a "continuum" model has been proposed by Slatkin [**1970**] and employed by Roughgarden [**1972**] and others in a number of

[10]Another possibility, or contributing factor, is dispersion (cf. §2.6). All the individuals in a given cohort do not age synchronously as we have assumed; some reach breeding age before others. The effect of this is to spread the reproductive age classes and destroy the synchrony within a given generation. This could have the effect of washing out the limit cycle after awhile. However, in light of the experimentally verified differential fecundity due to selection this is probably a secondary effect.

ecological contexts. Here we shall briefly describe the approach; a more complete description can be found in Rocklin and Oster [**1975**].

For expository purposes we shall concern ourselves initially with the evolution of the density function for the case when the generations do not overlap. Therefore, we can ignore the age distribution and regard time as a discrete variable, so that the equation of motion can then be written as a difference equation relating $n_t(\xi)$ to $n_{t+1}(\xi)$, where ξ is the phenotypic trait under consideration:

$$n_{t+1}(\xi) = \Psi[n_t(\xi)]. \tag{5-1}$$

After we determine the form of the functional $\Psi[\cdot]$ we can easily reintroduce the continuous dependence on age and time.

5.3. We shall summarize all of population genetics in a "scattering kernel" $L(\cdot)$ as follows.

Let

$$L(\xi|\xi_1, \xi_2) = \text{distribution of offspring phenotypes arising from matings of parental types } \xi_1 \text{ and } \xi_2 \tag{5-2}$$

(cf. Figure 28). $L(\cdot)$ is a conditional probability, $\int_0^\infty L(\cdot)\, d\xi = 1$, which contains a phenomenological description of the transmission of quantitative ("metric") traits by sexual reproduction. Like a constitutive relation its functional form must be derived from the mechanics of Mendelian inheritance. Alternatively, it can be measured empirically. For example, many quantitative traits tend to be distributed normally about the parental mean: $\bar{\bar{\xi}} = (\xi_1 + \xi_2)/2$, so that $L(\cdot)$ can be considered a function of a single variable $\zeta = \xi - \bar{\bar{\xi}}$. Other genetic processes such as heritability and dominance deviations can be included by suitably modifying the form of $L(\cdot)$ (Slatkin and Lande [**1975**]; Rocklin and Oster [**1975**]).[11]

To write the equation of motion for $n_t(\xi)$ we must compute the frequency

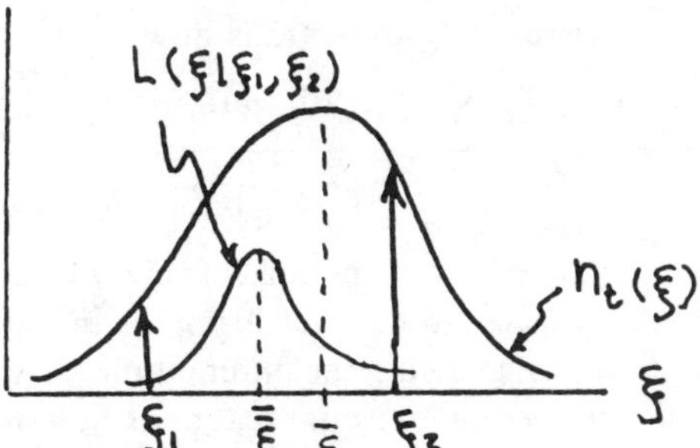

FIGURE 28

[11]Heritability, h^2, measures the tendency of the offspring distribution to regress to the population mean $\bar{\xi} = \int \xi n(\xi)\, d\xi / \int n(\xi)\, d\xi$. Thus we can redefine ζ as the convex combination $\zeta = \xi - (h^2\bar{\xi} + (1 - h^2)\bar{\bar{\xi}})$. Dominance deviations tend to make the offspring distribution asymmetric.

of matings between parental types ξ_1 and ξ_2. If mating is random with respect to the trait then, in a large population, we expect the number of matings to be the product of the parental frequencies: $p_t(\xi_1)p_t(\xi_2)$, where $p_t(\xi) = n_t(\xi)/N_t$. Thus in the absence of selection and assuming no differential fertility between mating pairs, the equation of motion is

$$p_{t+1}(\xi) = \int\int L(\xi|\xi_1, \xi_2)p_t(\xi_1)p_t(\xi_2)\, d\xi_1\, d\xi_2 \tag{5-3}$$

or, in terms of the density $n_t(\xi) = N_t p_t(\xi)$,

$$n_{t+1}(\xi) = \frac{r}{N_t}\int\int L(\zeta)n_t(\xi_1)n_t(\xi_2)\, d\xi_1\, d\xi_2 \tag{5-4}$$

where $r = N_{t+1}/N_t$ is the net reproductive rate, which we have assumed is independent of ξ. Assortative mating and differential fertility can be included into the model by defining a "mating function" $\phi(\xi_1, \xi_2)$ (Rocklin and Oster [**1975**]) defined to be the mean number of offspring arising from matings between parents of type ξ_1 and ξ_2.

$$p_t(\xi) = \frac{1}{r_t}\int\int L(\zeta)\phi(\xi_1, \xi_2)p_t(\xi_1)p_t(\xi_2)\, d\xi_1\, d\xi_2 \tag{5-5}$$

where r_t, the mean number of offspring per parent, is now given by

$$\int\int \phi(\xi_1, \xi_2)p_t(\xi_1)p_t(\xi_2)d\xi_1 d\xi_2.$$

Finally, selection is introduced by defining a survivorship function, $S_t(\xi)$ giving the fraction of newborn of type ξ which survive to reproduce. The equation of motion now becomes:

$$n_{t+1}(\xi) = \frac{\left(\int\int L(\xi|\xi_1, \xi_2)\phi(\xi_1, \xi_2)S_t(\xi_1)S_t(\xi_2)n_t(\xi_1)n_t(\xi_2)\, d\xi_1\, d\xi_2\right)}{\left(\int S_t(\xi)n_t(\xi)\, d\xi\right)} \tag{5-6}$$

The definition of the net reproductive rate is now

$$r_t = \frac{\left(\int\int \phi(\xi_1, \xi_2)S_t(\xi_1)S_t(\xi_2)n_t(\xi_1)n_t(\xi_2)\, d\xi_1\, d\xi_2\right)}{\left[\int S_t(\xi)n_t(\xi)d\xi\right]^2}. \tag{5-7}$$

We note that r_t may also be a function of the physiological variable x employed in §§2 and 3 to account for the variations in nutritional state *within* a particular generation. In general, the interaction between environmental and genetic forces is quite complex; our treatment here must be regarded as only a first approximation.

The equation for the total population is obtained by integrating over all ξ:

$$N_{t+1} = \frac{\left(\int\int \phi(\xi_1, \xi_2)S_t(\xi_1)S_t(\xi_2)n_t(\xi_1)n_t(\xi_2)\, d\xi_1\, d\xi_2\right)}{\left(\int S_t(\xi)n_t(\xi)d\xi\right)}. \tag{5-8}$$

5.4. To go further we must introduce specific assumptions about mating and survivorship. The trait we are concerned with here is the genetic component of the dependence of fecundity on protein. Recall that the birthrate function $b(\cdot)$ contained parameters, one of which controlled the threshold for egg-laying (Figure 7b). This parameter, heretofore considered fixed, we must now regard as an average with respect to $p_t(\xi)$. A complete study of the coupling between genetic and demographic dynamics can be found in Ipaktchi, Oster and Auslander [**1976**]. Here we shall only indicate how the changing genetic structure can drive the system through successive bifurcations by altering the effective parameter values (Oster, Ipaktchi and Rocklin [**1976**]).

5.5. For illustrative purposes we shall make the following reasonable assumptions about $S_t(\cdot)$ and $\phi(\cdot)$.

$$S_t(\xi) = \hat{S}(\xi)e^{-N_t/\hat{K}}, \tag{5-8a}$$

$$\phi(\xi_1, \xi_2) = B/(1 + \gamma|\xi_1 - \xi_2|). \tag{5-8b}$$

Substituting into the equation of motion we obtain for the equation for the total population

$$N_{t+1} = N_t e^{r_t(1 - N_t/K)} \tag{5-9}$$

where the effective net reproductive rate is given by

$$r_t = \ln B + \ln \left[\frac{\iint \frac{d\xi_1 d\xi_2}{1 + \gamma|\xi_1 - \xi_2|} \hat{S}_t(\xi_1)\hat{S}_t(\xi_2)p_t(\xi_1)p_t(\xi_2)}{\int \hat{S}(\xi)p_t(\xi)\, d\xi} \right] \tag{5-10}$$

and $K_t = \hat{K}r_t$.

Equation (5-9) is a discrete-time logistic model of the type encountered in §4; its behavior under variations in r_t is, by now, familiar (May and Oster [**1975**], Chapter 4). But now, as we see from (5-10), the effective value of r is a functional of the distribution of phenotypes, $p_t(\xi)$. In Figure 29 is shown the effect on the bifurcation structure of the variability of the population as measured by the variance, σ_t^2, of $p_t(\xi)$.

One feature of the dynamics of equation (5-8) is worth mentioning. The speed of the system's response to disturbances from equilibrium is faster the greater the genetic variability, as measured by the variance, σ_L^2, of the reproduction kernel (why is this to be expected?). Thus we would expect that a highly inbred strain of flies (i.e., low σ_L^2) would sustain a stable limit cycle longer than the wild type.

Acknowledgements. §4 of this paper is an informal account of research carried out with my colleagues D. Auslander and J. Guckenheimer under the auspices of NSF grant BMS 74-21240.

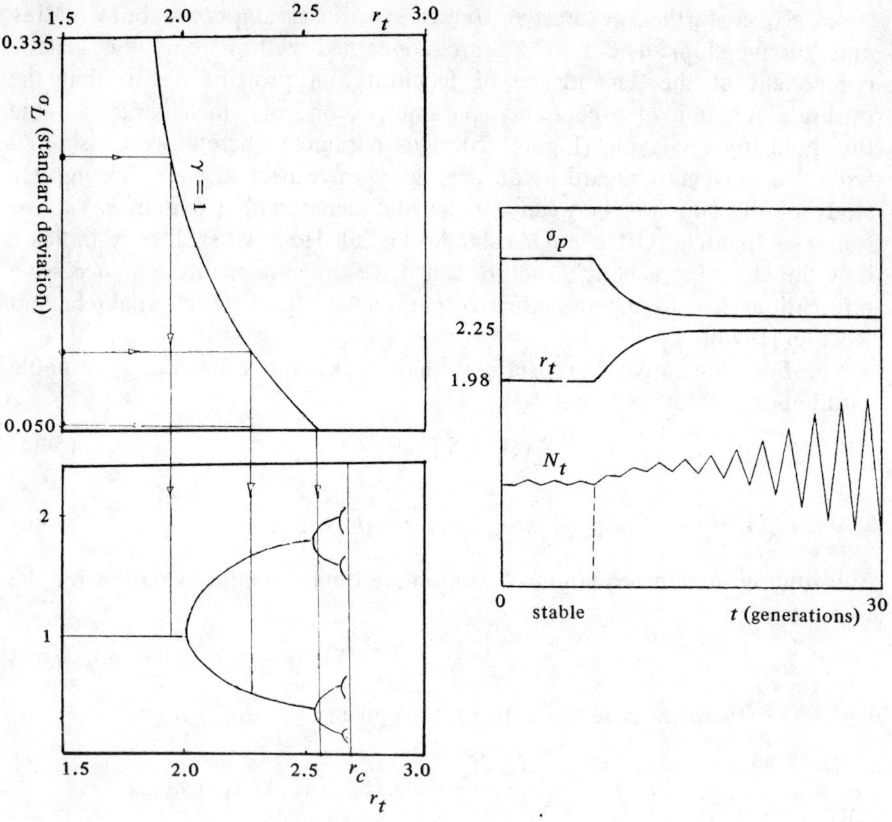

FIGURE 29

REFERENCES

D. Auslander, J. Guckenheimer and G. Oster (1976) (to appear).

D. Auslander, G. Oster and C. Huffaker (1974), *Dynamics of interacting populations*, J. Franklin Inst. **297**, 345–376.

R. Courant and D. Hilbert (1966), *Methods of mathematical physics*. Vol. II, Wiley, New York.

J. F. Crow and M. Kimura (1970), *An introduction to population genetics theory*, Harper & Row, New York. MR **42** #8944.

J. Frauenthal (1975), *A dynamic model for human population growth*, Center for Population Studies, Harvard University.

A. G. Frederickson, (1971), *A mathematical theory of age structure in sexual populations: random mating and monogamous marriage models*, Math. Biosci. **10**, 117–143.

J. Guckenheimer, G. Oster and A. Ipaktchi (1976), *The dynamics of density dependent population models*, J. Math. Biol. (to appear).

M. Gurtin and R. MacCamy (1974), *Non-linear age-dependent population dynamics* Arch. Rational Mech. Anal. **54**, 281–300.

L. Hood, J. Campbell and S. Elgin (1975), *The organization, expression and evolution of antibodies and other multigene families*, Ann. Rev. Genetics **9**.

A. Ipaktchi, G. Oster and D. Auslander (1976) (to appear).

A. Jacquard (1974), *The genetic structure of populations*, Springer-Verlag, New York.

D. Jones (1975), *Application of catastrophe theory to ecological modelling* (to appear).

N. Keyfitz (1968), *Introduction to the mathematics of populations*, Addison-Wesley, Reading, Mass.

A. M. Krall (1970), *The root locus method: A survey*, SIAM Rev. **12**, 64–72. MR **41** #5078.

C. J. Krebs (1972), *Ecology*, Harper & Row, New York.

H. L. Langhaar (1972), *General population theory in the age-time continuum*, J. Franklin Inst. **293**, 199–214. MR **46** #8688.

P. H. Leslie (1945), *On the use of matrices in certain population mathematics*, Biometrika **35**, 183–212. MR **7**, 465.

______ (1948), *Some further notes on the use of matrices in population mathematics*, Biometrika **35**, 213–245. MR **10**, 386.

______ (1959), *The properties of a certain lag type of population growth and the influence of an external random factor on a number of such species*, Physiol. Zool. **32**, 151–159.

T.-Y. Li and J. A. Yorke (1975), *Period three implies chaos*, Amer. Math. Monthly (to appear).

A. Lopez (1967), *Asymptotic properties of a human age distribution under a continuous net fertility function*, Demography **4**, 680–687.

S. Mac Lane (1968), *Geometrical mechanics*. I, II, Lecture Notes; Department of Mathematics, University of Chicago.

J. Marsden, D. Ebin and A. Fisher (1972), *Diffeomorphism groups, hydrodynamics and relativity*, Proc. 13th Biennial Seminar of the Canadian Mathematical Congress, J. R. Vanstone (editor), Canadian Math. Soc.

J. Marsden (1973), *The Hopf bifurcation for nonlinear semigroups*, Bull. Amer. Math. Soc. **79**, 537–541.

______ (1974), *Elementary classical analysis*, Freeman, San Francisco, Calif.

J. Marsden and M. McCracken (1976), *The Hopf bifurcation*, Lecture Notes in Math., Springer-Verlag, New York.

R. May (1972), *Limit cycles in predator prey communities*, Science **177**, 900–902.

______ (1973), *Time delay versus stability in population models with two and three trophic levels*, Ecology **54**, 315–325.

______ (1974), *Biological populations with nonoverlapping generations: stable points, stable cycles, and chaos*, Science **186**, 645–647.

______ (1975), *Stability and complexity in model ecosystems*, 2nd ed., Princeton Univ. Press, Princeton, N. J.

R. May and G. Oster (1975), *Bifurcations and dynamics complexity in simple ecological models*, Amer. Natr. (to appear).

A. J. Nicholson (1957), *The self adjustment of populations to change*, Cold Spring Harbor Symposia on Quantitative Biology **22**, 153–173.

Z. Nitecki (1971), *Differentiable dynamics*, M.I.T. Press, Cambridge, Mass.

G. Oster (1974), *Stochastic behavior of deterministic models*, SIMS Research Application Conference, July 1–5, Alta, Utah, Predicting the Response of Ecosystems to Perturbations.

G. Oster and Y. Takahashi (1974), *Models for age specific interactions in a periodic environment*, Ecological Monographs **44**, 483–501.

G. Oster, D. Auslander and T. Allen (1975), *Deterministic and stochastic effects in population dynamics*, J. Dyn. Sys. Meas. and Control (to appear).

G. Oster and J. Guckenheimer (1976), *Bifurcation Behavior of Population Models*, The Hopf Bifurcation, J. Marsden, M. McCracken (editors).

G. Oster, A. Ipaktchi and S. Rocklin (1976), *Genetic structure and bifurcation behavior of population models*, Theoret. Population Biology (to appear).

B. Parlett (1970), *Ergodic properties of populations*. I. *The one sex model*, Theoret. Population Biology **1**, 191–207. MR **48** #3529.

E. Pianka (1974), *Evolutionary ecology*, Harper & Row, New York.

J. H. Pollard (1973), *Mathematical models for the growth of human populations*, Cambridge Univ. Press, New York.

S. Rocklin and G. Oster (1975), *Competition between phenotypes*, J. Math. Biol. (to appear).

J. Roughgarden (1972), *Evolution of niche width*, Amer. Natr. **106**, 683–718.

D. Ruelle and F. Takens (1971), *On the nature of turbulence*, Comm. Math. Phys. **20**, 167–192. MR **44** #1297.

E. Seneta (1973), *Non-negative matrices*, Allen and Unwin, London.

S. Shahshahani (1975), *A new mathematical framework for the study of linkage and selection* (to appear).

J. Sinko and W. Streifer (1967), *A new model for age-size structure of a population*, Ecology **48**, 910–918.

M. Slatkin (1970), *Selection and polygenic characters*, Proc. Nat. Acad. Sci. U.S.A. **66**, 87–93.

M. Slatkin and R. Lande (1975), *Niche width in a fluctuating environment–density independent model*, Amer. Natr. (to appear).

S. Smale (1965), *Diffeomorphisms with many periodic points*, Differential and Combinatorial Topology (A Sympos. in Honor of Marston Morse), Princeton Univ. Press, Princeton, N.J., pp. 63–80. MR **31** #6244.

A. Streiffer and C. Istock (1973), *A critical variable formulation of population dynamics*, Ecology.

F. Takens (1973), *Unfoldings of certain singularities of vector fields: generalized Hopf bifurcations*, J. Differential Equations **14**, 476–493. MR **49** #4024.

R. Thom (1975), *Structural stability and morphogenesis*, Addison-Wesley, Reading, Mass.

K. Tognetti and A. Mazanov (1970), *A two-stage population model*, Math. Biosci. **8**, 371–378.

E. Trucco (1965), *Mathematical models for cellular systems: the von Foerster equation*. I, Bull. Math. Biophys. **27**, 285–303, II, **27**, 449–471.

G. H. Weiss (1968), *Equations for the age structure of growing populations*, Bull. Math. Biophys. **30**, 427–435.

DEPARTMENT OF ENTYMOLOGY, UNIVERSITY OF CALIFORNIA, BERKELEY, CALIFORNIA 94720

Lectures in Applied Mathematics
Volume 16, 1977

Amoeboid Motions

Garrett M. Odell

The cytoplasmic streaming that occurs in amoeboid pseudopodium extension is a fascinating biological phenomenon from which an interesting mathematical problem can be extracted. This article comprises a case study of continuum modeling of this biological event.

1. A statement of policy. Mathematical equations are conspicuously absent in the first *n* pages; the "small parameter" of my article is the ratio of equations to prose. I want to defend the reason for this at the outset.

Virtually no biological problem is simple. Even a minor feat performed by the humblest organism is so complicated that the construction of mathematical models of a biological phenomenon is guaranteed to be folly unless the construction begins with a genuine familiarity with the biological raw material at hand, and unless the construction is made, from the ground up, out of concrete, biologically realistic, building blocks. Scientific literature already contains an ample supply of sophisticated mathematical studies of biological phenomena consisting of three paragraphs of biological anecdotes, followed by an onslaught of equations whose connection to the biological problem is left largely to the reader's imagination. I hope this paper does not add to that abundance.

In most biomathematical problems, the central difficulty is concentrated in the question, "what *is* the problem?". This is a trite remark, but it highlights the fact that an engineer or mathematician who wishes to employ his skills to explore biological phenomena simply must be willing to spend an inordinate effort discovering what the problem is.

My purpose in the first part of this article is to identify a meaningful mathematical problem embedded somewhere in the long, careful, and difficult study experimental biologists have made of the question, "how does an amoeba move?". My primary aim here is to assemble a sufficiently coherent and realistic biological sketch (more accurately, a caricature) of this question that a reader who is not biologically oriented will be able to evaluate meaningfully the mathematical modeling presented in the second half of this essay. (Mathematical sophistication is not the aim.)

2. Why should anyone care how an amoeba moves? Almost all biological cells *move* at some phase of their life cycle. The various distinct mechanisms

AMS (*MOS*) *subject classifications* (1970). Primary 92A05, 92–02, 76Z99, 76S05; Secondary 35K55, 34B15.

that have evolved to create motion present fundamental scientific puzzles. Some cells, some creatures, move by thrashing whiplike appendages: flagella and cilia. Sperm is an example. Most highly evolved multicellular organisms have muscular contractile machinery (which will be described below).

The mechanisms just mentioned are highly specialized. Flagella are organelles that thrash to generate thrust in a fluid. Period. That is their entire function. Mammalian muscle cells do little else other than contract when stimulated. Fully one-half of the dry weight of a mammalian striated muscle consists of contractile proteins.

The living fluid-like contents (cytoplasm) of most biological cells has the intrinsic capability to move–usually to circulate around within a stationary cell. Ultimately, all life functions are performed by cytoplasm and membranes, so, in a primitive motile cell like an amoeba, only a small fraction of the cytoplasmic bulk is devoted to motile machinery (in sharp contrast to muscle cell specialization). The capability to stream is a one-line entry in a vast catalog of the remarkable abilities of cytoplasm, but an entry worthy of study. There is good evidence that muscle evolved as an amplification and refinement of the fundamental ability of cytoplasm to move.

Some organisms, devoid of muscle, flagella, cilia, spasmonemes and any other obvious movement organelles, use the ability of cytoplasm to create motion to propel themselves about. Some examples are: (i) amoebae, (ii) macrophages (amoeba-like white blood cells that crawl everywhere through your tissue spaces seeking to destroy microbial invaders, dead tissue, etc.), (iii) fibroblasts (cells that move within your tissue to spin out 'fibers' to repair damage).[1]

Amoebae are ideal objects for the study of cytoplasm's ability to move because their cytoplasm streams so dramatically, because they are easy to culture in large numbers and keep alive on the microscope stage, and because some amoebae are huge as individual cells go. Also, amoeboid motion is, and has been, studied because one glance through a microscope at a living, moving amoeba elicits a lasting wonder. Because an amoeba seems so devoid of permanent structures, yet is as much an animal as any other, one feels that its cytoplasm carries in it, unembellished by superfluous structure, some of the basiç axioms of living stuff.

3. What is an amoeba? An amoeba is a single-celled protozoan animal sufficiently important to modern experimental biologists to merit publication in 1973 of a 628-page monograph entitled *The Biology of Amoeba* (see Jeon (editor) [**1973**]). In this superb book, in a review article by Komnick, Stockem, and Wohlfarth-Botterman [**1973**], and in a crucial experimental study by Taylor, Condeelis, Moore, and Allen [**1973**], can be found everything about amoebae that any but an obsessed reader might care to know.[2]

[1]The motile mechanisms of amoebae, macrophages, and fibroblasts are dissimilar in detail.

[2]I will make no effort to survey the experimental literature or to assign credit for various experimental discoveries. When possible, I will confine citations to various chapters of *The Biology of Amoeba*.

The references just cited will provide not only the catalog of presently known facts about amoebae, but also the flavor of a long and vigorous controversy surrounding the question: How does an amoeba move?

The kinematics of a typical performance of amoeboid motion consists of the extension of one pseudopodium (false foot) or several. That is, the cell cytoplasm is caused to flow from the main cell body (called the *uroid*) out into a tube-like projection (called a pseudopodium). With the uroid attached to some substrate, the extending pseudopodium may select a new point of attachment on the substrate. The old attachment point can then release, and the cytoplasm can flow into the attached pseudopodium. The new attachment can now serve as home base from which new pseudopodial explorations can be launched.

Some amoebae are monopodial, consisting essentially of a single ever-changing pseudopodium. Others manage a number of pseudopodia (perhaps ten) simultaneously and, seemingly, independently. A conspicuous example of the latter case is the giant amoeba *Chaos carolinensis* ("giant" means the diameter is as much as several mm; most other amoebae are tens of μm in diameter). Particularly beautiful photographs of *Chaos* amoebae can be seen in R. D. Allen's article in *The Biology of Amoeba*. I will eventually restrict attention to this species.

A highly deformable membrane surrounds an amoeba's cytoplasm and nucleus. The amoeba consumes particles of food by engulfing them. The process is called phagocytosis, and results in a food item enclosed within the amoeba inside of a newly formed vacuole whose membrane used to be a piece of the external membrane. The membrane has an external mucuous coat which may play a role in attaching the amoeba onto the substrate on which he moves (see Komnick et al. [**1973**]). The cytoplasm is a heterogeneous soup of proteins, vacuoles, vesicles, mitochondria, etc.; in short, a substantial subset of the items to which molecular biology textbooks are devoted. In addition to streaming about to form pseudopodia, the cytoplasm must perform all of the life functions of the amoeba.

It is generally agreed that the cytoplasm can undergo a reversible phase transition. The peripheral layer of the cytoplasm is usually in a semirigid gel state (called plasmagel or ectoplasm) while the interior cytoplasm is more or less fluid (called plasmasol or endoplasm). The two states are distinguished by differences in index of refraction and by the fact that the ectoplasm does not deform significantly while the endoplasm easily deforms (to flow). Ectoplasm continually changes to endoplasm and vice-versa; any portion of cytoplasm can assume either state.

A pseudopodium consists of a semirigid cylindrical ectoplasmic tube through which endoplasm flows. (Refer to Figure 1.) At the tip of the tube, endoplasm gels to augment the extending tube.

To fix mechanical ideas, a typical *Chaos* pseudopodium is about 150 μm in diameter, as long as 800 μm, and shows stream velocities of about 8–30 μm/sec with a *plug* velocity profile in the endoplasmic stream. *N.B.* The velocity profile is *not* parabolic as in Poiseuille flow of a viscous liquid in a tube. Clearly, inertial forces are not important. Infrequently two streams of endoplasm move in opposite directions in the *same* pseudopodium at the same time. The backwards stream is usually much smaller than the forward stream. It is still not clear what pressure force must be applied in opposition to an advancing pseudopodium to arrest its advance, but the order of magnitude is 1 cm of water. These data may be found in Allen [**1973**].

4. A challenge to continuum field theory. In caricature, an amoeba consists of a glob of material (enclosed in a membrane whose *mechanical* role is probably not very important) which generates a coherent streaming deformation on its own volition, overpowering externally applied tractions. From this simplification emerges an enticing problem in continuum mechanics modeling: Develop a field theory to characterize the active mechanical performances this (living) material exhibits.

5. Some proposed models. The problem of how an ameoba moves has attracted a good deal of attention. There are, or have been, roughly four main ideas. A survey of these theories, and a list of their originators and proponents may be found in Allen [**1973**] or Komnick et al. [**1973**]. Figure 1 illustrates roughly the kinematics involved in all models.

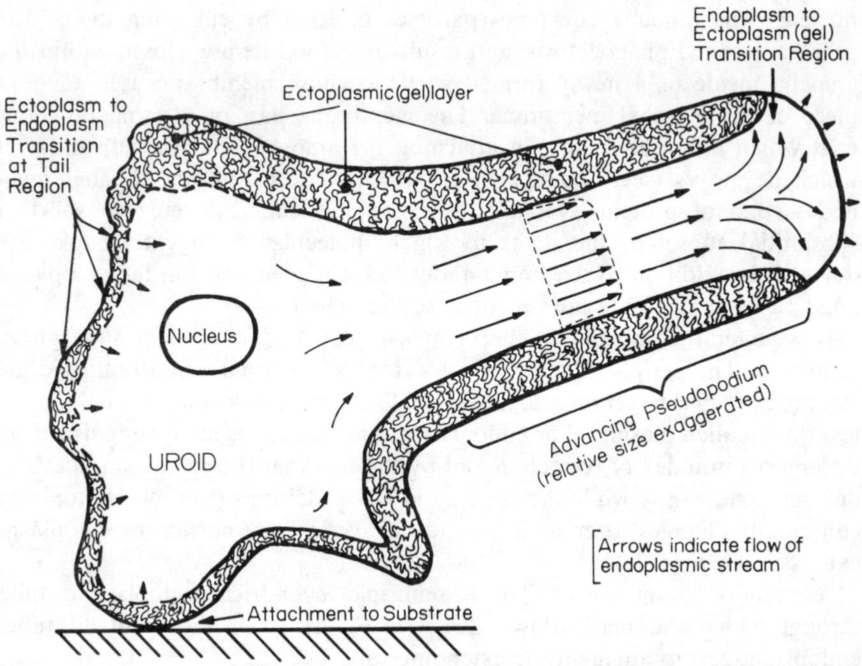

FIGURE 1

The hydraulic contraction model (or tail contraction model) posits a contraction of the ectoplasmic gel layer surrounding the (rear) uroid. This elevates the hydrostatic pressure in the uroid. A weakened region of the periphery then bulges out and the endoplasm is squeezed out to form a pseudopodium. As the endoplasm reaches the tip of the extending pseudopodium, it gels to form more of the ectoplasmic tube. According to this theory the hydrostatic pressure in the endoplasm should decrease from uroid to pseudopodium tip. This is the theory found in most modern biology texts.

The surface tension theory posits a variation of surface tension coefficient along the interface between membrane and external medium (with surface tension locally controlled by the amoeba). If, for example, surface tension is higher at the north pole of a fluid drop than at the south pole, the tangential gradient of surface tension will cause a flow of the interface from the south pole to the north pole. With the internal viscous fluid adhering to the flowing membrane an internal flow of fluid is driven concurrently. The corresponding continuum mechanics problem can be stated very clearly, and is mathematically fascinating. Mathematics aside, a drop of mercury in a flat dish of nitric acid can be made to chase a particle of chemical (potassium dichromate) which, as it dissolves and diffuses in the acid, lowers the surface tension on the part of the mercury drop interface nearest the particle of potassium dichromate. After a frantic chase, the mercury drop usually catches, and even engulfs, the potassium dichromate. In 1900 the problem of amoeboid motion was believed solved by this elegant analog. It turns out that the flow patterns of the interfaces are entirely different for such surface tension driven flows and actual amoeboid motions.

The active shear theory holds that some kind of "oars" affixed to the inner boundary of the ectoplasmic gel tube row the endoplasm, generating an active shear at the boundary between endoplasm and ectoplasm to propel the endoplasm toward the pseudopodium tip. This conjecture has been set down mathematically as a continuum mechanics boundary value problem, and studied by Subirana [**1970**]. One would intuitively expect that a viscous liquid driven down a round pipe by active shear applied at the boundary–*driven against an adverse pressure gradient*–would have a parabolic velocity profile with minimal forward speed at the axis and maximal forward speed at the active shear boundary. Subirana obtained that result rigorously. Measured velocity profiles in *Chaos. c.* amoebae show plug flow with maximum speed at the tube center, and a sharp (boundary layer) decrease of speed to zero at the sol-gel interface. This disagreement between theory and experiment casts doubt on the relevance of the active shear theory.

Allen's "fountain zone" or "frontal" contraction model (see Allen [**1973**]) sounds, at first, mechanically suspicious. According to this model pseudopodium extension is caused by a pull from the front instead of a push from behind. Allen's idea is that, due to submicroscopic fibers threading through the endoplasm, the endoplasm can support tension as can a viscoelastic fluid. At the tip region, where endoplasm flows out of the gel tube, everts, and gels to form additional gel tube, at the site where the endoplasm rounds the corner, a contraction occurs which generates tension to *pull* the viscoelastic column of flowing endoplasm forward.

One or two days of collaboration with R. D. Allen convinced H. L. Frisch and myself that his theory was essentially correct. We set about mathematizing his ideas. The major mathematical content of this article centers on a presentation of the mathematical model we developed (see Odell and Frisch [**1975**]). Toward the end of this article some new generalizations of that model are given. I will precede presentation of our mathematical field theory model by: (i) a minimal biological primer on the nature of the molecular machinery underlying the cytoplasmic streaming seen in pseudopodium extension, and (ii) a sketch of experimental evidence that seems to me to support Allen's frontal contraction model and render untenable the other models listed above.

Before leaving the topic of "some proposed models", I want to sketch a mathematically elegant study by Finlayson and Scriven [**1969**] entitled *Convective instability by active stress*. This study was inspired by films (made by Professor Noburo Kamiya) of gyrations and streaming exhibited by isolated strands of living *Physarum polycephalum* cytoplasm.

Finlayson and Scriven (hereafter referred to as F & S) set the following abstract question: Is it possible to write down a proper constitutive theory for an incompressible isotropic fluid whose stress tensor depends upon scalar concentration fields of chemicals diffusing in and convected by the fluid, in such a way that a quiescent state of the fluid, when infinitesimally perturbed, can generate[3] kinetic energy at a sufficient rate to drive a convective instability? (There are many kmown affirmative answers *if* surface-tension tractions or externally applied body forces are available; the point of this study was to do it without those.) The answer is a complicated "yes".

The topic of this book is continuum modeling. I am including a sketch of the study by F & S to illustrate a diffucult decision that must be made by anyone who sets out to construct a continuum model of an observed phenomenon, a decision regarding the mode of attack, that must be made at the outset. The approach of F & S is to begin with an abstract continuum constitutive theory of hopefully broad enough generality to characterize, as a special case, the phenomenon under study. Then a series of increasingly restrictive formal assumptions are made (in the form of *particular choices of constitutive equations*) to whittle away generality and home in on a particularized set of field equations whose solutions exhibit the desired qualitative behavior. This formal process has great power. At the least, one generates a catalog of models that *cannot* work, and, if the process succeeds, one obtains a class of models that can conceivably do the job. The power comes at a stiff price, however. At the end, one is faced with the extremely difficult task of attaching the formal theory to the underlying physics. That is, one must find, within the physical phenomenon, mechanisms that endow the formal constitutive choices with concrete reality.

[3]from available chemical potential energy...ATP?

Counterpoint to the above approach is the theory I shall present later. Instead of starting with general constitutive theories, I shall try to assemble what are, hopefully, reliable notions of what is really happening as regards force generation at the molecular level into field equations which constitute at least a stylized portrait of the phenomenon of cytoplasmic streaming. This approach has its uncertainties, too. The microscopic details are so complicated that one must simplify by discarding everything one feels is not essential. In this process lies a real danger of discarding some central feature of the physics, and thus constructing a superfluous model.

F & S did the following. Let ρ and $\mathbf{v}$ be mass density and velocity fields of the fluid. With the fluid incompressible, the mass balance equation is $\partial v_i / \partial x_i = 0$. The momentum balance equation without body forces is

$$\rho \left\{ \frac{\partial v_i}{\partial t} + v_j \frac{\partial v_i}{\partial x_j} \right\} = \frac{\partial}{\partial x_j} T_{ij}$$

where T_{ij} are cartesian components of the stress tensor. (Standard vector and tensor calculus notation with the Einstein summation convention is understood.)

If the fluid were Newtonian viscous fluid, its stress tensor would be:

$$\mathbf{T} = -p\mathbf{I} + \mu\mathbf{D}$$

where p is hydrostatic pressure, μ is viscosity, and $\mathbf{D}$ is the strain rate tensor. Such a fluid *dissipates* to heat the kinetic energy it starts with, and kinetic energy supplied by externally applied surface tractions. Clearly it is incapable of self-initiating streaming motions.

F & S studied the case when $\mathbf{T} = -p\mathbf{I} + \mu\mathbf{D} + \mathbf{T}_a$, where $\mathbf{T}_a$ is an "active" stress tensor. They supposed that $\mathbf{T}_a$ depended (via a constitutive law) upon the gradients of two scalar concentration fields, say ∇c and ∇b. c and b are two undeclared chemicals; that is, their relationship to real underlying physics is left vague. These substances diffuse in the fluid and, through "reactions", are produced or consumed locally. Thus, the balance law for c, say, is:

$$\partial c / \partial t + v_i \partial c / \partial x_i = D \nabla^2 c + R(x),$$

where D is diffusivity of c, and R is its local production rate density. b has an exactly similar balance law.

If the fluid is isotropic, how can $\mathbf{T}_a$ depend upon ∇c and ∇b? F & S explain in detail how no linear constitutive relation can satisfy the principle of material frame indifference (see Truesdell and Noll [**1965**] for an exposition on this "principle"). They also demonstrate that the following two classes of constitutive laws do satisfy this principal.

I. *Symmetric deviotoric active stress tensor*:

$$(\mathbf{T}_a)_{ij} = \lambda \left\{ -\frac{2}{3} \left[\frac{\partial c}{\partial x_k} \frac{\partial b}{\partial x_k} \right] \delta_{ij} + \frac{\partial c}{\partial x_i} \frac{\partial b}{\partial x_j} + \frac{\partial c}{\partial x_j} \frac{\partial b}{\partial x_i} \right\}.$$

II. *Assymmetric active stress*:

$$(\mathbf{T}_a)_{ij} = -k \begin{bmatrix} 0 & \omega_3 & -\omega_2 \\ -\omega_3 & 0 & \omega_1 \\ \omega_2 & -\omega_1 & 0 \end{bmatrix}$$

where the ω_i are components of the vector $\omega = \nabla c \times \nabla b$.

F & S provide examples of boundary value problems, for each case above, in which a convective instability can grow spontaneously from an initially quiescent state, that is, where the "active" part of the stress tensor, $\mathbf{T}_a$, can lead to a greater rate of kinetic energy production than the simultaneous dissipation rate through the Newtonian dissipative stress. Qualitatively, this is clearly what cytoplasm must do to stream. In this sense, the ingenious study by F & S may have *something* to say about cytoplasmic streaming.

Precisely *what* it has to say is not clear, and cannot be made clear until concrete physical analogs of the symbolic chemicals b and c are identified in cytoplasm, and until the biologicial mechanism of force generation underlying the constitutive relation exhibited above for $\mathbf{T}_a$ is similarly identified.

6. A minimal primer on the nature of actin and myosin. Before developing a mathematical model of cytoplasmic streaming, we turn for a brief look at the likely microstructure underlying the phenomenon.

It is now believed that two types of protein molecules, actin and myosin, constitute the nuts and bolts of the mechano-chemical transduction machinery in nonmuscular contractile cells. It is certain that those two molecules serve that role in striated muscle cells (mammalian skeletal muscle is striated muscle, e.g.). Since these two molecules underlie the model to be presented later, I want to put them in broad biological context by making the following simplified recitation of the current dogma on how the striated muscle actin-myosin system functions. This story can be found in most modern physiology texts (see Davson [**1970**] or Copenhaver [**1971**], e.g.).

A skeletal muscle fiber is composed of numerous parallel myofibrils bundled together. Each myofibril is a linear sequence of many sarcomeres (each about 2 μm in diameter, 3.5 μm in length). One sarcomere, shown schematically in Figure 2, consists, schematically, of a cylinder to each of whose ends (the ends are called Z-bands) are attached a great number of actin filaments (called thin filaments, each 50 Å in diameter, 1 μm long). The actin filaments run, parallel to each other and parallel to the axis of the sarcomere cylinder, from the Z-bands toward the center of the sarcomere. Each actin filament consists of two *helically wound* strands of globular protein subunits.

Centered in the sarcomere is another group of filaments, the myosin thick filaments (diameter = 100 Å, length = 1.65 μm) arranged as shown in Figure 2. These thick filaments are aggregates of myosin molecules. A myosin molecule (molecular wt. = 460,000, length ~ 1200 Å) consists, roughly, of a two-strand helix with a "claw-type" f-fragment attached onto one end of each strand. The helical strand parts of myosin molecules aggregate into a bundle

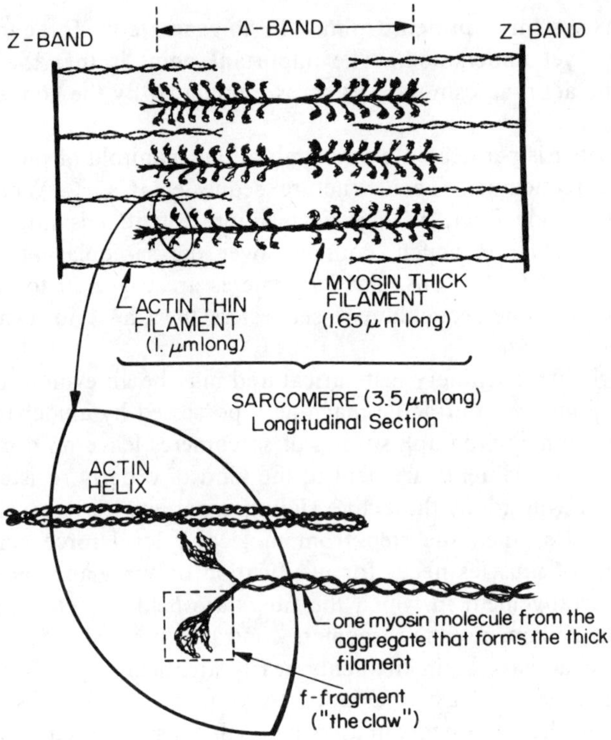

FIGURE 2

to form the thick filament, leaving the "claws" sticking out radially (the claws are about 40 Å long). The insert in Figure 2 depicts one piece of actin filament and a single myosin molecule. These *f*-fragment claws can latch onto special parts of the actin rods and pull them, causing the actin rods to slide toward the center of the myosin filaments. On opposite ends of the thick filament, the claws face in opposite directions. When the myosin claws pull the actin rods, a pair of opposite *Z*-bands is pulled together, generating the tension that results in a muscle twitch when an ensemble of sarcomeres execute the same performance in unison.

The triggering mechanism is fascinating, and is understood (see Huxley [**1969**]). Along each groove of the actin double helix lies another long chain protein called tropomyosin. Its exact position in the groove is controlled by still another protein (globular) called troponin. Troponin beads are spaced intermittently along the actin helix. When Ca^{++} ion is available (10^{-6} Molar Ca^{++} will suffice), the troponin binds it, then shifts its position on the actin helix, allowing the tropomyosin in turn to shift within the actin helix groove. With the position of tropomyosin so altered, the myosin claws can grapple the actin (whereas the initial configuration of the tropomyosin prevented this). When sufficient ATP is present to serve as chemical fuel, the claws start pulling on the actin which pull the *Z*-bands together. (The exact mechanism by which this system of proteins splits ATP to generate the force to "work the claws" is not yet understood.) The important point is that the contractile activity of the actin-myosin-etc. system is modulated by the concentration of Ca^{++} ion.

A muscle fiber is permeated by a membranous manifold of pipes called the sarcoplasmic recticulum. This structure sequesters Ca^{++}. When a nerve stimulates a muscle fiber, an electrical action potential is initiated on the muscle cell membrance, which, coursing over the sarcoplasmic recticulum, causes a dump of Ca^{++} ions. The sarcomeres are triggered to twitch once, then the sarcoplasmic recticulum re-sequesters the Ca^{++} to await the next action potential.

The remarkable machinery just caricatured may be an evolutionary refinement of the primitive contractile machinery possessed by amoebae.

Careful electron micrograph studies of sarcomeres leave no doubt that the actin and myosin filaments are held in the kind of orderly register indicated in Figure 2. If you admit the active sliding capability claimed of actin and myosin molecules, then the step from molecular level force generation to gross motions of muscles needs for justification only a glance at the regular geometrical arrangement in which the fibers are held. If actins are made to slide past thick filaments, the sarcomere contracts, period.

Some amoebae have actin filaments nearly identical to yours, and myosin molecules that are at least analogous to yours.[4] But, there is no orderly arrangement of these filaments in amoeba analogous to the sarcomere pattern just sketched. The actin and myosin are dispersed in the cytoplasm. With no such orderly structure, it is a great mystery how the molecular-level force generation by filaments sliding past each other can result in the sort of globally coherent streaming flow of cytoplasm involved in pseudopodium extension. The goal here is a mathematical exploration of that mechanical mystery.

For evidence that amoebae do in fact possess actin and myosin, see Thomas Pollard's [**1973**] review article in *The Biology of Amoeba*, and Taylor, Condeelis, Moore, and Allen [**1973**]. Briefly, actin has been purified from the cytoplasm from the multinucleate slime mold amoeba *Physarum*, from a small soil amoeba *Acanthamoeba*, and from *Dictyostelium* (the slime mold amoeba that biomathematicians know and love). These amoeboid actins are very similar to mammalian actin.

Myosins from different organisms differ. Rabbit myosin has a molecular weight of 460,000. *Acanthamoeba* myosin weighs in at 140,000. *Physarum* myosin has a molecular weight of 458,000. A general trait of primitive myosin molecules is a lesser tendency (than mammalian) to aggregate into thick filaments.

Thompson and Wolpert [**1963**] and Wolpert et al. [**1964**] made a pool of crude cytoplasmic extract from *A. Proteus.* The pool (devoid of membranes)

[4]Almost *all* organisms have similar actin, while there are great variations in myosins.

exhibited dramatic streaming motions. (The motion pictures are very dramatic.) Electron micrographs of this preparation show many actin rods interspersed with rare "thick" filaments (160 Å wide and 5000 Å long, presumably myosin with side bridges). When such thick filaments are centrifuged out, the pool of cytoplasm will not stream.

This is the kind of evidence that leaves little doubt that amoebae run on actin-myosin engines (the fuel is ATP). Very recently, Condeelis [**1975**] has purified both actin and myosin from single strands of cytoplasm of amoeba *Chaos c.*

7. Conjectures on how actin and myosin are arranged in amoebae. From now on, I want to fix attention upon pseudopodium extension in the giant amoeba *Chaos c.*

What is the distribution and geometrical arrangement of actin and myosin fibers in *Chaos*? This question has not been answered definitively due to the violent reactions *Chaos* has to any attempt to 'fix' him for electron microscope studies. *Chaos* resists quick permeation by the noxious fixation chemicals and, presumably, so alters his labile internal structure during fixation that the E. M. pictures are bland and unrevealing.

R. D. Allen has made polarized light microscope birefringence studies of live moving *A. Chaos* (too complicated to describe here in detail, see Allen [**1973**]) which indicate that submicroscopic fibers (actin?) lie oriented parallel to the long axis of an extending pseudopodium all along the column of endoplasm as it streams through the ectoplasmic tube. There is evidence that the volume fraction occupied by these fibers increases toward the tip of the pseudopodium. In the ectoplasmic gel itself, birefringence evidence indicates a patchwork quilt pattern of oriented fibers.

When a pseudopodium halts its advance, the birefringence gradually decays away indicating, perhaps, a gradual randomizing of the fiber orientation. During fixation of *Chaos* for E. M. preparation, the birefringence similarly decays.

8. Experimental evidence supporting the frontal contraction model. A clear exposition of this model together with supporting experimental evidence can be found in Allen [**1973**]. A few centrally important experiments and observations simultaneously make Allen's model tenable and present challenges to the other proposed models which, in my opinion, they cannot meet.

First, in *Chaos c.*, which moves chaotically, often extending several pseudopodia at the same time it is retracting several others, it is difficult to imagine how a manipulation of the hydrostatic pressure in the uroid could simultaneously sqeeze out and suck in. The hydraulic contraction model has been embellished to allow contractions at any site on the periphery of the amoeba. But then there is the observation, already mentioned, that infrequently, in a single pseudopodium, two distinct streams of cytoplasm move simultaneously in opposite directions. To my knowledge, proponents of the hydraulic contraction model have not suggested more embellishments to cover this kind of streaming.

Second, Allen discovered (accidentally by adjusting the microscope objective lens right through a preparation!) that a piece of pseudopodium, isolated in a short fractured length of glass capillary tube, could continue to exhibit a fountain streaming pattern similar to the flow observed in the normal extension of pseudopodia (see Allen [**1973**]). There was no tail region to contract! The cytoplasm streamed up the axis of the capillary tube and back along the walls. The implication of this extraordinarily lucky observation is that all the machinery for pseudopodial cytoplasmic streaming exists in each piece of cytoplasm. No globally imposed pressure gradient was necessary.

Third, Taylor et al. [**1973**] achieved *in vitro* isolation and control of membrane-free cytoplasm extracted from a single *Chaos* amoeba. They concocted three different chemical media, "a relaxation solution", a "contraction solution", and a "flare solution". The principal chemical *differences* between these solutions were variations in Ca^{++} and ATP concentration. Relaxation solution had 1 mM ATP and less than 10^{-7} M Ca^{++}; contraction solution had no ATP and 10^{-6} M Ca^{++}; flare solution had $\frac{1}{2}$ mM ATP and 7×10^{-7} M Ca^{++}. When bathed in relaxation solution, the cytoplasm relaxed and spread out into a puddle–no active streaming. Contraction solution caused the cytoplasm to contract into a ball, gel, and cease motion.[5] Flare solution elicited a remarkable behavior: loops of cytoplasm flared out into the bathing medium from the periphery of the pool of cytoplasm. A flare consisted of fibrils of cytoplasm moving rapidly outward (at from 19 to 43 μm per second) to some point where the cytoplasm assumed a semirigid state (ectoplasm), forming a sort of gel strut sticking out from the central pool of cytoplasm. This strut moved either not at all, or slowly back into the pool.

There were, here, no membranes, and this is the reason for the claim early in this paper that the membrane was probably not very active in the extension of pseudopodia. There was, of course, no tail to contract. Since there was no membrane the preparation was subject to quick fixation for electron microscope studies. The pictures showed thin (actin?) filaments dispersed in the loops of cytoplasm fixed in the act of flaring, and they showed thick filaments which were recognizable as bipolar myosin aggregates (about $\frac{1}{2}$ μm long, with bare central regions).[6] There was also an increase in birefringence locally in the cytoplasmic loops while they were flaring.

---

[5]I have de-emphasized the role of ATP. It is the common energy currency in biological cells. Extensive studies with striated muscle actin and myosin show that if Ca^{++} is present in sufficient concentration, but ATP is absent, the myosin "claws" lock onto actin, locking a muscle into a rigid state. Rigor mortis is one manifestation of this behavior. The behavior of amoeboid cytoplasm in the high Ca^{++}-low ATP contraction solution is thus exactly what one would expect of an assembly of actin and myosin filaments in such solution. In striated muscle actin-myosin systems, an abundance of ATP and an absence of Ca^{++} leads to dissociation of the myosin and actin molecules and, thus, relaxation. The same thing happens, evidently, in *A. Chaos* cytoplasm.

[6]Notice in Figure 2 the way the stylized thick myosin filaments are drawn with no cross bridge "claws" in the central part of each fiber. This bare central region is characteristic of all striated muscle myosin, and evidently is seen as well in amoeba *Chaos*.

By switching from one medium to another, etc., the pool of naked cytoplasm could be made to cycle repeatedly through the behavior appropriate for each solution.

When strands of cytoplasm plucked from relaxing medium were immersed in a medium with calibrated concentrations of Ca^{++}, they contracted, with the average speed achieved during contraction, V, depending upon the concentration of calcium ion $[Ca^{++}]$. This formula can be easily extracted from their paper:

$$V = \max\{0, 50.8 + 7.8 \log[Ca^{++}]\} \ \mu\text{m/sec}.$$

A strand of cytoplasm could be seen to reduce its length to 30% of its initial length when Ca^{++} was added to relaxing solution bathing the strand (to raise the calcium ion concentration to 10^{-3} M).

With all of the above as foundation, we turn to the construction of a field theory model of the phenomenon at hand. This model is intentionally oversimplified, the general target being a mathematical model with the fewest parts that embraces schematically the underlying contractile machinery, the modulation by Ca^{++} of the intensity of its action, and the global phenomenon of pseudopodium extension. The attachment of the mathematical entities in the next sections to the biological entities in the preceding sections will be left, in many particulars, as an exercise for the reader. This is somewhat different, I claim, from leaving the attachment process to the reader's imagination.

9. An intuitive description of the mathematical model.[7] We suppose that the column of endoplasm flowing up the ectoplasmic tube in pseudopodium extension consists of a mixture of contractile fibers and a viscous bathing medium, with the fibers oriented in the flow direction. The fibers are assumed to contract, reducing their length when triggered to do so by a trigger chemical, c.[8] We assume that there is a source of trigger chemical located at the tip of the pseudopodium.

It will turn out that the strength of this source determines the speed of the pseudopodium. We leave adjustment of the source strength to the discretion of the amoeba. Just how the control of internal Ca^{++} might be accomplished by an amoeba is not yet known. It is a topic of active experimental investigation.

c moves in the fluid part of the cytoplasm by diffusion and convection. For simplicity, it will be supposed that the trigger chemical and the fibers undergo a first order chemical reaction (resulting in a fiberbound version of c called B) and that each unit of trigger chemical that binds with the fibers causes the

[7]The material in §§9–14 is taken largely from Odell and Frisch [**1975**].

[8]These "contractile fibers" are to be thought of as assemblies of actin filaments threading through the cytoplasm, linked together by myosin thick filaments. When c is present the thick filaments can make the actin filaments slide to increase the degree of overlap and hence exhibit "contraction" of the "fibers".

fibers to shorten a specified amount. When and where this fiber-bound chemical, B, reaches a certain specified concentration per fiber, we shall assume that the cytoplasm forms a gel then and there. Wherever the mount of B per fiber is below a specified threshold, we assume that the fibers exist in a disaggregated state. As soon as the amount of B per fiber exceeds that threshold, the fibers link up and become capable of contraction.

We shall assume that the aggregated contractile fibers are attached near the pseudopodium tip and run continuously back to the point where the aggregation threshold is just met. As c reacts with the fibers along their entire aggregated length, each part of each fiber will shorten. With the fibers attached near the tip to the rigid outer gel tube, the fibers will move, all along their length, toward the tip. Their motion will exert a drag upon the bathing fluid and try to drag it forward. The fluid *will* move forward (to extend the pseudopodium) unless a sufficiently strong adverse pressure gradient is set up in the fluid (with pressure highest toward the tip of the pseudopodium). I will assume that there is some linear relationship between the forward speed of the pseudopodium tip and the pressure drop from inside the tip to outside the tip (that is, a certain pressure drop is needed to roll out the gel forming at the tip, and to overcome hydrodynamic drag opposing motion of the pseudopodium through the external fluid medium). With the fibers pulling forward and establishing a pressure excess inside the tip, the endoplasm column will move forward, too, at that speed which satisfies the above linear relationship.

I shall assume that there is an abundance of ATP present in the cytoplasm to fuel the contractile (sliding) process.

We shall confine attention to the formation of a pseudopodium and leave for future analysis the problem of how, at the rear end of the amoeba, the ectoplasmic gel is dissolved to recruit new endoplasm. This might be brought about by local secretion of some enzyme that strips off and sequesters c.

We now set down the contents of this section mathematically.

10. General balance laws. In this continuum theory we shall keep track of the following fields (bold face quantities are vectors or second order tensors, everything else is scalar, and all fields depend upon position $\mathbf{x}$ and time t):

f = the volume fraction of fibers ($1 - f$ is the volume fraction of fluid assuming that whatever is not fibers is fluid);

$\mathbf{u}$ = the velocity field of the fluid constituent;

$\mathbf{w}$ = the velocity of the fiber constituent;

P = the hydrostatic pressure in the fluid;

$\mathbf{e}$ = the vector orientation field of the fibers;

c = the concentration of trigger chemical in amount per unit volume of fluid (so $(1 - f)c$ is the amount per unit volume of space);

B = the concentration of the fiber-bound version of c in amount per unit volume of space (so B/f is the amount per unit volume of fiber).

The meaning and derivation of balance equations are discussed thoroughly by Lee A. Segel in this volume, so I shall just set out the appropriate balance laws for the fields defined above without detailed explanation.

Fiber conservation (*no creation of fibers*):

$$\partial f/\partial t = -\operatorname{div}(f\mathbf{w}). \tag{10.1}$$

Fluid conservation:

$$\partial[(1-f)]/\partial t = -\operatorname{div}[(1-f)\mathbf{u}]. \tag{10.2}$$

Conservation of c:

$$\partial[(1-f)c]/\partial t = -\operatorname{div}\{-D \operatorname{grad} c + (1-f)\mathbf{u}c\} - s\cdot f\cdot c. \tag{10.3}$$

Conservation of B:

$$\partial B/\partial t = -\operatorname{div}(\mathbf{w}B) + s\cdot c\cdot f. \tag{10.4}$$

REMARK. The $s\cdot f\cdot c$ terms in the last two equations represent the loss rate of c and the production rate of B due to the hypothesized reaction

$$c + f \xrightarrow{s} f + B \quad (s \text{ is the rate constant}).$$

c exists and moves only in the fluid; its convection velocity is $(1-f)\mathbf{u}$. D is the diffusivity of c. The fiber-bound B is convected at the fiber velocity $\mathbf{w}$ and does not diffuse.

For simplicity we shall make only a *kinematic* description of how the fibers move (this will be improved in §15). We shall confine attention to the part of flowing endoplasm where we can assume that the fibers lie straight and parallel to the pseudopodium axis (refer to Figure 3); that is, we shall ignore the interesting and difficult details of how the fibers and fluid round the corner to evert and gel at the pseudopodium tip.

Now $\mathbf{w}$ has only one component, w, in the x direction (x-axis parallel to pseudopodium axis). If $L(t)$ is the distance between two marked points on a contractile fiber, then $-(1/L)(dL/dt)$ is the strain rate of shortening of the fiber segment. A moment's thought will suffice to identify:

$$\partial w/\partial x = (1/L)(dL/dt)$$

where the left-hand side is evaluated at the position of the center of the "marked" fiber segment just considered.

The creation rate density of B is $s\cdot c\cdot f$, so the creation rate density of B per volume of fiber is $s\cdot c\cdot f/f = s\cdot c$. By making the strain rate of shortening of fiber segments proportional to this creation rate of B per fiber (which is the rate trigger chemical is bound per fiber) we obtain this equation for w:

$$\partial w/\partial x = -(\gamma f_0 s)c. \tag{10.5}$$

I have called the proportionality constant γf_0 (where f_0 is the fiber volume fraction in the cytoplasm back in the uroid) for later convenience.

(10.5) is the only equation so far whose applicability is restricted to one-dimensional geometry. It is important to realize that, for a different reason, (10.5) represents physical nonsense. It says that the fibers' rate of contraction is proportional to the local concentration of trigger chemical *regardless of what forces might be opposing fiber contraction*. I am therefore supposing here that whatever forces may be required to slide filaments and produce "contraction" can be generated by the actin-myosin system. This assumption is made for simplicity. The direction toward a more realistic model accounting for tension generation in the fibers will be indicated in §15.

We now account for fluid momentum balance, but instead of using the conventional sort of momentum balance equation, we use Darcy's law for flow in a porous medium suitably generalized (see for example, Whitaker [**1970**] for some discussion of this theory).

In its simplest form, Darcy's law states that for (vanishingly small local Reynolds number) flow of a viscous fluid through a porous medium (think of water oozing through a packed bed of sand), the fluid goes instantaneously and only where the pressure gradient pushes it against the resistance to flow presented by the porous medium. Generalized to the present case where the fibers constitute the porous medium, and those fibers are moving at velocity **w**, Darcy's law is:

$$\mathbf{u} - \mathbf{w} = -(1/\mu)\mathbf{K}(f, \mathbf{e})\,\mathrm{grad}\, P. \tag{10.6}$$

Here μ is the viscosity of the fluid, and **K** is the permeability tensor of the porous matrix of oriented fibers. This tensor field, which is anisotropic because flow perpendicular to oriented rods encounters greater resistance than flow parallel to the rods, will be exhibited explicitly in the following section.

(10.6) says all of the following things: (i) if the pressure gradient vanishes, then the fluid constituent moves faithfully with the fiber constituent, $\mathbf{u} = \mathbf{w}$; (ii) if the porous medium matrix does not move, i.e., $\mathbf{w} = 0$, then the fluid velocity is a linear function of the pressure gradient, with fluid velocity parallel to the pressure gradient when and only when the pressure gradient is parallel to one of the eigenvectors of **K**; (iii) when the underlying porous matrix of fibers is moving with local velocity **w**, then to an observer riding along on a fiber, case (ii) holds locally.

(10.6) is *not* a momentum balance equation. Inertial terms are conspicuously absent. The absence of inertial terms declares the approximation that the viscous resistive forces opposing flow through the porous medium so completely overwhelm fluid inertia that, were the pressure gradient suddenly abolished, flow through the medium would cease instantly. The reader is invited to calculate any Reynolds number that comes to mind based upon the data for filament size and flow speeds given at the end of §3 and to draw his own conclusions about the importance of inertia in this problem.

(10.6) is fraught with theoretical difficulties if one looks closely. What about boundary conditions? If one uses that equation for flow driven down a cylindrical pipe full of porous medium, how can the no-slip boundary condition at the pipe walls be met? The necessary boundary layer theory will not be addressed in this article. I do not know of any *rigorous* derivation of Darcy's equation from a *real* momentum balance law. We shall use it anyway because it is empirically established that it works for problems like water flow through packed soil beds, etc.

One theoretical bright spot is that (10.6) does satisfy the principle of material frame indifference cited in §5. If it did not have this property, it would be worthless.

11. The permeability tensor. Low Reynolds number flow of viscous liquid in an infinite universe pierced by a regular array of parallel, solid, straight, circular cylinders is discussed in Happel and Brenner [**1965**] (hereafter referred to as H & B). Only approximate solutions to the Stokes equations are obtained. From these come the following expressions for the components of $\mathbf{K}(f, \mathbf{e})$ relative to a coordinate system $(\mathbf{i}_1, \mathbf{i}_2, \mathbf{i}_3)$, where $\mathbf{i}_1$ coincides with $\mathbf{e}$ which is parallel to the lengths of the cylinders, and $\mathbf{i}_2$, $\mathbf{i}_3$ are any unit vectors forming an orthonormal triad with $\mathbf{i}_1$:

$$\begin{bmatrix} k(f) & 0 & 0 \\ 0 & \gamma(f) & 0 \\ 0 & 0 & \gamma(f) \end{bmatrix}$$

where

$$\frac{k(f)}{a^2} = \frac{1}{8(1-f)}\left(\frac{\pi}{4f}\right)\left\{-2\ln\left(\frac{4f}{\pi}\right) + 4\left(\frac{4f}{\pi}\right) - \left(\frac{4f}{\pi}\right)^2 - 3\right\}, \tag{11.1}$$

$$\frac{\gamma(f)}{a^2} = -\frac{1}{8(1-f)}\left(\frac{\pi}{4f}\right)\left\{\ln\left(\frac{4f}{\pi}\right) + \frac{1-[4f/\pi]^2}{1+[4f/\pi]^2}\right\}, \tag{11.2}$$

where a is the radius of the cylinders. To extract these expressions from H & B, the following identifications are needed: With b in H & B denoting the distance between the axes of neighboring cylinders, the volume fraction of cylinders is $f = \pi a^2/(2b)^2$. (11.1) and (11.2) are valid asymptotically as a/b approaches 0, thus for volume fractions much less than 1. H & B's "superficial velocity" is equivalent to $(1 - f)\mathbf{u}$ in this article.[9]

$k(f)$ and $\gamma(f)$ are given above by ungainly expressions. Since those expressions are approximations, there is not much point in using them if they can be "fit" by simpler expressions. A least squares fit of $K(f)$ for $0.0004 \leqslant f \leqslant 0.1$ by the expression $k(f)/a^2 = mf^n$ yields:

$$K(f)/a^2 \approx 0.1 f^{-4/3}. \tag{11.3}$$

A least squares fit of $\gamma(f)$ by a similar expression over the domain $0.0009 \leqslant f \leqslant 0.1$ yields[10]

$$\gamma(f)/a^2 \cong (2/3)K(f)/a^2. \tag{11.4}$$

These fits are good to about $\pm 20\%$. The details are given in Table 1.

---

[9]This point was missed in Odell and Frisch [**1975**]. Thus the right-hand sides of equations (A3) and (A4) in Appendix A of that paper should be multiplied by $1/(1 - f)$. Since $f \ll 1$ in applications, this makes no real difference.

[10]The domains over which the fits are made differ slightly. I am being sloppy in the middle to be neat at the end.

f	$k(f)/a^2$ from (11.1)	approx. by $.1f^{-4/3}$	% error
.0004	2987.	3393.	+13.6
.0006	1859.	1976.	+6.3
.0008	1324.	1346.	+1.7
.001	1016.	1000	−1.5
.002	440.5	396.8	−9.9
.004	186.8	157.5	−15.7
.006	111.9	91.7	−18.
.008	76.9	62.5	−19.
.01	57.3	46.4	−19.
.02	22.2	18.4	−17.
.03	12.4	10.7	−14.
.04	8.07	7.31	−9.4
.05	5.70	5.43	−4.8
.06	4.25	4.26	0.
.07	3.29	3.47	+5.2
.08	2.62	2.90	+10.6
.09	2.13	2.48	+16.3
.10	1.76	2.15	+22.3

$\gamma(f)/a^2$ from (11.2)	approx. by $.066f^{-4/3}$	% error	$\gamma(f)/k(f)$
1616.	2239.	38.6	
1011.	1304	+29.	.54
723.3	888.7	+23.	.55
556.8	660.0	+18.	.55
244.6	261.9	+7.1	.56
105.5	104.0	−1.4	.56
63.8	60.5	−5.1	.57
44.4	41.2	−7.0	.58
33.4	30.6	−8.2	.58
13.4	12.2	−9.2	.60
7.65	7.08	−7.5	.62
5.07	4.82	−4.8	.63
3.64	3.58	−1.6	.65
2.76	2.81	+1.9	.65
2.16	2.28	+5.8	.66
1.74	1.91	+10.0	.66
1.43	1.64	+14.5	.67
1.19	1.42	+19.3	.68

What is a rough estimate for the fiber volume fraction for amoebae? This is not known experimentally. Probably $f < 0.05$. f_0 is left as a parameter in the model, awaiting experimental determination.

With the above data fits (made for convenience only), we have the permeability tensor in component form relative to the basis $(\mathbf{i}_1, \mathbf{i}_2, \mathbf{i}_3)$ as

$$\mathbf{K}(f, \mathbf{e})_{ij} = 0.1 f^{-4/3} a^2 \begin{bmatrix} 1 & 0 & 0 \\ 0 & \frac{2}{3} & 0 \\ 0 & 0 & \frac{2}{3} \end{bmatrix} \tag{11.5}$$

where a is the filament radius, and $\mathbf{i}_1 = \mathbf{e} =$ fiber orientation direction (which, note, may vary with position and time in a general application). $\mathbf{i}_2$ and $\mathbf{i}_3$ are any vectors that form an orthonormal triad with $\mathbf{e}$.

Note that (11.4) indicates that the resistance to flow perpendicular to the rods is $3/2$ the resistance to parallel flow.

12. Specialization to the simplest interesting case of the general balance laws. First, we make the geometrical assumption used already to derive (10.5): that fiber orientation, $\mathbf{e}$, and velocity, $\mathbf{w}$, are everywhere parallel to the x-axis which is the axis of the extending pseudopodium. Further, we assume that nothing varies perpendicular to this direction, so that all vector fields, all gradients, have only x-components. This means we give up for the moment the difficult task of characterizing how fibers and fluid round the corner at the tip, and it means that we do not make the endoplasm flow adhere to the inner wall of the gel tube (that is, we do not entertain the boundary layer theory needed to do this).

Second, we attach the origin of our coordinate system to the (moving) tip of the pseudopodium. (We are free to make this Galilean coordinate system transformation because all of our balance laws satisfy the principle of material frame indifference. Since there are no inertial terms in any equation, we can do this even if the pseudopodium tip is accelerating arbitrarily.) We shall treat the cases in which, seen from this moving coordinate system, nothing changes with time. A physical interpretation will be given later. Now, all quantities depend only on x.

The coordinate origin, $x = 0$, will be located by the declaration that it is the point where the endoplasm first gels to form ectoplasm. The rest of the problem occurs in the domain $x < 0$. From §9, the gelation condition is that B/f reaches a specified threshold. Naming that threshold B_0/f_0, we have the boundary condition

$$B(0)/f(0) = B_0/f_0. \tag{12.1}$$

The assumption that the amoeba controls the strength of a source of c at the tip becomes

$$[\text{flux of } c|_{x=0} = [-D dc/dx + u(1-f)c|_{x=0} = -r_0 \tag{12.2}$$

with r_0 left to the amoeba's discretion.

We shall suppose that, at $x = 0$, the fibers and fluid move together at the same speed, which should be the case if the cytoplasm has become gel there.[11] Let Q_0 denote the common speed at $x = 0$. With the gel tube taken to be

[11]This condition can be relaxed to allow the fluid to ooze through the gel in the neighborhood of the tip. The price in analytical complication is stiff.

cylindrical, Q_0 is the mass-averaged velocity of the endoplasm for all x. We have

$$w(0) = u(0) = Q_0. \tag{12.3}$$

B is the fiber-bound version of c. Let us assume that all the c that leaves the origin (by convection and diffusion in the fluid) returns to the origin in the bound form, and that the only source for B is the c that leaves the origin. Thus

$$r_0 = -\left[\text{flux of } c|_{x=0} = +\left[\text{flux of } B|_{x=0} = w(0)B(0) = Q_0 B(0). \tag{12.4}$$

Let V' be the speed of the pseudopodium tip and let Q_0' be the speed of the endoplasm flowing up the gel tube, both as seen from the laboratory coordinate system. If A_1 and A_2 are the cross-sectional areas of the entire pseudopodium and the lumen of the gel tube, respectively, then global mass balance requires

$$A_1 V' = A_2 Q_0'. \tag{12.5}$$

In the moving coordinate system attached to the tip, the rigid gel tube will appear to move backwards at the speed V', while the speed of the endoplasm will be Q_0:

$$Q_0 = \left[A_1/A_2 - 1\right] \cdot V'. \tag{12.6}$$

If we take far ambient external pressure in the medium to be 0, then the pressure drop from inside the tip to ambient is $P(0)$. Taking the force this pressure head can exert to roll out the gel at the tip and overcome hydrodynamic drag on the tip to be proportional to the forward speed of the tip, we obtain

$$P(0)A_1 = HV', \tag{12.7}$$

where H is some proportionality constant that says how hard it is to roll out the forming gel. (12.6) and (12.7) combine to yield the boundary condition

$$P(0) = \hat{D} Q_0, \tag{12.8}$$

where $\hat{D}$ is a constant for a given pseudopodium of a given amoeba.

Aside. It is appropriate to think, here, about global momentum balance. The aggregated fibers described in §9 are attached to the rigid gel tube at the tip and contract. These fibers will be under tension. To understand what must go on at the tip to make this model work, imagine some demons, standing upon the farthest edge of the gel tube, and pulling on ropes that run backward into the core of fluid cytoplasm. The demons can exert a tension on the ropes to pull them forward (thus bringing all the cytoplasm forward to form more gel tube) if and only if they have a rigid gel tube to stand upon, to which they can transmit the reaction force. The reaction force transmitted by the demons' feet to the gel tube must push the gel tube backwards (as seen in the coordinate system attached to the tip). To complete this analogy, imagine that each incremental amount of cytoplasm rolling out gels around the demons, trapping them and incorporating them into the forming gel tube.

Each new incremental advance of the pseudopodium requires a new troop of demons.

We need more boundary conditions. As stated in §9, it is assumed that the fibers remain disaggregated and unable to contract until B/f exceeds some threshold. Naming the threshold $(\varepsilon B_0/f_0)$, that is, a small number ε ($\ll 1$) times the gelation threshold, we have the point $\hat{x}$, at which gelation occurs, located by

$$B(\hat{x})/f(\hat{x}) = \varepsilon B_0/f_0. \tag{12.9}$$

Continuous, aggregated fibers extend from $x = \hat{x} < 0$ forward to $x = 0$. (Refer to Figure 3.)

$P(0)$, through equation (12.8), determines pseudopodium speed (Q_0 remains to be determined). Equation (10.6) is the only balance law involving $P(x)$, and only the *gradient* of P appears there. Thus, the differential equations will determine P only up to an arbitrary additive constant. Because the actual value of $P(0)$ is important, that arbitrary constant is troublesome. In short, we must specify P *somewhere* in the amoeba in order to find $P(0)$. We do not want to rely upon squeezing by the uroid to elevate the hydrostatic pressure above ambient. We enforce the constraint that the hydrostatic pressure (which will be highest at $x = 0$) must fall to ambient *somewhere* between $\hat{x}$ and 0. Then at $\hat{x}$, the pressure will be ambient or below, so that no additional forces are needed to get to endoplasm moved up to $\hat{x}$, from which point fiber traction can pull it the rest of the way to the tip. The constraint is the existence of some L such that

$$\hat{x} \leqslant -L \leqslant 0 \quad \text{and} \quad P(-L) = 0. \tag{12.10}$$

For a sketch of how the pressure field comes out, refer to Figure 3.

13. The balance laws restricted to the steady-state case. With the restrictions made in the last section, the balance equations of §10 collapse to the seventh order system of nonlinear ODE's:

$$d(fw)/dx = 0, \tag{13.1}$$

$$d[(1-f)u]/dx = 0, \tag{13.2}$$

$$u - w = -\frac{0.1\,a^2}{\mu}\, f^{-4/3}\frac{dp}{dx}, \qquad (13.3)^{12}$$

$$\frac{d}{dx}\left(D\frac{dc}{dx} - (1-f)uc\right) - s\cdot c\cdot f = 0, \tag{13.4}$$

$$-d(Bw)/dx + s\cdot c\cdot f = 0, \tag{13.5}$$

$$dw/dx = -(\gamma\cdot s\cdot f_0)c. \tag{13.6}$$

The above set of equations holds for $\hat{x} < x < 0$. In the domain $x < \hat{x}$, no fiber shortening occurs, and only the reaction chemistry is important. For $x < \hat{x}$, where filaments are disaggregated, we suppose that the fiber and fluid

[12] The material of §11 has been used to obtain this equation from (10.6).

constituents go at the same speed, which by global mass conservation, must be the (still-unknown) speed Q_0 of equations (12.3) and (12.8). For $x < \hat{x}$, c and B are determined by (13.4) and (13.5) with u replaced by Q_0 and f replaced by f_0.[13] The resulting equations are linear, and can be easily solved in terms of real exponentials in x. The coefficient of the exponential term that grows as $x \to -\infty$ is set equal to zero for obvious reasons.

It remains to patch the solutions in $x < \hat{x}$ to the solutions in $x > \hat{x}$. $\hat{x}$ is a point at which the fields B, f, and w must be discontinuous. The patching is done by jump conditions, namely: the fluxes of f, B, and c must be continuous at $\hat{x}$ (since nothing accumulates at $\hat{x}$). Moreover, c must be continuous at $\hat{x}$ because c is a diffusing substance.

14. A description of the boundary value problem and its solution. The statement of the boundary value problem for steady-state motion (seen from the pseudopodium tip) has just been completed. The answer to the problem consists of the fields $c(x)$, $f(x)$, $B(x)$, $P(x)$, $u(x)$, $w(x)$, and the values of $\hat{x}$ (the location of filament aggregation), Q_0 (the speed of the endoplasm), $-L$ (the location where pressure falls to ambient). These unknowns are determined by the six differential equations (13.1)–(13.6) (a seventh order system); the five boundary conditions at $x = 0$, (12.1)–(12.4) and (12.8), the constraint (12.10); boundary conditions at $\hat{x}$ consisting of (12.9) and the jump conditions described at the end of the preceding section.

Although it is not obvious, this turns out to be a well-posed problem. I have used my space and effort here to construct and motivate the mathematical model, by far a more important matter (in the context of this volume) than the details of how to solve the resulting boundary value problem. These details can be found in Odell and Frisch [**1975**], in particular a mathematical detail of overriding importance: a rigorous proof that solutions to the boundary value problem exist and are unique.

This section concludes with a remark on a single critical manipulation of the solution strategy, then a description of the solutions.

To put the above boundary value problem in form for analytical treatment, one writes all equations in terms of scaled dimensionless variables. This reduces the number of essential parameters. In this problem, scaling the variables plays a much more important role than that. In the above problem, $\hat{x}$ is an unknown and $\hat{x}$ is the point where a number of nonlinear boundary conditions must be applied. That is an inconvenient place for an unknown to appear. Thus, in introducing dimenionless variables, x is scaled by $\hat{x}$ to get (a dimensionless version of) $\hat{x}$ shifted to appear *only* in the coefficients of the differential equations.

By introducing suitable variable transformations, the above boundary value problem can be put into the form of a single, nonlinear, second order differential equation, with boundary conditions imposed at two fixed points.

---

[13]The constant f_0 was introduced some sections back as fiber volume fraction in the cytoplasm back toward the uroid. It is one parameter of the model.

There are *three* boundary conditions, and one integral constraint, but the problem is not ill-posed because there is one unknown parameter ($\hat{x}$) in the coefficients of the differential equation, and there is a dimensionless version of the L that appears in equation (12.10) which can be adjusted.

It turns out that only five dimensionless parameters are important:

$$\varepsilon,\ f_0,\ \tilde{B} = \gamma B_0,\ \tilde{D} = \frac{0.1 a^2 \hat{D}}{f_0^{4/3}} \left(\frac{s}{D} \right)^{1/2},\ \tilde{Q} = \frac{r_0 (1 - f_0)}{B_0 (sD)^{1/2}}.$$

The first two have been defined. $\tilde{B}$ is a measure of the total contraction strain a fiber undergoes before it turns to gel. $\tilde{D}$ is a measure of the resistance the gel at the tip offers to being rolled out.

In the solution process, an early result is that the speed Q_0 is determined by the c-source strength, r_0, simply as $Q_0 = r_0 / B_0$. $\tilde{Q}$ is a dimensionless version of this speed. $X = -\hat{x}(s/D)^{1/2}$ is the dimensionless version of the length over which fibers are aggregated.

Solutions exist and are unique in the following sense. If $\tilde{Q}$, $\tilde{B}$, f_0, and ε are prescribed and positive with $\varepsilon < 1$, then there exist a unique X and unique dependent variable fields that satisfy all of the boundary conditions except an integral constraint that comes from the condition (12.10). All solutions for sufficiently small $\tilde{Q}$ can satisfy this integral constraint: for $\tilde{Q}$ exceeding a certain value (depending upon B, f_0, ε, and $\tilde{D}$) the solutions are not valid. That is to say, the theory predicts a speed limit for pseudopodium extension.

Typical solution fields are shown in Figure 3. They are representative of the qualitative features of *all* solutions for all parameter values. Fibers outrun fluid ($w > u$) in all of the domain $\hat{x} < x < 0$. According to (13.6), the local fiber contraction strain rate is proportional to c. $c(x)$ is seen to increase rapidly toward the tip (refer to Figure 3). Thus the "contractile activity" is sharply concentrated near the tip of the pseudopodium, justifying the name "frontal contraction model" for this mathematical transcription of Allen's biophysical theory of the same name.

Looking at $w(x)$ in Figure 3, we see that fiber speed is greatest near $\hat{x}$, slowing as the tip is approached. Particles can be seen in living amoebae executing the same kinematic performance–a hint of agreement between model and experiment if those particles were somehow hooked to the fibers.

Fiber volume fraction, $f(x)$, is seen to increase from $\hat{x}$ toward the tip, and, while fiber volume fraction is high for $x < \hat{x}$, the fibers are disaggregated there. This seems to be in agreement with the birefringence studies, referred to earlier, that indicated an increase in some kind of filament orientation density from tail to tip.

In the domain $\hat{x} \leqslant x \leqslant 0$, hydrostatic pressure increases in the direction of the pseudopodium tip. It is a prediction in exact opposition to the hydraulic contraction model's. Local pressure measurement inside a moving amoeba is an exceedingly difficult technical problem. Nearly any kind of probe inserted into an amoeba insults the amoeba. It reacts by surrounding the probe with hard gel. This pressure field prediction may not be testable for this reason,

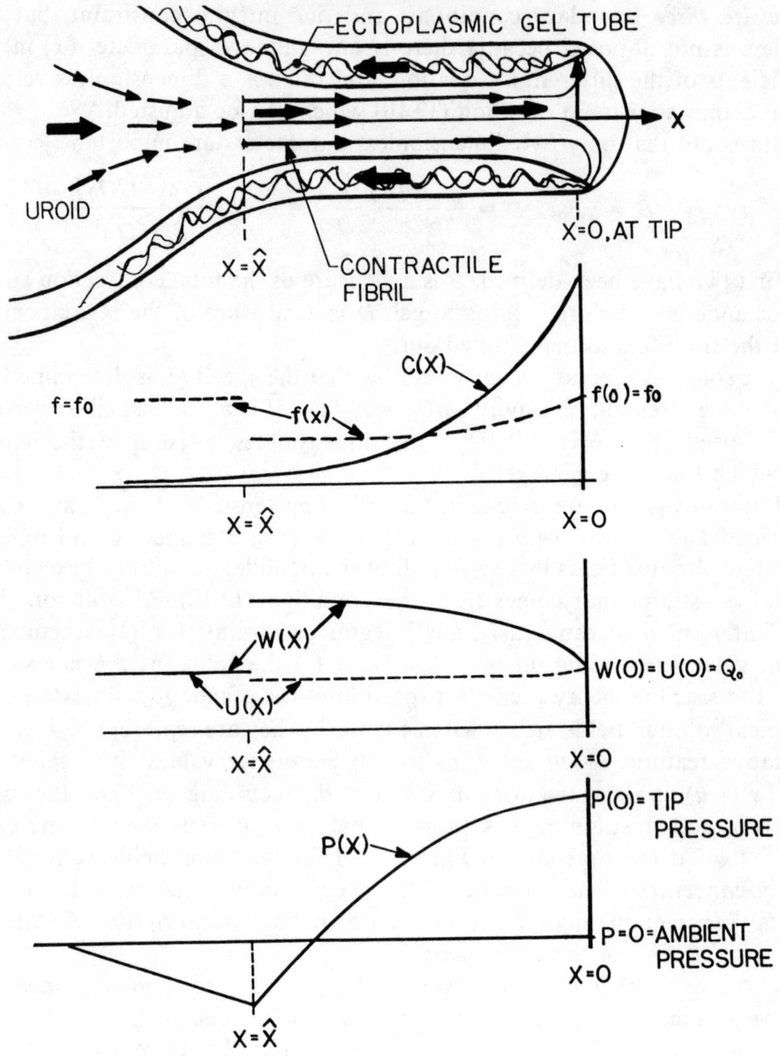

FIGURE 3[14]

but the possibility exists in principle. Note that measurements of the effects of externally applied pressure forces upon pseudopodium motion have little to say about the details of the internal pressure field.

Dimensionless speed, $\tilde{Q}$, and dimensionless length, X, turn out to be related as shown in Figure 4. This is a typical case for fixed $\tilde{B} = 0.5$, $\varepsilon = 0.025$, and $f_0 = 5\%$. Other cases are plotted in Odell and Frisch [**1975**].

The message in Figure 4 is that the longer they are, the slower they go. For fixed $\tilde{B}$, f_0, and ε, *there is a maximal possible pseudopodium length predicted.* It

[14]Reprinted from The Journal of Mechanochemistry and Cell Motility, vol. 4, pp. 1–13, with permission of Gordon and Breach, Publishers.

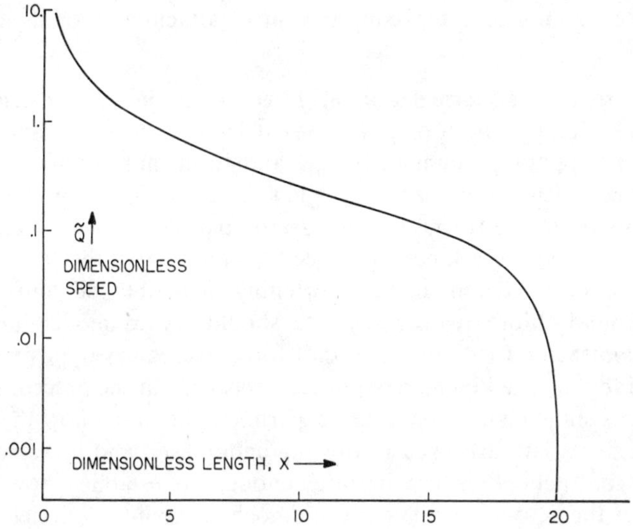

FIGURE 4

must be remembered that these solutions are for a steady-state problem as seen from the tip, hence an artificial problem because amoebae start their pseudopodia and stop them. Therefore, comparison of these predictions with experiment is uncertain.

Recall that high speed (high $\tilde{Q}$) solutions are disallowed as explained above. From Figure 4, this means that the steady-state version of too-short pseudopodia are disallowed. This raises the interesting question of whether or not pseudopodia can start up (in a time-varying version of the problem presented here) from zero length. The time-varying problem, of course, involves full partial differential equations, and is analytically formidable. A numerical study of this question is in progress, but no results are available yet.

The interpretation of the steady-state problem solved here is as follows. Riding along at the pseudopodium tip, *nothing* changes with time. The solid gel tube moves backward; the endoplasm column moves forward; the fluid of the external medium flows past the tip, rearward, at the same speed as the gel tube. Somewhere in the tail end of the amoeba, the ectoplasmic gel must be dissolved to recruit new endoplasm to flow forward. The physical interpretation is that of a continuously streaming fountain pattern of amoeboid cytoplasm "swimming" through a viscous medium. This is none too realistic in the sense that amoeba do not ordinarily do this. Nevertheless, it is the simplest caricature of what amoebae do. This model corresponds almost exactly to the situation described by Taylor et al. [**1973**]. Refer to their Figure 9. In that figure, they show a point of attachment of the gel to a substrate. This point of attachment would be seen to move backwards when viewed from the coordinate system attached to the tip. But, as regards the mechanics of internal streaming of cytoplasm, a point of attachment, or its absence, is not relevant.

15. A more general formulation of fiber orientation and mechanics. To approach a characterization of exactly what happens as fibers and filaments roll out at the pseudopodium tip to gel and form more of the ectoplasmic tube, we need field equations that allow arbitrary fiber orientation, and arbitrary motion of the fibers, not necessarily parallel to their orientation.

Equation (10.5) gave a kinematic specification of how the fibers move by postulating a specified amount of shortening of a fiber per unit of trigger chemical bound. More realistically, we should try to model how a fiber generates contractile force and how that force overcomes opposing forces to cause contraction. The kinematic approach was used in the preceding sections because of its simplicity, and because experimental data on how the actin and myosin filaments are arranged in flowing endoplasm, and how they interact to cause "contraction" is not definite enough to establish how the force dynamics of the acto-myosin system in amoebae should be modeled.

In this section we develop the formalism needed for the tasks cited in the preceding paragraphs. (This material is heretofore unpublished to my knowledge.)

It is a straightforward matter to trace through the development of the permeability tensor, as described in §11, but this time to find what forces are exerted upon the fibers, perpendicular and parallel to their orientation, commensurate with a specified difference between fluid and fiber velocities.

Let $\mathbf{F}(\mathbf{x})$ denote the force per unit length exerted upon the fiber segment at $\mathbf{x}$ by the bathing viscous fluid due to a difference between fluid and fiber velocities at $\mathbf{x}$. (We shall supress the t independent variable; in general, everything is understood to depend upon t.)

It is a worthwhile exercise to use the material in Happel and Brenner [**1965**] and the material in §11 to produce the following relationship:

$$\mathbf{u} - \mathbf{w} = \frac{1}{\mu} \frac{4f}{\pi^2 a^2} \mathbf{K}(f, \mathbf{e}) \cdot \mathbf{F}, \tag{15.1}$$

where a is the filament radius. $\mathbf{K}$ is the same permeability tensor that was produced in §11.

We now relate the force given in (15.1) to the radius of curvature, $R(\mathbf{x})$, of the fiber, and the tension the fiber at $\mathbf{x}$ bears, $T(\mathbf{x})$. In Figure 5, the solid curve depicts a fiber. s measures arc length along the fiber. $\boldsymbol{\tau}(s)$ and $\mathbf{n}(s)$ are the unit tangent and normal vectors to the fiber. The Frenet formulae (see for example, McConnell, [**1957**]) have:

$$d\boldsymbol{\tau}/ds = \mathbf{n}/R. \tag{15.2}$$

It is another worthwhile exercise to write down the conditions that both normal and tangential components of net force on an infinitesimal segment of fiber vanish (because inertial effects are being ignored). These conditions

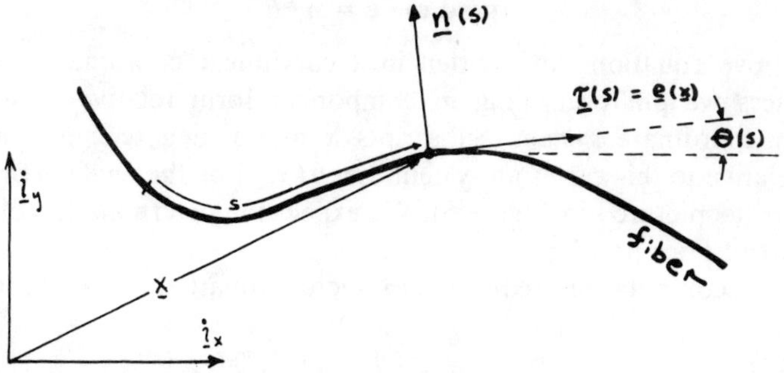

FIGURE 5

yield, first, a relationship between applied normal force, tension in the fiber, and radius of curvature,[15]

$$\mathbf{F} \cdot \mathbf{n} = T/R; \tag{15.3}$$

and, second, the condition that the tangential applied force component, $-\mathbf{F} \cdot \boldsymbol{\tau}$, equal the directional derivative of the tension in the direction of the fiber orientation (i.e., in the direction of the tangent vector field, $\boldsymbol{\tau}$). (Note that $\boldsymbol{\tau}$ and $\mathbf{e}$ coincide at each point; they depend upon different arguments.) Thus:

$$\mathbf{F} \cdot \boldsymbol{\tau} = F \cdot \mathbf{e} = -\mathbf{e} \cdot \operatorname{grad} T(\mathbf{x}). \tag{15.4}$$

I have restricted attention to the two-dimensional case, which suffices to illustrate all essential ideas. For the three-dimensional case, there is another normal-component force balance equation, and another Frenet formula involving curve torsion to use.

Now we use (15.1) and the component form of $\mathbf{K}(f \cdot \mathbf{e})$, relative to a basis oriented with the fibers, given in equation (11.5), to obtain:

$$\begin{aligned} \mathbf{u} - \mathbf{w} &= \frac{4f}{\mu\pi^2 a^2}\left[0.1 f^{-4/3} a^2\right]\left\{(\mathbf{F} \cdot \mathbf{e})\mathbf{e} + \tfrac{2}{3}(\mathbf{F} \cdot \mathbf{n})\mathbf{n}\right\} \\ &= \frac{0.4}{\mu\pi^2} f^{-1/3}\left\{(\mathbf{F} \cdot \mathbf{e})\mathbf{e} + \tfrac{2}{3}(\mathbf{F} \cdot \mathbf{n})\mathbf{n}\right\}. \end{aligned} \tag{15.5}$$

Taking the dot product of (15.5) with $\mathbf{n}$ then $\mathbf{e}$, we find $\mathbf{F} \cdot \mathbf{n}$ and $\mathbf{F} \cdot \mathbf{e}$, which, when combined with (15.3) and (15.4) yields

$$\frac{T}{R} = \frac{3}{2}\left(\frac{\mu\pi^2}{0.4}\right) f^{1/3}(\mathbf{u} - \mathbf{w}) \cdot \mathbf{n}, \tag{15.6}$$

$$\mathbf{e} \cdot \operatorname{grad} T = -\left(\frac{\mu\pi^2}{0.4}\right) f^{1/3}(\mathbf{u} - \mathbf{w}) \cdot \mathbf{e}. \tag{15.7}$$

In terms of the field $\mathbf{e}(\mathbf{x})$, the Frenet formula (15.2) becomes

[15] I am assuming the fibers have no intrinsic (elastic) resistance to bending.

$$(\text{grad } \mathbf{e}) \cdot \mathbf{e} = \mathbf{n}/R. \tag{15.8}$$

The above equations are written in a curvilinear coordinate system. For definiteness we put everything in component form relative to some fixed cartesian coordinate system. Subscripts x and y denote components of a vector relative to this cartesian system. Let $\theta(x, y)$ be the angle between $\mathbf{e}$ and the x-axis (as pictured in Figure 5). We expect the determination of this field to be part of later problems.

An easy computation reduces the vector equation (15.8) to the single equation

$$\frac{1}{R} = \cos\theta \frac{\partial\theta}{\partial x} + \sin\theta \frac{\partial\theta}{\partial y}. \tag{15.9}$$

We use (15.9) to eliminate R in (15.6), and obtain from (15.6) and (15.7) these two equations

$$T(x, y)\left[\cos\theta \frac{\partial\theta}{\partial x} + \sin\theta \frac{\partial\theta}{\partial y}\right]$$
$$= \frac{3}{2} \frac{\mu\pi^2 f^{1/3}}{0.4}\{-(u_x - w_x)\sin\theta + (u_y - w_y)\cos\theta\}, \tag{15.10}$$

$$\cos\theta \frac{\partial T}{\partial x} + \sin\theta \frac{\partial T}{\partial y}$$
$$= -\frac{\mu\pi^2 f^{1/3}}{0.4}\{(u_x - w_x)\cos\theta + (u_y - w_y)\sin\theta\}. \tag{15.11}$$

Let us now count unknowns and equations. With all fields depending upon (x, y, t) we have as unknowns in a general problem u_x, u_y, w_x, w_y, f, c, B, P, T, θ, that is, *ten* fields to determine. For balance laws we have: the four (scalar) conservation equations (10.1)–(10.4) for fibers, fluid, c, and B; the two components of the generalized Darcy's law (10.6); the two scalar equations (15.10) and (15.11) above relating fiber tension and its gradient to fiber geometry and fluid slip.

That makes eight equations. We want to dispense with equation (10.5) which determined the fiber motion *kinematically*. We should replace (10.5) by an equation that relates the local tension in the fibers to the rate of binding of trigger chemical per fiber, or something similar.[16] When we do that, we obtain a set of *nine* scalar differential balance laws to be used to determine the ten unknown fields listed above. This is a strong clue that another differential equation is needed and has been overlooked.

The discovery of the missing ingredient is left as a problem for the interested reader. Several clues have been dropped in preceding sections of this article.

16. Conclusion. This case study ends in an incomplete form because research on this problem is in a preliminary stage and still in progress. In its

[16]Mathematically, (10.5), generalized, would be interchangable with this proposed equation. (10.5), generalized, is $\mathbf{e} \cdot \text{grad}(\mathbf{w} \cdot \mathbf{e}) = -(\gamma f_0 s)c$.

present incomplete form, it must be regarded as merely a *plausibility argument* that the kind of field theory model of streaming cytoplasm presented here, involving a mixture of contractile fibers (composed of actin filaments linked together and moved by myosin thick filaments) and a viscous bathing fluid in which a locally secreted trigger chemical moves, has a chance of characterizing the biological phenomenon to which it is addressed.

Protein molecular filaments which may now be imagined to have the ability to slide past each other or contract, and which may be proved experimentally to have this ability in the future, seem to be ubiquitous in biological cells. I hope that the continuum field theory whose beginnings have been presented in this article, or some variant of that theory, will prove to have as ubiquitous a scope of application.

ACKNOWLEDGEMENT. I would like to acknowledge the careful correction of errors and substantial assistance toward making my sentences intelligible that Karen Wood contributed to the writing of these notes. Lee A. Segel made valuable suggestions about how to structure the article. Robert D. Allen either personally communicated, or aimed me at the literature containing, much of the biological information upon which this work was founded, much of which was his own research. Douglas L. Taylor kindly communicated crucial experimental results before their publication. Conversations with Thomas D. Pollard helped in sorting out how actin and myosin filaments, disperse in cytoplasm, might collaborate to produce organized "fiber contraction". I wish to thank Virginia Steffen for producing careful typescript from marginally legible manuscript.

Most of the author's research in this study was supported by the National Science Foundation under grant MPS 73-08922.

REFERENCES

R. D. Allen (1973), *The biology of amoeba*, edited by K. W. Jeon, Academic Press, New York, London, Chap. 8.

J. Condeelis (1975), Ph. D. Dissertation, Biology Department, S.U.N.Y. at Albany.

H. Davson (1970), *A textbook of general physiology*. Vol. II, 4th ed., Williams and Wilkens, Baltimore, Chap. 5.

W. M. Copenhaver, R. P. Bunge and M. B. Bunge (1971), *Bailey's textbook of histology*, Williams and Wilkens, Baltimore, 16th ed., Chap. 8.

B. A. Finlayson and L. E. Scriven (1969), *Convective instability by active stress*, Proc. Roy. Soc. London A **310**, 183.

J. Happel and J. Brenner (1965), *Low Reynolds hydrodynamics with special applications to particulate media*, Prentice-Hall, Englewood Cliffs, N. J., p. 392, MR **33** #3562.

H. E. Huxley (1969), *The mechanism of muscular contraction*, Science **164**, 1356.

K. W. Jeon (editor) (1973), *The biology of amoeba*, Academic Press, New York and London.

H. Komnick, W. Stockem and K. C. Wohlfarth-Botterman (1973), *Cell motility: Mechanisms in protoplasmic streaming and amoeboid movement*, International Review of Cytology **34**, 169.

A. J. McConnell (1957), *Application of tensor analysis*, Dover, New York, Chap. XIII, MR **19**, 1074.

G. M. Odell and H. L. Frisch (1975), *A continuum theory of the mechanics of amoeboid pseudopodium extension*, J. Theoret. Biology **50**, 59.

J. A. Subirana (1970), *Hydrodynamic model of amoeboid movement*, J. Theoret. Biology **28**, 111.

D. L. Taylor, J. S. Condeelis, P. L. Moore and R. D. Allen (1973), *The contractile basis of amoeboid movement*. I. *The chemical control of motility in isolated cytoplasm*, J. Cell Biology **59**, 378.

C. M. Thompson and L. Wolpert (1963), *The isolation of motile cytoplasm from amoeba proteus*, Experimental Cell Research **32**, 156.

C. A. Truesdell and W. Noll (1965), *The non-linear field theories of mechanics*, V. III/3 of Handbuch der Physik, Springer-Verlag, Berlin and New York, p. 44.

S. Whitaker (1970), *Advances in theory of fluid motion in porous media*, Flow Through Porous Media, A Sympos. Washington, D. C., American Chemical Society, p. 31.

L. Wopert, C. M. Thompson and C. H. O'Neill (1964), in *Primitive Motile Systems in Cell Biology* (R. D. Allen and N. Kamiya, editors), Academic Press, New York, p. 79.

Department of Mathematical Sciences, Rensselaer Polytechnic Institute, Troy, New York 12181

Lectures in Applied Mathematics
Volume 16, 1977

Earthquake Sources

Leon Knopoff and John O. Mouton

I. **Introduction**. The association of earthquakes with sudden motions on geological faults was first reported by Reid [**1910**] as a consequence of the observations of major sudden relative displacements on the San Andreas fault in connection with the 1906 San Francisco earthquake. Over a portion of the fault approximately 450 km in length, points on opposite sides of the fault that had been next to one another only moments before the earthquake were, after the shock, found to be displaced an average of 4 meters!

The identification of the earthquake source meant that it was possible to consider the recording of seismograms as a simple passive network consisting of a source, a transmitting medium and detector, in series. The theory of the linear seismograph has been known for some time; in any case, response curves of complex magnification could be measured in lieu of calculation. The theory of wave propagation through a linear elastic medium as complex as the earth is, with all its heterogeneity–and even anisotropy in some instances–has kept the attention of theoretical seismologists for the first three-quarters of this century. Basically, these problems are easily described, even though the execution may be complicated. Problems of sphericity, layered media, the surface of the earth, the core of the earth, etc., can be, and have been, dealt with. The equations of linear elasticity are well known and solutions to problems in elastic wave propagation are becoming more and more commonplace.

However, the specification of the source, and especially the time dependence of the earthquake source, has not been a subject of much concern until recently. Usually, the time dependence at the source is modeled to be impulsive in time, which in part reflects the impression that the duration of the earthquake is short. However, for the San Francisco earthquake, if one takes a velocity of rupture as being of the order of 3 km/sec, the fault took roughly $2\frac{1}{2}$ minutes to tear from one end to the other (assuming that the faulting was unilateral), a time which is not short by seismological standards. Nevertheless, earthquakes of the size of the 1906 event occur only once every five years or so somewhere in the world. Most of the seismic activity is concentrated in more frequent, smaller earthquakes. In Southern California, for example, an average of two earthquakes per year occur that have the capability of causing damage, were they to occur in populated areas (of course, most of them do not). One can guess that earthquakes at this threshold of size have a fault length of about 1 km or less; hence their source

AMS (*MOS*) *subject classifications* (1970). Primary 73N10.

spectral energy will be concentrated in much higher frequencies and, at distances of some thousands of kilometers, the events can indeed be treated as impulsive in time and as points in space. Of course, the theoretician recognizes that the solution to the wave-propagation problem for a source-time function $h(t)$ can be obtained by a simple convolution once the result has been obtained for impulsive time dependence. The nature of the source time function remains both a theoretical as well as an experimental subject of uncertainty.

Until recently, the subject of the source-time dependence was not considered to be too important for all but the largest earthquakes. Seismographs used in observation were recording in frequency bands outside the range governed by the length of small and moderate sized earthquake ruptures; only for the largest earthquakes did the approximation that the sources are impulsive in time break down. Recent interest in the problems of earthquake prediction have aroused interest in extending the frequency range of recording and thereby studying the source-time function.

In these notes, we propose to give exposition to the problems of the initiation, extension and ultimate cessation of motion on an earthquake fault. We shall assume that the forces between atoms are perfectly linear, i.e., Hookean elasticity applies, except at the moment of rupture. Thus, the elastodynamic field radiated is calculable by all the relationships of ordinary elastic wave theory. These relationships are easily summarized in terms of a representation theorem. For a homogeneous, isotropic medium, the wave equation of elasticity is

$$\alpha^2 \nabla\nabla \cdot \mathbf{U} - \beta^2 \nabla \times \nabla \times \mathbf{U} - \frac{\partial^2 \mathbf{U}}{\partial t^2} = \mathbf{F}(\mathbf{x}, t)$$

where $\mathbf{u}$ is the displacement of a particle, α and β are the compressional and shear elastic wave velocities, and $\mathbf{F}$ is the density of body accelerations per unit volume. A solution to this equation is (Knopoff [**1956**])

$$\begin{aligned} 4\pi \frac{\partial^2 \mathbf{U}}{\partial t^2} = &- \int_V \frac{[K_\alpha F_r]}{\alpha^2} \mathbf{r}_1 \, d\tau + \int_V \frac{\langle K_\beta F_r \rangle}{\beta^2} \mathbf{r}_1 \, d\tau \\ &+ \int_V \frac{[L_\alpha \mathbf{F}]}{\alpha^2} \, d\tau - \int_V \left\{ \frac{\langle L_\beta \mathbf{F} \rangle}{\beta^2} + \frac{\langle \partial^2 \mathbf{F}/\partial t^2 \rangle}{\beta^2 r} \right\} d\tau \\ &+ \int_S [K_\alpha \nabla \cdot \mathbf{U}](\mathbf{r}_1 \cdot d\mathbf{s})\mathbf{r}_1 - \frac{\alpha^2}{\beta^2} \int_S \langle K_\beta \nabla \cdot \mathbf{U} \rangle (\mathbf{r}_1 \cdot d\mathbf{s})\mathbf{r}_1 \\ &- \int_S [L_\alpha \nabla \cdot \mathbf{U}] \, d\mathbf{s} + \frac{\alpha^2}{\beta^2} \int_S \left\{ \langle L_\beta \nabla \cdot \mathbf{U} \rangle + \frac{\langle \nabla \cdot (\partial^2 \mathbf{U}/\partial t^2) \rangle}{r} \right\} d\mathbf{s} \\ &+ \int_S \left\{ \left\{ \frac{[\partial^3 \mathbf{U}/\partial t^3]}{\alpha r} - \frac{[\partial^2 \mathbf{U}/\partial t^2]}{r^2} \right\} \cdot d\mathbf{s} \right\} \mathbf{r}_1 \end{aligned} \tag{1.1}$$

$$-\int_S\left\{\frac{r}{\beta^2}\langle L_\beta\partial^2\mathbf{U}/\partial t^2\rangle\times d\mathbf{s}\right\}\times\mathbf{r}_1$$

$$-\frac{\beta^2}{\alpha^2}\int_S\{[K_\alpha\nabla\times\mathbf{U}]\times\mathbf{r}_1\cdot d\mathbf{s}\}\mathbf{r}_1+\int_S\{\langle K_\beta\nabla\times\mathbf{U}\rangle\times\mathbf{r}_1\cdot d\mathbf{s}\}\mathbf{r}_1$$

$$-\frac{\beta^2}{\alpha^2}\int_S[L_\alpha\nabla\times\mathbf{U}]\times d\mathbf{s}$$

$$+\int\left\{\langle L_\beta\nabla\times\mathbf{U}\rangle+\frac{\langle\nabla\times(\partial^2\mathbf{U}/\partial t^2)\rangle}{r}\right\}\times d\mathbf{s}$$

where the operators K and L are defined as

$$[K_\alpha]=\frac{1}{r}[\partial^2/\partial t^2]+\frac{3\alpha}{r^2}[\partial/\partial t]+\frac{3\alpha^2}{r^3}[\ ],$$

$$[L_\alpha]=\frac{\alpha}{r^2}[\partial/\partial t]+\frac{\alpha^2}{r^3}[\ ];$$

the square bracket implies a retardation of time inversely proportional to the velocity of compression waves and the angular bracket implies a retardation of time inversely proportional to the velocity of shear waves.

A second version of this result is given by de Hoop [1958]

$$U_i(\mathbf{x},t)=\int_V G_{ij}(F_j)dV+\int_S c_{jkpq}G_{ij}(\partial U_p/\partial\xi_q)n_k\,dS$$

$$+\partial/\partial x_q\int_S c_{jkpq}G_{ip}(U_j)n_k\,dS \tag{1.2}$$

where, for a homogeneous isotropic medium,

$$c_{ijpq}=\rho(\alpha^2-2\beta^2)\delta_{ij}\delta_{pq}+\rho\beta^2(\delta_{ip}\delta_{jq}+\delta_{jp}\delta_{iq})$$

and where the infinite medium Green's function is

$$G_{ij}(\phi)=\frac{1}{4\pi\rho}\left\{\frac{\partial^2}{\partial x_i\partial x_j}\left(\frac{1}{r}\int_0^\infty\phi(\boldsymbol{\eta},t-r/\alpha-v)v\,dv\right.\right.$$

$$\left.-\frac{1}{r}\int_0^\infty\phi(\boldsymbol{\eta},t-r/\beta-v)v\,dv\right)$$

$$\left.+\frac{1}{\beta^2}\delta_{ij}\frac{\phi(\boldsymbol{\eta},t-r/\beta)}{r}\right\},$$

$$r^2=(x_i-\eta_i)(x_i-\eta_i).$$

In the above expression, the summation convention applies. The coordinate $\boldsymbol{\eta}$ is a point in the fault surface. The first version can be derived from the second by differentiation. The integration in either of these two expressions gives the motion at a point interior to a volume V which is bounded by a surface S.

We can now formulate the radiation from the moving fault as an integral equation, which leads to an initial value problem for the seismogram at a point distant from the fault surface. Let the surface S be a surface of two sheets; one sheet is at infinity which, by a causality condition, will contribute nothing to the surface integrals; the other sheet surrounds the fault at a very small distance from it. Let it be assumed that no body forces act at the time of the earthquake. Then the motion at a remote point is given by the surface integrals in either of the above expressions, integrated over the fault surface itself, and where $\mathbf{U}$ has become the jump in the displacement vector across the fault. The initial value problem referred to above involves the *a priori* specification of the jump $\mathbf{U}(\boldsymbol{\eta}, t)$ across the fault surface and the subsequent evaluation of the integrals. The problem of the solution of an integral equation, as is well known, has been reduced to the evaluation of some integrals.

The problem of this solution to a realistic model of faulting has been reduced in this formulation to one of devising the jump in displacement $\mathbf{U}(\boldsymbol{\eta}, t)$ with some artistry. The literature is relatively full of different attempts to design this function; unfortunately, the results differ so widely that one can only conclude that the states described by the assumptions are physically unrealizable. For example, the assumption referred to above is

$$\mathbf{U}(\boldsymbol{\eta}, t) = \overline{\mathbf{U}}(\boldsymbol{\eta})H(t),$$

with $\overline{\mathbf{U}}$ the final relative displacement after the earthquake motion has terminated and $H(t)$ the Heaviside unit step function. The principal objection to this formulation is that it is often written $\overline{\mathbf{U}}(\boldsymbol{\eta}) \simeq \mathbf{U}^{(s)}(\boldsymbol{\eta})$ where $\mathbf{U}^{(s)}$ is the solution to the corresponding problem of static elasticity. In the static problem, the final state of stress on the fault is specified; the solution to the corresponding set of stresses or displacements then follows. In the dynamic problem, one does not in general know the final state of either stress or displacement, except as the dynamical problem itself is solved.

There are other variations on the representation theorems quoted above. One version is

$$U_k(\mathbf{x}) = \int_V G_{ki}(\boldsymbol{\xi}, \mathbf{x}) F_i(\boldsymbol{\xi})\, dV + \int_S \{ G_{ki}(\boldsymbol{\xi}, \mathbf{x}) \tau_{ij}(\mathbf{U}) - U_i(\boldsymbol{\xi}) \tau_{ij}(G_k) \} n_j \, dS \tag{1.3}$$

where τ_{ij} are the elastic stresses corresponding to the displacements $\mathbf{U}$:

$$\tau_{ij} = c_{ijpq} U_{p,q},$$

i.e.,

$$\tau_{ij} = \rho(\alpha^2 - 2\beta^2) U_{k,k} \delta_{ij} + 2\rho\beta^2 U_{i,j}$$

for an isotropic medium. In the expression (1.3), U_i is the Fourier transform of the displacement and G_{ij} is the Fourier transform of the Green's function (Hudson and Knopoff [**1964**]).

Finally, we note that an arbitrary set of discontinuous displacements across a fault will give the identical seismogram at a remote point as the solution to a problem in which body forces are applied to an unfaulted medium. The body force density equivalents are (Burridge and Knopoff [**1964**])

$$F_p(\mathbf{x}, t) = -\int_S \left\{ [U_i(\xi, t)] \nu_j c_{ijpq}(\xi) \frac{\partial}{\partial x_q} \delta(\mathbf{x}, \xi) \right\} dS_\xi.$$

For tangential displacements along a fault, the equivalent body forces are a suite of double couples, with the same time dependence.

Considerable progress has been made in the solution of the problems of rupture dynamics for some special cases. In particular, analysis of self-similar problems have been made with much success (Willis [**1973**], Burridge and Willis [**1969**], Kostrov [**1964**]). In self-similar configurations, the crack extends itself in such a way that its shape is always preserved. This assumption constrains the problem significantly. In particular, the velocity of extension of the crack is one of the results of the solution to the general problem. However, self-similar problems are convenient to solve since they permit the introduction of homogeneous functions into the formulation in some cases.

An important class of problems, which in general are not solved by the assumption of self-similarity, concern the problem of the growth of a fracture surface under the conditions of brittle fracture. In this case, a characteristic length must be introduced which represents the property of cohesion at the crack tip. Very little progress has been made to date on the extension of cracks into materials that fail by brittle fracture; these problems, however, are very important for the largest class of earthquakes, namely those which occur at shallow depths in the earth's crust. In the remainder of this discussion, we will present an exposition of the one-dimensional problems of fracture in the presence of cohesion, for which a complete solution exists, an example of a study of the two-dimensional problem of fracture in the absence of cohesion and an example of the use of an energy criterion to study the extension of a crack for a case in which cohesion is present, but self-similarity is not present.

II. **One-dimensional problem**. Discussions of the nature of the dynamical process of rupture in solids usually center on the shear fracture of previously unfractured materials (Kostrov [**1964**], Broberg [**1960**], Kostrov and Nikitin [**1967**], [**1970**]). To study the problem of the propagation of a tear on an earthquake fault during an earthquake, it is undoubtedly more appropriate to study the dynamics of fracture on a pre-existing fracture surface in a solid. We can imagine that an earthquake fault is prevented from moving by the presence of a static shear friction along the fault. The normal forces keeping the two blocks which abut the fault closed are derived from hydrostatic stresses, or some approximation thereto. The static shear friction is the product of the normal stress and the coefficient of static friction. While rupture is in progress, the motion is inhibited by the presence of a dynamic friction, which is less than the static friction. Only Burridge and Halliday [**1971**] have considered the effect of friction on the process of rupture on a pre-existing fault in a continuum. The examples in the literature show that the mathematical problems describing the tearing of a crack in a continuum are exceedingly complex.

The difficulty with models of tearing of a continuum is that a singularity of stress is found in the vicinity of the crack tip, varying as $r^{-1/2}$ with distance r from the tip. This singularity would appear to endow an advancing crack with the ability to overwhelm the finite breaking strength of the material in front of the crack; if this is indeed the case, then the static friction or breaking strength can have no influence whatever on the speed of rupture. This conclusion has been reached by Burridge and Halliday.

A remarkable diversity of opinion exists in the literature concerning the appropriate treatment of this singularity. Some authors moderate the effect of the singularity in stress by introducing an ancillary notion, such as a zone of plastic deformation surrounding the crack tip (Kostrov [**1964**], Dugdale [**1960**]), or a finite crack thickness which effectively blunts the crack tip, at least in the static case (Griffith [**1920**], Starr [**1928**]). Burridge and Halliday take the implications of the singularity at face value, and find that static friction influences the propagation of cracks only at the moment of onset of fracture, when the nucleation of the crack occurs.

Barenblatt [**1962**] proposed that cohesive forces play an important role in the description of the stress field in the neighborhood of the crack tip. In Barenblatt's model, a transition zone exists between the unfractured part of a crystal and the fractured part; in this transition zone, the bonds holding atoms together steadily weaken between the fractured and unfractured parts (Figure 1a). On a macroscopic basis, an impulse of force is found, spread over the interval of the transition zone, since a stress gradient is a force density and the transition connects two regions each of uniform stress (Figure 1b). The consequence of Barenblatt's model is that the stress exterior to the crack falls off as $r^{1/2}$ instead of as $r^{-1/2}$ in the Griffith model. Thus, the cohesive forces may be imagined to have introduced a stress singularity at the crack tip which varies as $r^{-1/2}$ and which is perfectly adjusted to cancel the singularity resulting from the relief of the prestress by the fracture.

The relative dimensions of the two functions of distance mean that the cohesive forces at the crack tip have a characteristic length over which these forces are deployed (Figure 1). The presence of a characteristic length at the crack tip means that the two- or three-dimensional problems cannot be analyzed as a self-similar crack; the one-dimensional problem can be so analyzed in certain special cases. However, in all cases, the conditions on the distributions of static and dynamic frictions as well as of external driving stress on the fault must be very unusual to guarantee self-similarity. We prefer in these discussions to carry the general problem as far as possible.

The problems of the mathematical theory of the propagation of cracks,

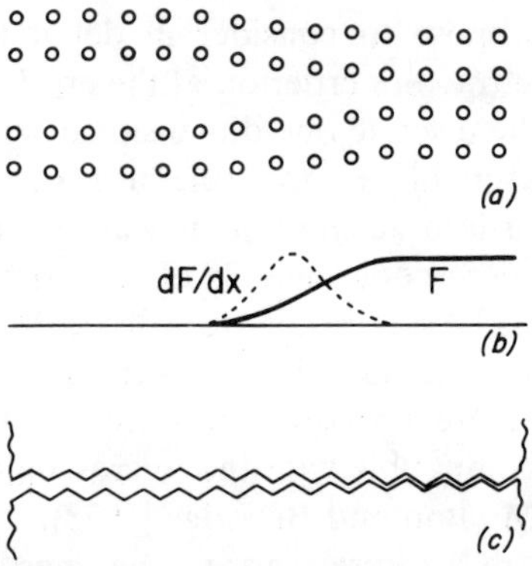

FIGURE 1. (a) Fracture of a crystalline material. (b) The forces per particle F across the interface. The stress is localized near the crack tip. (c) The fracture of a fault under friction. The tensile character of the fractures shown in (a) and (c) is for the purpose of illustration; the scheme applies also to shear fractures.[1]

whether in the presence of a pre-existing fault or not, seem to be complicated by the presence of some mathematical features which would appear to obscure the physical processes which govern the process of tearing. We believe that the physical processes are rather simple and we have sought a way to strip off the baroque mathematics to lay bare the physics beneath. Much of the complex mathematics of the dynamics of rupture of a fault in a continuum appears to be associated with the two- or three-dimensionality of the continuum. Accordingly, we investigate below the properties of the tearing of a one-dimensional continuous system in which friction plays a prominent role. We have found that, although the mathematics of the one-dimensional crack analog with friction is much simpler than the other cases, additional arguments must be presented to avoid the pitfall of merely postulating a differential equation and proceeding to its solution. The reasons for this would seem to be that, although our familiarity with the nature of the influence of friction is large when we study the motion of discrete particles, our intuition is not as strong when we consider the case of the continuum.

One way to approach the problem of the one-dimensional continuum is to consider it as the limiting case of the problem of a one-dimensional discrete array, with the limit being taken as the particle spacing vanishes. The discrete array has been considered by Burridge and Knopoff [**1967**] and many of the features of focal processes are reproduced in their description of the faulting in such a medium. It is not our intention to consider the dynamical fracture criterion from the point of view of atomic processes, although such criteria are certainly of interest to metallurgists and others. Instead, we use the continuum limit of the discrete array to derive a suitable set of differential equations and edge conditions.

---

[1] Reprinted from "The dynamics of a one-dimensional fault in the presence of friction" by L. Knopoff, J. O. Mouton and R. Burridge, Geophysical Journal of the Royal Astronomical Society **35** (1973), 169–184, with permission of the publisher.

In this section, we propose to consider in detail the role of friction in formulating a reasonable fracture criterion at the crack tip. We will show that the continuum limit of the discrete one-dimensional array can lead to a result which is physically reasonable in the role it assigns to static friction if cohesive forces are taken into account. This will not involve us with *ad hoc* remedies or particular microscopic models of fracture in the construction of the continuum equations. As a result, it will be found that, in fact, the cohesion acts with singular strength at the crack tip as the tip moves along a continuous fault. A fracture criterion, or edge condition, will be derived which we believe to be compatible with the microscopic fracture mechanisms listed above (Knopoff, Mouton and Burridge [**1973**]).

Suppose the particles in a discrete array have mass M and are separated one from another by a distance a; each is coupled to its nearest neighbors by linear springs with force constant k. The left-most particle, or particle number one, is attached by such a spring to a rigid wall, and the entire semi-infinite array rests on a rigid substrate (see Figure 2). Initially, the particles are bound to the substrate by static friction, and this bond persists until the local breaking strength B_n is exceeded. After rupture from the substrate, particle n is impelled by a pre-existing "tectonic" force T_n, together with the internal spring forces and a restraining contribution from "dynamic friction" D_n. The equations of motion for the moving particles are given by

$$M\ddot{U}_n + k(2U_n - U_{n-1} - U_{n+1}) = T_n - D_n, \qquad \dot{U}_n > 0, \qquad (2.1)$$

where U_n is the displacement of the nth particle relative to its initial position.

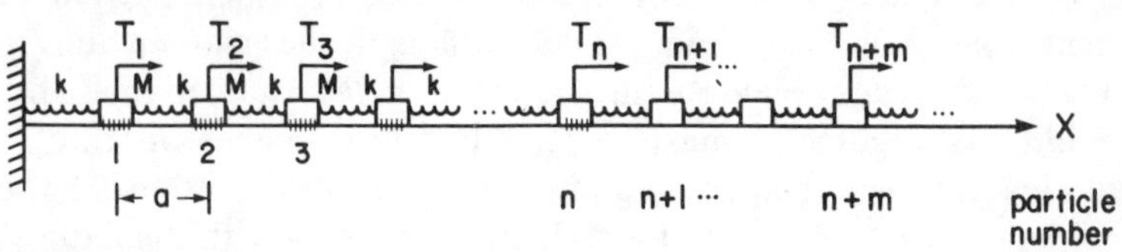

FIGURE 2. Schematic diagram of a discrete system of massive particles interconnected by springs to each other and to a source of tectonic stress, and resting on a plane under friction.[2]

Suppose that the first particle is about to rupture ($T_1 = B_1$) at time $t = 0$, and that thereafter it moves to the right until (a) it causes the next particle to rupture, or, failing that, (b) it comes to rest and is again frozen in place by static friction. If the tectonic forces T_n are sufficiently large relative to the breaking strengths B_n, a cascading sequence of ruptures can result which can be likened in a natural way with the propagation of a continuous crack.

We wish to describe the continuum limit of this model, by which we mean, a dynamical description in terms of the continuous variable x, the distance along the fault, as the separation a between the discrete particles approaches zero. In this limit, the local displacements $U_n(t)$ may be replaced by a

---

[2]See footnote 1, p. 227.

function $U(x, t)$ of two continuous variables. We scale the masses and spring constants as $M = \rho a$, $k = \mu/a$, in such a way that the quantities $\rho(x)$ and $\mu(x)$ remain finite. These two quantities are of course the mass density and the elastic modulus. Finally, since we want finite force densities ("stresses"), the quantities T_n and D_n are scaled so that for some function $f(x)$, usually called the stress drop,

$$\lim_{a\to 0} \frac{T_n - D_n}{a} = f(x), \qquad \lim_{a\to 0} na = x.$$

In the limit $a \to 0$, the equations resulting from a division of the equations of motion (2.1) by a, may be replaced by

$$\rho U_{tt}(x, t) - \mu U_{xx}(x, t) = f(x) \tag{2.2}$$

for μ independent of x. The subscripts t and x denote the corresponding partial derivatives. We recognize this result as the one-dimensional wave equation, with characteristic velocity $c = (\mu/\rho)^{1/2}$. In view of the subsequent discussion, it is important to note that the quantity $U(x, t)$ is not the usual particle displacement in the elastic wave equation; instead it is the jump in the displacement across the walls of the crack. We are taking $f(x)$ to be independent of t.

A dimensional analysis of the terms in equation (2.2) shows that the terms μU_{xx} and f are stresses. In the usual wave equation in three dimensions, which would formally look like equation (2.2), even though the wave function is presumed to be a function of only one coordinate (x) and time (t), the terms μU_{xx} and f are stress gradients. This anomaly is evidently due to the fact that we are describing the motion of a thin chain. Thus we see that the quantity μU_{xx} must be used to compute the stress drop in the case of the linear chain.

This seemingly disturbing feature of the calculation can be made more understandable if we imagine the chain to have a unit width in the fault plane, perpendicular to the direction of motion of the particles. Equation (2.2) still holds, but now the quantity μU_{xx} becomes a stress gradient since μ takes on the normal role of a shear modulus. The quantity f now becomes a stress drop per unit width of fault. Hence μU_{xx} is the correct operation to obtain the stress drop, in the case of the linear chain, or the stress drop per unit width of fault in the two-dimensional problem constrained so that motion takes place in one direction. We have replaced the seeming logical inconsistency by a physical one: by making the fault surface a plane, or strip of unit width, and requiring that all points in the plane with the same coordinate x move in the same way, we require that an infinite rupture velocity be present in the fault plane in the direction of the dimension of unit width. A more realistic problem cannot be one-dimensional.

As may have been expected, the breaking strength B_n, or its scaling as $a \to 0$, does not appear explicitly in the equation of rupture (2.2). This is reasonable since the breaking strength affects only the condition of tearing at the crack tip, if at all, and not the subsequent history of relative motion of the crack where the dynamical friction is important. The question of the scaling of B_n as $a \to 0$ is a point of some delicacy. As the development in the next section will show, B_n does not become a force density in the continuum limit, for the simple reason that it acts at a single point, namely the crack tip, and is therefore distinguishable on dimensional grounds from the other forces in the discrete theory. If the Burridge-Halliday model is appropriate, the strength of the singular force is zero; on the other hand, if cohesive forces on pre-existing cracks play much the same role as they do on the fracture of fresh material, the strength of the singularity is not zero.

Next, we determine how and whether static friction acts upon a continuous system, and derive a fracture criterion which governs the rupture velocity in our unilateral, one-dimensional fault. Some unavoidable complexities arise in the two- and three-dimensional cases which will amply justify our investigation of the simple one-dimensional model. We draw upon the method of Courant and Friedrichs [**1948**] who used the basic conservation laws to determine the physical conditions which arise in the neighborhood of moving discontinuities. We propose to follow the method of Courant and Friedrichs with the modification that we investigate continuity of momentum across the tearing edge for the discrete case before proceeding to the limit of the continuum.

Let the nth particle begin to move at time t. As the particle spacing a approaches zero, n increases so that $na \to \xi(t)$ for the continuum. Suppose that, at a slightly later time $t + \Delta t$, the $(n + m)$th particle has just begun to move. As $a \to 0$, m also increases to infinity. When Δt is small enough, we can write $\xi(t + \Delta t) \cong \xi(t) + \dot{\xi}(t)\Delta t$, where $\xi(t)$ is the coordinate of the tearing edge at the time t, by using the average time interval between successive ruptures, $\delta t = \Delta t / m$, and setting $\dot{\xi}\delta t = a$. We anticipate that we shall take the continuum limit $a \to 0$, followed by the independent limit $\Delta t \to 0$.

Restricting our attention to the subsystem consisting only of particles n, $n + 1, \ldots, n + m$, the total change in momentum ΔP of this subsystem during the interval $(t, t + \Delta t)$ is found to be

$$\Delta P = \sum_{j=n}^{n+m} M\dot{U}_j(t + \Delta t) \tag{2.3}$$

where $\dot{U}_j$ is the velocity of particle j, since all the particles in the subset are at rest at time t. Replacing M by $\rho a = \rho\Delta x$, this sum becomes an integral in the continuum limit

$$\Delta P = \int_{\xi(t)}^{\xi(t+\Delta t)} \rho U_t(x, t + \Delta t)\, dx.$$

For Δt small, the integral can be approximated as

$$\Delta P = \rho\dot{\xi}(t) U_t(\xi(t), t + \Delta t)\Delta t. \tag{2.4}$$

In this approximation we have used the value of the integrand at the lower limit $x = \xi(t)$, since we can anticipate the possibility of a discontinuity of $U_t(x, t + \Delta t)$ at the crack tip $x = \xi(t + \Delta t)$.

We invoke Newton's law in the form

$$\Delta P = \int_t^{t+\Delta t} F(\tau)\, d\tau, \tag{2.5}$$

where the force F is the sum of all external forces acting on the subsystem during the time interval $(t,\ t + \Delta t)$; we then take the limit $a \to 0$. The contributions to F fall naturally into three categories, which need to be discussed separately.

The first contribution, F_1, is that of the left-most spring acting on the subsystem; the influence of this force is felt throughout the interval $(t,\ t + \Delta t)$. In the discrete case this is

$$F_1 = k[U_{n-1}(\tau) - U_n(\tau)], \qquad t < \tau < t + \Delta t. \tag{2.6}$$

The corresponding contribution to the right-hand side of equation (2.5) in the continuum is then

$$\begin{aligned} \int_t^{t+\Delta t} F_1\, d\tau &= -\mu \int_t^{t+\Delta t} U_x(\xi(t), \tau)\, d\tau \\ &= -\mu U_x(\xi(t), t + \Delta t)\Delta t. \end{aligned} \tag{2.7}$$

We have once again chosen an approximation to the integral which avoids a possible discontinuity of the first partial derivatives of U at the crack tip.

Another contribution, F_2, is made up of the driving forces $T_j - D_j$ on the particles of the subsystem $j = n, n+1, \ldots, n+m$ subsequent to rupture. The jth particle ruptures at time $t + (j - n)\delta t$, and is driven by the constant external force $T_j - D_j$ from that time until the end of the time interval $t + \Delta t$. If Δt is sufficiently small that the physical properties of the fault do not vary from $j = n$ to $j = n + m$, we can approximate each of these forces by the single term $T_n - D_n$. The contribution to Newton's law, equation (2.5), is therefore

$$\begin{aligned} \int_t^{t+\Delta t} F_2\, d\tau &= (T_n - D_n) \sum_{j=n}^{n+m} (n + m - j)\delta t \\ &= \tfrac{1}{2} - m(m+1)(T_n - D_n)\delta t. \end{aligned}$$

This can be rewritten in terms of the rupture velocity on the discrete chain by setting $m = \Delta t/\delta t$, $\dot{\xi} = a/\delta t$. We get

$$\int_t^{t+\Delta t} F_2\, d\tau = \tfrac{1}{2}(\Delta t)^2 \dot{\xi}(T_n - D_n)/a. \tag{2.8}$$

In the limit $a \to 0$, this becomes

$$\int_t^{t+\Delta t} F_2\, d\tau = \tfrac{1}{2}(\Delta t)^2 \dot{\xi} f(\xi(t)) \tag{2.9}$$

where as before $f(x) = \mathrm{Lim}_{a \to 0}(T_n - D_n)/a$. Since (2.9) indicates F_2 is of second order in Δt, it may be neglected in equation (2.5) as $\Delta t \to 0$, since we shall retain only first order terms.

The last contribution, F_3, must be treated with great delicacy for it is here that breaking strength or the static friction must enter our discussion. F_3 arises from the force exerted by the substrate to resist each successive rupture. The jth particle ruptures at time $t + (j - n)\delta t$ $(j = n, n + 1, \ldots, n + m)$ and, accelerating from rest, compresses the spring between it and the $(j + 1)$st particle. The $(j + 1)$st particle, already subject to the tectonic force $T_{(j+1)}$, receives an additional spring force equal to $kU_j(\tau)$. The bond between the substrate and particle $(j + 1)$ supplies a resistance (oppositely directed force) equal in magnitude to $T_{(j+1)}$ initially; this increases to

$$T_{(j+1)} + kU_j(\tau) = B_{(j+1)}$$

at time $t + (j - n + 1)\delta t$; at the latter instant, rupture of the $(j + 1)$st particle occurs.

The force in the spring which connects the jth and $(j + 1)$st particles is just $-kU_j(\tau)$ in the time interval between the onset of motion of the two particles. This force is zero at the beginning of the interval and is $T_{(j+1)} - T_{(j+1)} - B_{(j+1)}$ at the end of the interval. The duration of the interval is δt. Here we assume, as in the case of equation (2.2), that the threshhold static friction is $B_{(j+1)}$ until the moment of rupture and, immediately after this instant, the frictional force is $D_{(j+1)}$; other models of the transition (in time) from static friction to dynamic friction are possible. The general behavior of the force in the connecting spring is depicted in Figure 3. The integral of this force is $-p(B_{j+1} - T_{j+1})\delta t$, for some constant p, $0 < p < 1$. No matter how small a is, it is unreasonable that p should assume either of the two extreme values: if $p = 0$, then the jth particle does not move until the critical instant for the $(j + 1)$st particle, while if $p = 1$, the jth particle reaches its terminal displacement at the instant of rupture. The contribution to the right-hand side of equation (2.5) of each of the forces in the succession of springs which connects the ruptured and nonruptured elements, is equal to

$$-p\sum_{j=n}^{n+m}(B_{j+1} - T_{j+1})\delta t = -p(B_n - T_n)\Delta t \tag{2.10}$$

where once again we have assumed (B_j, T_j) constant across the subset of particles.

We can now pass easily to the limit $a \to 0$. We set

$$p(B_n - T_n) = g(na).$$

Below we comment on the differences in the scaling of the forces $(B_n - T_n)$ and $(T_n - D_n)$. The continuum limit of the contribution from (2.10) is then

$$\int_t^{t+\Delta t} F_3\, d\tau = -g(\xi(t))\Delta t. \tag{2.11}$$

We collect the contributions (2.4), (2.7) and (2.11). Newton's law (2.5) gives us

$$\rho U_t(\xi(t), t + \Delta t)\dot{\xi}(t)\Delta t = -\mu U_x(\xi(t), t + \Delta t)\Delta t - g(\xi(t))\Delta t. \tag{2.12}$$

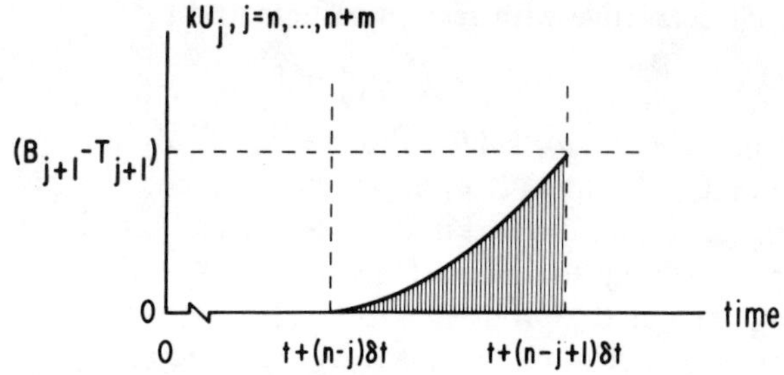

FIGURE 3. Force in the jth coupling spring connecting the jth particle which is already in motion and the adjacent $(j + 1)$st particle which is not yet in motion, during the time interval between the two onsets of motion.[3]

As $\Delta t \to 0$, we obtain the following *edge condition*:

$$U_x(x, t) + \frac{\dot{\xi}}{c^2} U_t(x, t) = \frac{-1}{\mu} g(x) \quad \text{at } x = \xi(t)^-, \tag{2.13}$$

where the symbol $\xi(t)^-$ implies that x is located infinitesimally to the left of the crack tip $\xi(t)$.

The conditions for tearing of a brittle crack are associated with the presumed time independence of $f(x)$ and $g(x)$. Physically, this means that the friction falls instantaneously from $B(x)$ to $D(x)$. In practical cases, time-dependent relaxation of the frictions occurs, but that is clearly beyond the scope of these elementary considerations.

It also will not have escaped attention that we have assumed that the parameter p is taken to be independent of the velocity of rupture $\dot{\xi}$. From (2.13) and (2.16),

$$0 \leqslant \dot{\xi} \leqslant c \quad \text{for } g(x) > 0,\ u_x(\xi) < 0.$$

The fact that p is bounded, $0 < p < 1$, suggests that p is probably independent of $\dot{\xi}$ except possibly for $\dot{\xi}$ close to zero. We have not proved this assertion.

There are some remarks that should be made here which are of particular importance in view of equation (2.13). Let us define the notion of a jump discontinuity in a function $F(x)$ at the point x: let

$$[F(x)] = F(x^+) - F(x^-).$$

Since $U(x, t) \equiv 0$ for $x > \xi(t)$, we can rewrite (2.13) as

$$[\mu U_x] + \dot{\xi}[\rho U_t] = g(\xi(t)). \tag{2.14}$$

Quantities such as $[\mu U_x]$ are thus the left-right jump discontinuity in the stress across the crack. The continuity of U itself across the crack tip is expressed by

$$[U] = 0. \tag{2.15}$$

[3]See footnote 1, p. 227.

We take a partial derivative with respect to time and get

$$[U_t] + \dot{\xi}[U_x] = 0. \tag{2.16}$$

Using the fact that $dF(x)/dx = [F(x)]\delta(x - x_0)$ at a point of discontinuity $x = x_0$, we see that (2.14) and (2.16) imply the existence of a point force (i.e., a delta function source) at $x = \xi(t)$ in the wave equation (2.2); the strength of this source is proportional to $g(\xi(t))$. Thus the effect of the breaking strength is a singular one, as noted above.

From the discussion relative to equations (2.2) and (2.11) we see that the force $T_n - D_n$ in the discrete case becomes the quantity $af(x)$ in the limit as $a \to 0$, while the force $B_n - T_n$ in the discrete case becomes the quantity $g(x)/p$ in the limit as $a \to 0$. Thus the two types of forces have different limits; the first becomes the body force density f, i.e., a one-dimensional stress, while the second remains a force g (i.e., product of stress and length). Hence the effect of the static friction is to produce a singular term in the stress gradient in the differential equation (2.2) with the singularity at the tearing edge. The singularity can be deleted from the differential equation and reintroduced in the boundary condition (2.14). We note that the $r^{-1/2}$ singularities found in the two-dimensional case do not arise in the one-dimensional problem; these terms now act as delta functions at the crack tip. The boundary condition (2.14) is our fracture criterion.

Our fracture criterion is in fact the counterpart, for the dynamical problem in one dimension, of the model of cohesion described by Barenblatt, but in this case applied to the problem of a pre-existing crack. In contrast to Barenblatt's treatment, we have taken the cohesive forces to be zero outside in an infinitesimal region in the vicinity of the tip of the fault. Because of this latter assumption the mathematics has been simplified over that which we might expect in the case of cohesive forces distributed over a finite interval; we do not believe that this assumption limits the extent of the inferences we can draw from the theory. Thus the properties of the static friction in the discrete case, taken to the continuum limit, can now be described as having two parts: one is the ordinary static friction to be observed in the unfaulted region far from the crack tip and the other is the singular cohesive force at the crack tip. Although both properties are manifestations of the same physical process, namely the static friction, we refer to them separately since they act differently on the dynamics of the crack. In fact, the friction in the unfaulted region has no influence on the rupture dynamics of the one-dimensional crack; thus it suffices henceforth to discuss only the cohesion as the only aspect of the static friction pertinent to this problem. In the subsequent section we shall use "cohesive forces", "cohesion" or "breaking strength" interchangeably.

We can imagine friction to be modelled by roughness on the interface (Figure 1c). The shear stresses in the fractured region have been large enough to allow the sawteeth to be relatively free from one another; in the unfractured region, the sawteeth are firmly enmeshed. In the fractured region, because of stresses tending to close the crack, the sawteeth tend to bounce off one another while the surfaces are in relative motion; this is the model analog of dynamical friction. In a transition zone, the teeth are able to slide a little over one another but the surfaces are not completely free. Thus in the fractured region, a small or zero force per unit length is needed for relative motion across the interface; in the unfractured region a relatively large force per unit length is required. The force and stress distributions are, once again, as shown in Figure 1b. There is an analogy with the cohesive forces in the crystalline case (Figure 1a).

It can be shown that the rate at which energy is absorbed at the crack tip is $\gamma = \frac{1}{2} g U_t$. This quantity is once again singular: in the discrete case this quantity is zero, but in the limit as the particle spacing $a \to 0$, this becomes finite in view of the singular nature of the concentrated force g at the edge. The quantity γ must represent a loss of energy in terms not accountable by mechanical effects; we propose that this represents thermal vibrations outside the low-frequency spectrum of the seismic signal, which are set up as the bonds of static friction are broken. Thus the static friction has a strong similarity to the cohesive forces arising in the tearing of fresh crystalline matter. We cannot be more specific than this; if we were able to specify precisely the atomic nature of the rupture of a crack in the presence of static friction, we could both specify the parameter p and indicate the mode of dissipation of the energy at the tip of the crack. As we have indicated above, we believe it is unlikely that p is either zero or one and hence it is unlikely that g is zero. To state this another way, we cannot conceive of static friction without the presence of cohesive forces.

We seek solutions to the one-dimensional unilateral rupture problem (2.2) subject to the conditions (2.14) and (2.15). The relation giving the position of the crack tip, $x = \xi(t)$, is determined by finding solutions to

$$(1/c^2)U_{tt} - U_{xx} = (1/\mu)f(x), \qquad 0 < x < \xi(t), \tag{2.17}$$

subject to the boundary conditions

$$U(0, t) = U(\xi(t), t) = 0 \tag{2.18}$$

and the fracture criterion

$$U_x(\xi^-, t) + (\dot{\xi}/c^2)U_t(\xi^-, t) = -(1/\mu)g(\xi). \tag{2.19}$$

When $g(x)$ is a linear function of x, $g(x) = \alpha x$, and $f(x)$ is a constant, $f(x) = f$, we can verify trivially that

$$U(x, t) = (f/2\mu)x(vt - x), \qquad \xi(t) = vt, \tag{2.20}$$

is a solution, provided the (constant) velocity v of the crack tip satisfies

$$v = c(1 - 2\alpha/f)^{1/2}. \tag{2.21}$$

The solution (2.20), (2.21) has several properties: the maximum speed of rupture is the velocity of sound c; this occurs when cohesion is absent, $\alpha = 0$. For values of $\alpha = dg/dx$ which are greater than $f/2$, rupture does not take place. The particle velocity increases with increasing distance from the point of initiation of the crack; once a particle starts to move, it continues to move with constant velocity as long as the conditions of constancy of f and dg/dx are preserved.

A somewhat wider class of solutions may be found having the general form

$$U(x, t) = \tfrac{1}{2}x\{at + bt^2 - h(x)\}. \tag{2.22}$$

The boundary conditions (2.18) imply

$$h(\xi) = at + bt^2 \tag{2.23}$$

and, to satisfy the wave equation (2.17), we must have

$$\{xh(x)\}'' = 2f(x)/\mu - 2bx/c^2. \tag{2.24a}$$

Finally, the edge condition (2.19) demands

$$\xi h'\left\{1 - \frac{a^2 + 4bh}{h'^2c^2}\right\} = \xi h'(\xi)\left(1 - \frac{\dot{\xi}^2}{c^2}\right) = \frac{2}{\mu}\, g(\xi). \tag{2.24b}$$

These solutions admit a variable rupture velocity $\dot{\xi}(t)$ and a variable particle velocity as well. Equations (2.24a) and (2.24b) require that the functions $g(x)$ and $f(x)$ have some relationship to each other, through $h(x)$ as a parametric representation. We do not propose to pursue the matter of more general solutions further in this paper.

We have considered the solution given in equations (2.20) and (2.21) corresponding to constant stress drop $f(x)$ and linearly increasing breaking strength $g(x) = \alpha x$; the result obtained was that of a constant rupture velocity for $\alpha < f/2$. If tearing is interrupted by an unbreakable barrier at $x = L$, $t = t_0$, how does this system heal, i.e., come to rest because of the *condition* $g(L) = \infty$? In the discrete case, one can easily envision that the particles pile up one against each other at the right side of the fault; each in turn comes to rest and is again frozen by static friction which we suppose acts whenever a particle comes instantaneously to rest. Similarly in the continuous case, a particle ceases to move when its velocity U_t drops to zero. The sequence of successive stopping is initiated at the barrier and proceeds leftward at a velocity equal to the velocity of sound. The configuration at the time the barrier is encountered is

$$U(x, t_0) = fx(vt_0 - x)/2\mu, \qquad vt_0 = L. \tag{2.25}$$

The healing wave arrives at $x < L$ at time $t = t_0 + (L - x)/c$; at that time, $U(x, t)$ is frozen at the value

$$\overline{U}(x) = \tfrac{1}{2} fx(L - x)(1 + v/c)/\mu, \qquad t > t_0 + (L - x)/c. \tag{2.26}$$

An alternative description of the extinction process is to imagine that the barrier at $x = L$ is a source of leftward travelling waves which serve to constrain the particle at $x = L$ to zero velocity for all time. Since the velocity in the forward travelling wave is $fLv/2\mu$ at this point, a wave of particle velocity $-fLv/2\mu$ travelling in the $(-x)$-direction is excited. This wave should satisfy the homogeneous wave equation with wave velocity c. However such a backward travelling wave of constant amplitude of particle velocity would cause each point $x < L$ to acquire a negative particle velocity. Instead, each particle comes to rest at the time of arrival of the backward travelling wave, without reversal of sign. (The dynamical friction acts in a direction opposed to the particle motion; we assume that the dynamical friction is large enough in absolute value to preclude the possibility of backswing.) This process is evidently nonlinear. Thus although the process of tearing is described by a dynamical condition, which involves a linear equation (2.17) subject to edge condition on a moving boundary (2.19) (and hence produces a nonlinear dynamical condition), the process of healing is described by a linear kinematic condition which has a nonlinear effect on the dynamics.

As described above, the stress on the linear chain is given by the quantity μU_{xx}; this term is not a stress gradient. During rupture, the stress drop on this type of fault is simply f; this can be obtained by operating with μU_{xx} on (2.20). We can understand this result from the following argument; since the particle velocities are constant, there can be no force on the particles. Thus the force driving the motion must be equal and opposite to the dynamical friction which opposes the motion. Hence the force on each particle has dropped from the tectonic force to the dynamical friction; this is the classical stress drop. We note however that the constancy of particle velocity is special to this model.

After the shock is over, the stress drop must be given by μU_{xx} operating on (2.26). This result is

$$f(1 + v/c). \tag{2.27}$$

Thus, the final state of stress is uniform; the final stress drop is between 1 and 2 times the stress drop f, which is the difference between the the tectonic and the dynamical frictional stress. The overshoot is not negligible, and the final stress on the fault will depend strongly on the mode of healing.

Suppose, instead of encountering an unbreakable barrier, the crack tip encounters free space, that is, the "earthquake" starts below the surface of the earth and ruptures until it reaches the free surface. We continue to assume that the cohesion increases with distance from the point of initial tearing; thus the maximum cohesion is at the free surface. In this case, a wave with velocity c starts at the free surface at time $t = t_0$, $x = L = vt_0$, and travels back to the beginning of the fault at $x = 0$. This backward travelling wave increases the amplitude of the particle motion over and above that due to the initial tearing. Healing then commences from the origin, which is a point of infinite strength, and progresses to the end of the fault at $x = L$. The healing wave thus arrives at x at time $t = t_0 + (L + x)/c$.

To calculate the history of this event, we note that the boundary conditions in this case are $U_x(L, t) = 0$ and, since U is continuous at the reflected wave front,

$$[U(x, L/v + (L - x)/c)] = 0.$$

Since the wave reflected from the free end, $x = L$, must satisfy the homogeneous wave equation, i.e., it is of the form $F(x + ct)$, we find that the complete solution is

$$U(x, t) = \frac{fx}{2\mu}(vt - x)H(t - x/v) + \left\{ \frac{fLc\tau}{2\mu} - \frac{fvc\tau^2}{4\mu} \right\} H(\tau),$$

$$\tau \equiv t - L/v + (x - L)/c, \qquad t < L/v + (L + x)/c. \tag{2.28}$$

The final phase of the healing process, namely that which arrives as a reflection from the barrier at $x = 0$, arrives at a point on the fault at the time $t = L/v + (L + x)/c$. From (2.28) the velocity field in general is

$$U_t(x, t) = \frac{f}{2\mu}\left\{ xvH\left(t - \frac{x}{v}\right) + c(L - v\tau)H(\tau) \right\}, \qquad t < \frac{L}{v} + \frac{L + x}{c}, \tag{2.29}$$

and in particular, at the time of arrival of the final healing pulse it is

$$U_t(x, L/v + (L + x)/c) = (f/2\mu)(cL - vx). \tag{2.30}$$

Since this quantity decreases with increasing x, the reflection from the point $x = 0$ is sufficiently large to drive the point x to the condition of zero velocity and beyond; by the argument of the first case considered above, the system freezes with the displacement condition given by equation (2.28) at the time $t = L/v + (L + x)/c$. The final displacement is

$$\overline{U}(x) = \frac{fx}{2\mu}\left\{ \left(3 + \frac{v}{c}\right)L - \left(1 + \frac{v}{c}\right)x \right\}, \qquad t > \frac{L}{v} + \frac{L + x}{c}. \tag{2.31}$$

The stress drop in this case is $f(1 + v/c)$.

Thus the stress drop in the case in which the fault breaks the surface is the same as that for the case in which the fault collides with an unbreakable barrier. In either case the stress drop is greater than the classical stress drop f.

The energy released in the two cases is also of interest. Unfortunately we cannot make as strong a statement as we would like concerning the potential energy released by the shocks. This is because the change in potential energy depends on the state of prestress of the system. In point of fact, for the model described in Figure 2 and its continuum limit, the quantity T is the sum of the prestress P on the fault and the tectonic stress S obtained from external sources:

$$T = P + S. \tag{2.32}$$

For the purposes of mathematical convenience we have set the quantity $P = 0$ in equation (2.2). However, energy balance requires that

$$(PE)_{\text{before}} - (PE)_{\text{after}} + S\Delta U = D\Delta U, \tag{2.33}$$

namely that the difference between the potential energies of deformation before and after faulting, plus the work done by the tectonic forces in translation of. the system a distance ΔU, be equal to the frictional heat developed, where the latter quantity is the product of the dynamical friction and the displacement of each particle. Since we do not know the partition of T among terms P and S in (2.32) we cannot solve for

$$\Delta(PE) = (PE)_{\text{before}} - (PE)_{\text{after}}.$$

However, we see that, for the same partition of T into its P and S constituents, and for constant D, P and S along the fault, the change in potential energy will be proportional to the mean displacement on the fault. Thus ΔU becomes a measure of the energy released. The result is

$$\Delta(PE) = (fL^3/12\mu)\sigma(1 + v/c) \tag{2.34}$$

for the first model, in which healing takes place due to an encounter with an unbreakable obstacle. In the second model, in which healing takes place due to an encounter with a free surface, and later with the unbreakable origin, the potential energy released is

$$\Delta(PE) = (fL^3/12\mu)\sigma(7 + v/c). \tag{2.35}$$

In both cases the quantity σ is a constant stress given by

$$\sigma = \lim_{a\to 0}(D_n - S_n)/a.$$

Thus the event which breaks the surface releases considerably more potential energy than the event which does not, even through the two events have the same stress drop.

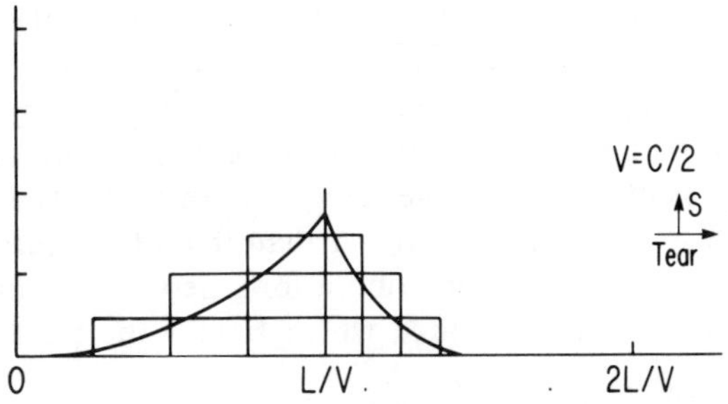

FIGURE 4. Theoretical seismogram (heavy line) for a velocity detector at great distance, broadside to a fault of finite length L and tearing with velocity $c/2$. The fault heals due to an unbreakable barrier at $x = L$. The particle velocities at points $L/4$, $L/2$, $3L/4$ and L as functions of time are also shown (light lines) relative to the onset of faulting as recorded at the seismograph.[4]

In Figure 4, we show the signal to be expected at a velocity detector located at a great distance from a fault of length L and tearing with velocity $c/2$. The

[4]See footnote 1, p. 227.

detector is broadside to the linear extent of the fault. Healing takes place due to an unbreakable barrier at $x = L$. The seismogram is the heavy line in the figure. The lighter lines show the particle velocity of individual points located at distances $L/4$, $L/2$, $3L/4$ and L from the origin of the fault; the origin of time for these latter motions has been shifted to allow the onset of rupture to coincide with the arrival of the signal at the distant recorder.

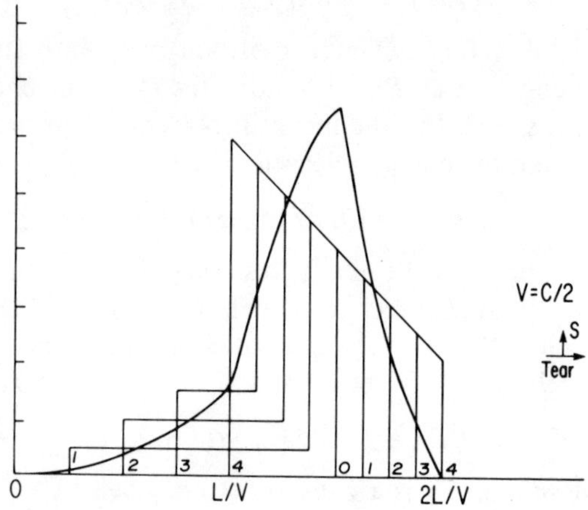

FIGURE 5. Same as Figure 4 except healing is due to a massless reflector at $x = L$.[5]

In Figure 5, the same calculation as the preceding is carried out, except that the fault heals after breaking the surface, according to the mechanism leading to the result of equation (2.28). The scales of Figures 4 and 5 are the same; thus it is readily seen that the net motion due to the second type of mechanism of healing is much greater than that due to the first, for the same length of fault. As might have been expected, the early history of the pulses of Figures 4 and 5 for the first L/v units of time is the same, since the tearing part of the process of faulting is the same in both cases. The duration of the impulse in the case of Figure 5 is greater than that of the case of Figure 4 by the amount L/c; hence we expect more long period spectral energy in the radiation spectrum from the source which breaks the surface than for the buried source.

In Figure 6 the calculation of Figure 5 is repeated except that the seismograph is located at great distance from the fault, along the extension of the fault in a direction opposite that of the initial tearing pulse. Figures 5 and 6 show that there is a considerable azimuthal dependence of the radiation pattern. In the case of Figure 6, a sharp wave front is apparent at the time $3L/2v$ after the arrival. This is due to a favorable interference of waves, corresponding to the Doppler effects that one might expect to arise.

We have formulated a fracture criterion for the one-dimensional crack in

[5]See footnote 1, p. 227.

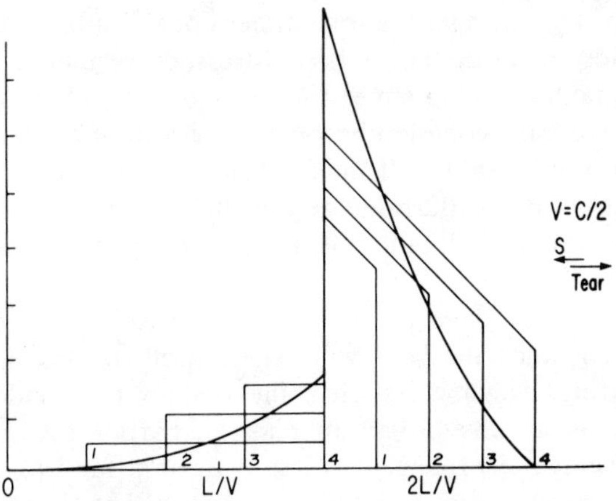

FIGURE 6. Same as Figure 5, except the seismograph (S) is end-on to the fault, in a direction opposite to that of the initial rupture.[6]

which the cohesion and the stress drop are both preeminent factors controlling the rupture speed. This conclusion is in contrast with the implications of the result obtained by Burridge and Halliday, namely, that the static friction has no quantitative effect on rupture speed in models in which there is no cohesive force.

Our model has assumed that the quantity $B(x)$ is not a function of time or of stress rate. This has been a matter of mathematical convenience for the purposes of the illustrations. We have shown that there are nonnegligible stresses radiated in advance of a moving crack tip, without the nuisance of finite *displacements,* which only complicate the dynamical picture. The fact that stresses can be propagated in advance of the crack tip while keeping $U = 0$ in this region is readily understood if one recalls that U is the jump in the displacement across the fault and not the displacement itself. This idealized version of the breaking strength can be modified in more general models with no more than a subtle change in interpretation of the edge condition, equation (2.14). In the present example, the quantity

$$\mu U_x + \rho \dot{\xi} U_t \tag{2.36}$$

is a dynamical quantity, evaluated just behind the advancing crack tip, at $x = \xi(t)^-$. The same quantity, in the present model, vanishes at $x = \xi(t)^+$, just ahead of the tear. In a more general case, the left-hand side of equation (2.14) will be the same as before; only the value of expression (2.25) for $x = \xi(t)^+$ will not be zero. We may also interpret (2.14) by stating that $g(\xi(t))$ is part of the total *stress* just ahead of the crack tip, so that the edge condition is

$$[\text{stress}] + \dot{\xi}[\text{particle momentum}] = 0. \tag{2.37}$$

[6]See footnote 1, p. 227.

The velocity of tearing of a one-dimensional fault in the presence of friction depends on both the classical stress drop and the cohesion. The extension of a fault can stop either because the classical stress drop becomes negative or because the cohesion becomes too great, or as we presume, due to a combination of both causes. If the fault stops because of reversal of sign of the classical stress drop, Burridge and Halliday have shown that the fault actually extends itself well into the region of negative stress drop before stopping.

Some special cases have also been considered. If the cohesion has a constant gradient, then the fault will extend itself indefinitely at a constant velocity of tearing, which is less than the velocity of sound. If this type of fault stops because it runs into an unbreakable barrier, the energy released is less and the average displacement is less than on a fault of the same length which stops because it intersects the free surface. The duration of the latter event is the greater of the two. The stress drop in both cases is the same and is greater than the classical stress drop f.

III. **Two-dimensional rupture on faults without cohesion**. As indicated above, the statement of the problem of the dynamical history of a two- or three-dimensional fault in the presence of cohesive forces has not yet been constructed mathematically. In this section, we reproduce Burridge and Halliday's [**1971**] analysis of the history of a two-dimensional crack under antiplane strain in the absence of cohesive forces. The two-dimensional problems, and also the three-dimensional problems, differ on several counts from the one-dimensional cases. First, there is the mathematical complexity of the Green's function; the Green's function for the wave equation in one dimension is considerably simpler to manipulate than the Green's function for the two-dimensional wave equation; convolutions of functions such as

$$\{\beta^2(t-t_0)^2-(y-y_0)^2-(x-x_0)^2\}^{-1/2}$$

are difficult to manipulate. Second, the two- and three-dimensional problems have the feature of elastic wave radiation to contend with: For one-dimensional problems, the motions are all confined to the fault line. For more-dimensional problems, the fault surface is imbedded in the elastic medium and the propagation of elastic waves can communicate stresses ahead of the tearing edge as well as reduce the complement of kinetic energy on the surface by radiation.

We consider the problem of fault history for a vertical fault in a half-space under antiplane strain (Figure 7). Consider the homogeneous, elastic half-space $y > 0$. The only admissible displacements are in the z-direction (the displacements and stresses are independent of z). Thus the problem is a scalar problem.

Faulting takes place on the plane $x = 0$; stresses τ_{xz} are applied to the system. When the shear stresses τ_{xz} on $x = 0$ become equal to the static friction, faulting initiates at some point $y = a$ (actually, the line $y = a$, $-\infty < z < \infty$). From then on, the static friction plays no role in determining

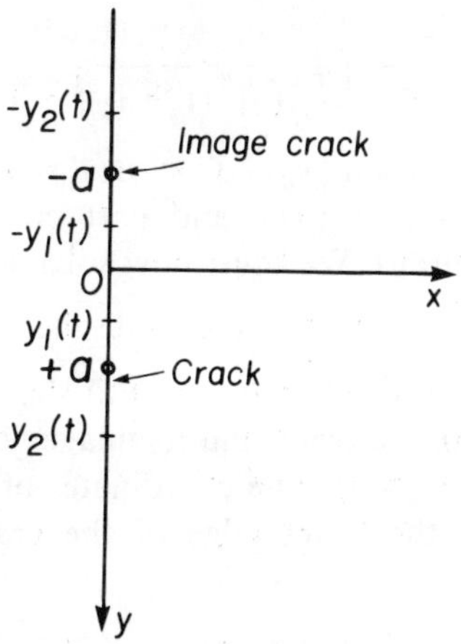

FIGURE 7. Geometry of a two-dimensional scalar fault. The fault nucleates at $(0, a)$ and extends in both positive and negative y-directions.[7]

the subsequent history of the fault. The extension of the fault takes place under the influence of the stress drop $f(y)$ which, as in the one-dimensional case, is the difference between the driving prestress $T(y)$ on $x = 0$ and the dynamic friction. If the driving stress is a constant for all depth $y > 0$ and the dynamic friction increases with depth because the normal stress $\rho g y$ increases with depth, the stress drop will be increasingly negative with increasing y for $y > b$, where b is the coordinate where $f = 0$.

Let the elastic displacement of a particle in the z-direction be $w(x, y, t)$. Thus initially $w = 0$ everywhere and w satisfies the scalar equation

$$\frac{1}{\beta^2} w_{tt} = w_{xx} + w_{yy}, \qquad y > 0, x \neq 0, \tag{3.1}$$

where β is the shear wave velocity. The stresses are

$$\tau_{zx}(x, y) = f(x, y) + \mu w_x, \qquad \tau_{zy}(x, y) = \mu w_y,$$

where μ is the shear modulus. On $y = 0$, the surface is traction-free so that $\partial w / \partial y = 0$.

The fault line $x = 0$ is divided into the part that has already torn and the part that has not. On the part that has torn,

$$\mu w_x = -f(y) \tag{3.2}$$

while on the part that has not yet torn, $w = 0$. Further, the effect of the free surface of the earth can be taken into account by considering the whole space $-\infty < y < \infty$, and considering the function $f(y)$ as an even function of y, causing cracks to be initiated at both $y = \pm a$. In fact, the antisymmetry of the stress permits us to consider the half-space $x > 0$, $-\infty < y < \infty$.

We can obtain an integral equation for the displacement on the crack surface

[7]Reprinted from "Dynamic shear cracks with friction as models for shallow focus earthquakes" by R. Burridge and G. S. Halliday, Geophysical Journal of the Royal Astronomical Society 25 (1971), 261–283, with permission of the publisher.

$$w(t_0, 0, y_0) = \frac{-\beta}{\pi} \iint_S \frac{w_x(t, 0, y)\, dt\, dy}{(\beta^2(t - t_0)^2 - (y - y_0)^2)^{1/2}} \tag{3.3}$$

where S is the triangular region $|y_0 - y| < \beta(t_0 - t)$. Part of S includes the crack surface where $w_x = -f(y)/\mu$ and part of S lies outside the crack where w_x is not given a priori. We must now solve (3.3) together with

$$w = 0 \quad \text{on } x = 0 \text{ outside the crack.}$$

If the upper edge of the crack extends until the crack runs into the free surface of the earth $y = 0$, the crack and its image will coalesce. Let the time of collision with $y = 0$ be $t = t_1$. The coordinates of nucleation of the crack are $y = \pm a$, $t = \bar{t}$. Let the lower edge of the crack be $y_2(t)$. The crack occupies

$$\bar{t} \leqslant t \leqslant t_1: \quad -y_2(t) \leqslant y \leqslant -y_1, y_1(t), \quad y_1(t) \leqslant y \leqslant y_2(t),$$
$$t \geqslant t_1: \quad -y_2(t) \leqslant y \leqslant y_2(t),$$

where $y_1(t_1) = 0$. Figure 8 shows the y–t space defined by the solution.

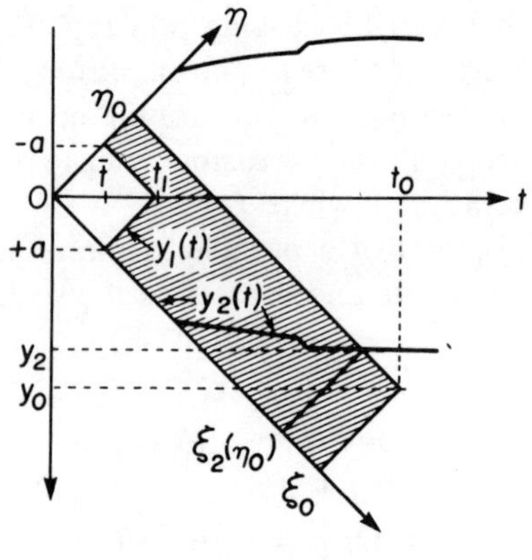

FIGURE 8. Locus of the rupture history of a hypothetical fault. The points of nucleation are at $(0, \pm a)$.[8]

Let the equations of the characteristics be $\xi = \beta t + y$, $\eta = \beta t - y$. Consider a point y_0 such that $y_0 > y_2(t_0)$. Then $w(0, y_0, t_0) = 0$. Then, by (3.3),

$$\int_0^{\eta_0} \frac{d\eta}{(\eta_0 - \eta)^{1/2}} \int_0^{\xi_0} \frac{w_x(\xi, \eta)\, d\xi}{(\xi_0 - \xi)^{1/2}} = 0 \tag{3.4}$$

where $w_x(\xi, \eta) = w_x(0, y, t)$. The region of integration is actually the shaded region of Figure 8 but, since $w_x = 0$ outside the shaded region, the integration can be written as taking place over the entire rectangle $0 \leqslant \xi \leqslant \xi_0$, $0 \leqslant \eta \leqslant \eta_0$, and has been so written in (3.4). The result (3.4) must hold for all upper

[8]See footnote 7, p. 243.

limits η_0 as long as $y_0(t_0) > y_2(t_0)$. Thus

$$\int_0^{\xi_0} \frac{w_x(\xi, \eta_0)\, d\xi}{(\xi_0 - \xi)^{1/2}} = 0, \qquad \xi_0 > \xi_2(\eta_0). \tag{3.5}$$

Thus, from (3.2),

$$\int_{\xi_2(\eta_0)}^{\xi_0} \frac{\mu w_x(\xi, \eta_0) d\xi}{(\xi_0 - \xi)^{1/2}} - \int_0^{\xi_2(\eta_0)} \frac{f(\xi, \eta_0) d\xi}{(\xi_0 - \xi)^{1/2}} = 0 \tag{3.6}$$

where we assume, for the moment, $\xi_1(\eta_0) = 0$.

If the second term is transposed to the right-hand side, we have an Abel integral equation whose solution is

$$\pi\mu w_x(\xi_0, \eta_0) = (\xi_0 - \xi_2(\eta_0))^{-1/2} \int_0^{\xi_2(\eta_0)} \frac{f(\xi, \eta_0)}{\xi_0 - \xi} (\xi_2(\eta_0) - \xi)^{1/2}\, d\xi. \tag{3.7}$$

This gives the stress beyond the crack tip in terms of an integral, as soon as the locus $\xi_2(\eta_0)$ is determined.

It is now possible to investigate the behavior of the shear stress just outside the crack tip. If we write

$$\frac{\{\xi_2(\eta_0) - \xi\}^{1/2}}{\xi_0 - \xi} = \frac{1}{(\xi_2(\eta_0) - \xi)^{1/2}} - \frac{\xi_0 - \xi_2(\eta_0)}{(\xi_0 - \xi)(\xi_2(\eta_0) - \xi)^{1/2}} \tag{3.8}$$

then (3.7) becomes

$$\pi\mu w_x(\xi_0, \eta_0) = (\xi_0 - \xi_2(\eta_0))^{-1/2} \int_0^{\xi_2} \frac{f(\xi, \eta_0)\, d\xi}{(\xi_2 - \xi)^{1/2}} - \int_0^{\xi_2} \frac{(\xi_0 - \xi_2) f(\xi, \eta_0)\, d\xi}{(\xi_0 - \xi)(\xi_2 - \xi)^{1/2}}. \tag{3.9}$$

Thus, as $\xi_0 \to \xi_{2+}$,

$$\pi\mu w_x(\xi_0, \eta_0) = \frac{1}{(\xi_0 - \xi_2(\eta_0))^{1/2}} \int_0^{\xi_2} \frac{f(\xi, \eta_0)\, d\xi}{(\xi_2 - \xi)^{1/2}} - \pi f(\xi_2, \eta_0) + O(1). \tag{3.10}$$

For $\xi_0 - \xi_2(\eta_0)$ small,

$$\xi_0 - \xi_2(\eta_0) \sim \frac{2\beta}{\beta - \dot{y}_2(t_0)} (y_0 - y_2(t_0)).$$

Thus, just ahead of the crack tip, the dominant term is

$$\pi\mu w_x(\xi_0, \eta_0) \sim \frac{1}{(y_0 - y_2(t_0))^{1/2}} \left\{ \left(\frac{\beta - \dot{y}_2(t_0)}{2\beta} \right)^{1/2} \int_0^{\xi_2} \frac{f(\xi, \eta_0)\, d\xi}{(\xi_2 - \xi)^{1/2}} \right\}. \tag{3.11}$$

The term in front shows the expected reciprocal square root dependence.

We can now determine the velocity of extension of the crack if we use the fracture criterion that the dynamical stress μw_x just ahead of the advancing crack must be large enough to overcome the cohesive forces. If the cohesive forces are zero, (3.11) vanishes and either

$$\dot{y}_2 = \beta \tag{3.12}$$

as postulated in (3.6), or

$$\int_0^{\xi_2(\eta_0)} \frac{f(\xi, \eta_0)_\xi}{(\xi_2(\eta_0) - \xi)^{1/2}} = 0. \tag{3.13}$$

This condition is both necessary and sufficient for the fracture condition to be satisfied. If (3.13) holds then, by (3.10), the stress just beyond the crack tip is $-f(\xi_2, \eta_0)$. Given the function $f(\xi, \eta)$, (3.12) and (3.13) give us the locus of the rupture edge $\xi_2(\eta_0)$; note that this locus may have jump discontinuities corresponding to jump discontinuities in f as a function of η_0.

We can now calculate the displacement. Inside the shaded region of Figure 9 where the velocities of extension of the crack are $\pm\beta$,

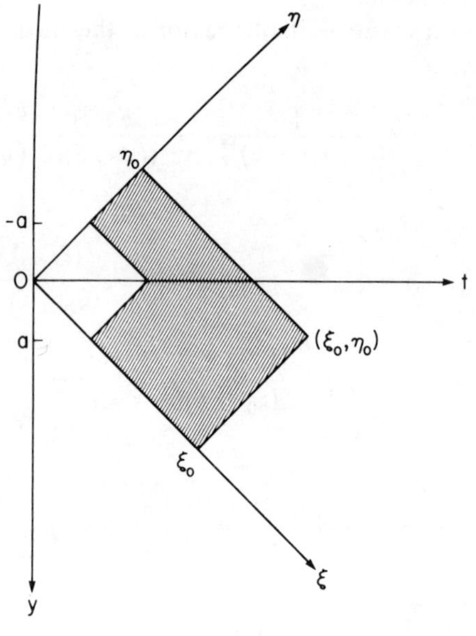

FIGURE 9. Rupture locus of a fault without cohesion. In the early part of the process, the fault tears with velocity $\pm\beta$.[9]

$$w(\xi_0, \eta_0) = -\frac{1}{2\pi}\int_0^{\eta_0} \frac{d\eta}{(\eta_0 - \eta)^{1/2}} \int_0^{\xi_0} \frac{w_x(\xi, \eta)\, d\xi}{(\xi_0 - \xi)^{1/2}} \tag{3.14}$$

from equation (2.6). In the shaded region $w_x = -f(\xi, \eta)/\mu$ so the integration is straightforward. In fact, anywhere within the rupture locus shown in Figure

[9]See footnote 7, p. 243.

10, the result of (3.14) applies since the contribution to the integrals from the regions outside the rupture locus vanishes (since $w = 0$ outside this region).

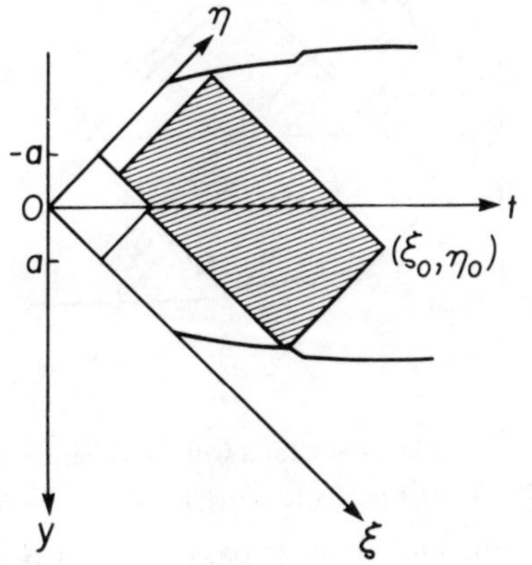

FIGURE 10. Rupture locus of a fault without cohesion. At a certain time, the speed of extension becomes less than β.[10]

The history of the crack can now be outlined (Figure 11). We assume $f(y)$ is positive for $-b < y < b$ and negative elsewhere. The crack nucleates at A and A'. The crack extends itself with velocity $\pm\beta$ along AC and AP. On AC, it overshoots the point B where $f(y) = 0$. At C, the crack slows down and extends itself according to equation (3.13) along CD. At D, a reflection from the intersection of the crack with the surface at P arrives, and causes the crack to resume sonic extension along DE. It slows again in accordance with (3.13) on EF and the outer edge ultimately comes to rest at F. In fact, $B'F$ is a characteristic. Motion in the wake of the extending crack ceases first at the point G; the crack then splits into two parts and ultimately all motion ceases with the extinction at S and F.

To find the point F where the crack edge first stops, $\dot{y}_2(t_0) = 0$, we write (3.13) in terms of t and y and get

$$\int_{l(t_0,y_0)}^{y_2(t_0)} \frac{f(y)\,dy}{(y_2(t_0) - y)^{1/2}} = 0$$

where l is the lower limit of integration. Differentiate with respect to t_0 and insert $\dot{y}_2(t_0) = 0$; we have

$$\frac{\dot{l}(t_0, y_0) f(l(t_0, y_0))}{(y_2(t_0) - l(t_0, y_0))^{1/2}} = 0.$$

Since the other factors are not zero, we have that F is defined by

$$f(l(t_0, y_0)) = 0.$$

[10] See footnote 7, p. 243.

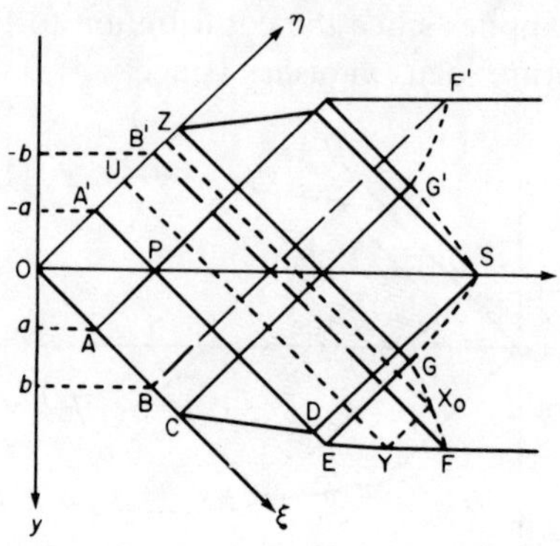

FIGURE 11. Rupture and healing locus of a fault without cohesion. Tearing begins at A and A'. Healing begins at G and G'. The last points to move are F, F' and S.[11]

Thus, the characteristic through T must pass through B', since $f(\pm b) = 0$.

Our last task is to compute the stopping locus, which is given by the condition $\partial w/\partial t_0 = 0$. Consider the point X_0 on the stopping locus GF. Construct the rectangle UZX_0Y. Then, from (3.14),

$$2\pi\mu w(\xi_0, \eta_0) = \int_0^{\xi_0} \frac{d\xi}{\xi^{1/2}} \int_0^{\eta_0 - \eta(Y)} \frac{f(\xi_0 - \xi, \eta_0 - \eta)}{\eta^{1/2}}\, d\eta$$

where $\eta(Y)$ is the η-coordinate of the point Y. Since $f(y)$, $\partial f/\partial t_0 = 0$. Thus, we need to compute only the contributions to $\partial w/\partial t_0$ from the limits ξ_0, $\eta_0 - \eta(Y)$. Thus

$$2\pi\mu \frac{\partial w}{\partial t_0} = \frac{\beta}{\xi_0^{1/2}} \int_0^{\eta_0 - \eta(Y)} \frac{f(0, \eta_0 - \eta)\, d\eta}{\eta^{1/2}} + \left(\frac{\beta - \partial\eta(Y)/\partial t_0}{\eta_0^{1/2}} \right) \int_0^{\xi_0} \frac{f(\xi_0 - \xi, \eta(Y))\, d\xi}{\xi^{1/2}}.$$

Changing to new variables,

$$2\pi\mu \frac{\partial w}{\partial t_0} = \frac{\beta}{\xi_0^{1/2}} \int_0^{\eta_0} \frac{f(0, \eta)\, d\eta}{(\eta_0 - \eta)^{1/2}} + \left(\frac{\beta - \partial\eta(Y)/\partial t_0}{\eta_0^{1/2}} \right) \int_0^{\xi_0} \frac{f(\xi, \eta(Y))\, d\xi}{(\xi_0 - \xi)^{1/2}}.$$

The second integral vanishes since Y lies on the crack edge, by (3.13). Thus $\partial w/\partial t_0 = 0$ at X_0 if

$$\int_U^Z \frac{f(0, \eta) d\eta}{(\eta_0 - \eta)^{1/2}} = 0.$$

For points on the locus GS, the condition is

[11] See footnote 7, p. 243.

$$\int \frac{f(\xi)d\xi}{(\xi_0 - \xi)^{1/2}} = 0$$

where the range of integration is on some part of the line $A'E$, determined by an appropriate inscribed rectangle.

IV. **The propagating Griffith crack.** In the one-dimensional problem, there are two methods for bringing the extension of a crack to a halt. In the first, we place high-barrier, cohesive forces in the path of the crack, so that the stresses in advance of the crack tip are incapable of tearing fresh material; this is an unbreakable object, at least unbreakable by the given crack. The second is to have the driving stress $f(x)$ become negative, through the expedient of allowing the dynamical friction to exceed the driving stresses; in this case, the crack "runs out of gas". In general, we presume that a combination of the two influences will act to bring the crack to a halt. In the two-dimensional case of the preceding section, in the absence of cohesive forces, the crack moved at sonic velocity, well past the point of zero driving stress $f(y)$ and ultimately came to rest by the second mechanism; it "ran out of gas".

We have already remarked that cohesive forces serve to moderate the $(r - vt)^{-1/2}$ singularity in the shear stress at the tip of the advancing crack in the two-dimensional case. But this is done at considerable expense in mathematical tractability of the problem. The presence of a new distance scale makes the analysis difficult. In fact, we know of no two-dimensional problems that have been solved which consider a fracture criterion of critical shear stress for extension of cracks; presumably, as in the one-dimensional case, when the critical shear stress in the neighborhood of the crack tip exceeds the cohesive forces, the crack extends. The location of the crack tip is difficult to define because of the extended zone over which this is found. The crack presumably extends itself at less than the sonic velocity. It becomes valuable to preserve the mathematical properties of the $(r - vt)^{-1/2}$ singularity associated with the cohesionless problem and to reconsider the effects of the finite extent of the crack tip by modeling this finite zone as a point. The impact of this reconsideration is obtained by considering a different fracture criterion. The Griffith [**1920**] fracture criterion provides such a model.

Under the Griffith fracture criterion, the transition zone is modeled as a point as before. The critical shear stress is no longer a criterion; instead, it is replaced by a critical flux of energy into the crack tip. Since the faulted solid has a lower energy than the unfaulted solid, the extension of the crack requires that there be a flux of potential (deformational prestress) energy through the crack tip into the crack. This energy is used to supply kinetic energy to the solid as well as do work in moving matter. This energy goes into creating new dynamic crack surface.

If the criterion of critical shear stress is appropriate for extension of a crack in the presence of cohesion, and if the cohesion is an absolute criterion based on the physical properties of materials, the specification of the energy flux through the crack tip is no longer an absolute physical criterion for fracture. This is most simply demonstrated through the result of §II for the relation between energy flux and cohesive strength $\gamma = \frac{1}{2} g U_t$. Clearly, if γ is the physical property controlling fracture, then the quantity g controls fracture only when moderated by the dynamical condition of motion on the crack, and vice versa. Our personal prejudices are to view the intractable critical stress criterion as being the physical member of the pair. Thus forewarned, we proceed to sketch a fracture problem using the Griffith criterion. We follow the treatment of Kostrov [**1966**] for the scalar problem.

As already seen in (3.11), the singularity in stress μw_x ahead of the crack tip is

$$\mu w_x(x_0, t_0) = \frac{k_2}{\pi\left(y_0 - y_2(t_0)\right)^{1/2}}$$

where

$$k_2 = \left(\frac{\beta - v}{\beta}\right)^{1/2} \int_{y_2(t_0)-\beta t_0}^{y_2(t_0)} \frac{f(y')\,dy'}{\left(y_2(t_0) - y'\right)^{1/2}}; \tag{4.2}$$

k_2 is called the stress intensity factor.

If a transformation is made to a moving coordinate system with origin at the edge of the crack, then

$$k = (1 - v/\beta)^{1/2} \int_0^t \frac{f(y')\,dy'}{(y')^{1/2}}, \qquad y' = y_2(t) - y, \tag{4.3}$$

where $y' > 0$ on the crack and $y' < 0$ on its extension; v is the velocity of the crack. We have dropped the subscript on k so that this last expression is valid for either edge of a finite crack.

Let it be assumed that the work done in the rupture process, in overcoming the cohesive forces, depends only on the speed of propagation v. This energy can be related to the coefficient of stress intensity by

$$E(v) = k^2/\pi\left(\beta^2 - v^2\right)^{1/2}. \tag{4.4}$$

As $v \to 0$, this becomes the Griffith fracture criterion for a static crack. The function $E(v)$ has to be given from some physical basis. For example, if we assume that the surface tension Σ is some constant of matter

$$2\Sigma = k^2/\pi\left(\beta^2 - v^2\right)^{1/2}.$$

Thus, an equation for the location of the end of the crack is

$$\int_{y_2-\beta t}^{y_2} \frac{f(y)\,dy}{(y_2 - y)^{1/2}} = (\pi E(v))^{1/2}\left(\frac{\beta + v}{\beta - v}\right)^{1/4}, \qquad v = \dot{y}_2, \tag{4.5}$$

with a similar equation for y_1.

References

G. I. Barenblatt, *The mathematical theory of equilibrium cracks in brittle fracture*, Advances in Appl. Mech., vol. 7, Academic Press, New York, 1962, pp. 55–129. MR **26** #7213.

K. B. Broberg, *The propagation of a brittle crack*, Ark. Fys. **18** (1960), 159–192. MR **22** #6210.

R. Burridge and G. S. Halliday, *Dynamic shear cracks with friction as models for shallow focus earthquakes*, Geophys. J. R. Astr. Soc. **25** (1971), 261–283.

R. Burridge and L. Knopoff, *Body force equivalents for seismic dislocations*, Bull. Seism. Soc. Amer. **54** (1964), 1875–1888.

——— *Model and theoretical seismicity*, Bull. Seism. Soc. Amer. **57** (1967), 341–371.

R. Burridge and J. R. Willis, *The self-similar problem of the expanding elliptical crack in an anisotropic solid*, Proc. Cambridge Philos. Soc. **66** (1969), 433–468.

R. Courant and K. O. Friedrichs, *Supersonic flow and shock waves*, Interscience, New York, 1948. MR **10**, 637.

A. T. de Hoop, *Representation theorems for the displacement in an elastic solid and their application to elasto dynamic diffraction theory*, Doctoral Thesis, Delft, 1958.

D. S. Dugdale, *Yielding of steel sheets containing slits*, J. Mech. Phys. Solids **8** (1960), 100–104.

A. A. Griffith, *The phenomenon of rupture and flow in solids*, Philos. Trans. Roy. Soc. London **A** **221** (1920), 163–198.

J. A. Hudson and L. Knopoff, *Transmission and reflection of surface waves at a corner*. 2. *Rayleigh waves (Theoretical)*, J. Geophys. Res. **69** (1964), 281–289.

L. Knopoff, *Diffraction of elastic waves*, J. Acoust. Soc. Amer. **28** (1956), 217–229. MR **17**, 1255.

L. Knopoff, J. O. Mouton and R. Burridge, *The dynamics of a one-dimensional fault in the presence of friction*, Geophys. J. R. Astr. Soc. **35** (1973), 169–184.

B. V. Kostrov, *An axisymmetric problem concerning the propagation of a normal fracture*, Prikl. Mat. Meh. **28** (1964), 644–652.

———, *Unsteady propagation of longitudinal shear cracks*, Prikl. Mat. Meh. **30** (1966), 1042–1049.

B. V. Kostrov and L. V. Nikitin, *The elasto-plastic crack under longitudinal shear*, Geophys. J. R. Astr. Soc. **14** (1967), 101–112.

———, *Some general problems of mechanics of brittle fracture*, Arch. Meh. Stosowanej **22** (1970), 749–776.

H. F. Reid, *The mechanics of the earthquake*, Calif. State Earthquake Commission Report, 2: The California Earthquake of April 18, 1906, Carnegie Institution of Washington, 1910.

A. T. Starr, *Slip in a crystal and rupture in a solid due to shear*, Proc. Cambridge Philos. Soc. **24** (1928), 489–500.

J. R. Willis, *Self-similar problems in elastodynamics*, Philos. Trans. Roy Soc. London Ser. **A** **274** (1973), 436–490. MR **49** #1877.

Institute of Geophysics and Planetary Physics, University of California, Los Angeles, Los Angeles, California 90024

Hughes Aircraft Company, Canoga Park, California 91304 (Current address of John O. Mouton).

BCDEFGHIJ–CM–898765432